常耀斌 郑智民 周贤波 著

大数据
架构之道与项目实战

清华大学出版社

北京

内 容 简 介

大数据和人工智能技术发展正当时，如何快速构建一个高水平的企业级大数据平台是撰写本书的出发点。

本书从总体技术要求出发，深入分析了全栈技术的各自优势和应用场景，传授了三十多种主流技术的架构设计、技术原理和集成方法。第1章介绍企业级大数据平台服务的总体设计，突出研究经典设计模式之美、吸纳分布式技术的精髓、深耕微架构的演变内涵。第2章～第9章是项目实战环节，介绍高并发采集、灵活转发、高可扩展海量存储、高并发海量存储、高可靠海量存储、实时计算、智能分析和自定义迁移等微服务，手把手传授架构设计和核心代码，让读者掌握商用微服务产品开发全流程。

本书封面贴有清华大学出版社防伪标签，无标签者不得销售。
版权所有，侵权必究。侵权举报电话：010-62782989　13701121933

图书在版编目（CIP）数据

大数据架构之道与项目实战/常耀斌，郑智民，周贤波著．—北京：清华大学出版社，2018
ISBN 978-7-302-51586-9

Ⅰ.①大⋯　Ⅱ.①常⋯ ②郑⋯ ③周⋯　Ⅲ.①数据处理-研究　Ⅳ.①TP274

中国版本图书馆 CIP 数据核字（2018）第 257183 号

责任编辑：杜春杰
封面设计：常雪影
版式设计：楠竹文化
责任校对：马军令
责任印制：丛怀宇

出版发行：清华大学出版社
　　　网　　　址：http://www.tup.com.cn，http://www.wqbook.com
　　　地　　　址：北京清华大学学研大厦 A 座　　邮　　编：100084
　　　社 总 机：010-62770175　　邮　　购：010-62786544
　　　投稿与读者服务：010-62776969，c-service@tup.tsinghua.edu.cn
　　　质量反馈：010-62772015，zhiliang@tup.tsinghua.edu.cn
印 装 者：三河市君旺印务有限公司
经　　　销：全国新华书店
开　　　本：185mm×260mm　　印　张：24.75　　字　数：509 千字
版　　　次：2018 年 12 月第 1 版　　印　次：2018 年 12 月第 1 次印刷
定　　　价：75.00 元

产品编号：081565-01

推荐序一

我国高度重视大数据在经济社会发展中的作用,十八届五中全会提出"实施国家大数据战略",国务院印发《促进大数据发展行动纲要》,全面推进大数据发展,加快建设数据强国。大数据技术和应用正处于创新攻关期,国内市场需求正处于爆发期,大数据产业面临前所未有的发展机遇。推动大数据平台建设和智能应用发展,加快传统产业数字化、智能化,是支撑国家战略的重要举措。当前我国正在推进供给侧结构性改革和服务型政府建设,加快实施"互联网+"行动计划,为大数据产业创造了广阔的市场空间,是我国大数据产业发展的强大内生动力。

我国大数据产业具备了良好基础,目前正在内蒙古创建国家大数据综合试验区,为产业发展奠定了坚实的基础研发和商业环境,但是发展机遇和挑战并存,主要挑战有四点:一是大数据资源的开放程度和大数据质量不高,难以被快速抽取和挖掘利用;二是技术创新能力有差距,我国在分布式计算架构、大数据处理等核心技术研究上没有长时间积累,尤其是目前在大数据的开源技术上没有太多的国家影响力,我们要向谷歌等顶级互联网公司学习优秀的研发经验和创新技术;三是大数据应用的普适性存在差距,目前的大数据应用在部分垂直领域有了较深的发展,但是绝大多数领域还是空白;四是大数据人才难以满足企业发展需求,大数据基础技术研究、产品研发和业务应用等各类人才短缺,难以满足发展需要。

加强大数据技术和通用服务产品研发应以行业垂直应用为导向,推动产品和解决方案研发及产业化,创新技术服务模式,形成可以商用的、完备的技术产品体系。大数据技术,从产品体系研发上说,包括大数据采集、传输、存储、管理、处理、分析、应用、可视化和安全等关键技术;从基础技术和技术引擎上说,包括大规模异构数据融合、集群资源调度、分布式文件系统、通用计算框架技术、流计算、图计算等;从前沿技术创新上讲,包括机器学习、深度学习、认知计算、区块链和虚拟现实等。总之,打造大数据核心产品,建立完善的大数据工具型、平台型和系统型产品体系,形成成熟大数据解决方案,是大数据产业化的重点工作方向。

本书作者是我的研究生,他于2007年获得北京邮电大学计算机专业硕士学位,曾任职于中国电科，参与并负责一体化指挥平台的核心服务研发工作，历时2年研制了全军一体化指挥作战平台，在业内取得了良好声誉。2009年加入华为技术，担任云平台高级技术专家和架构师，主导设计了国内首个企业公有云平台，负责云计算和大数据核心技术研究工作，获得华为技术"金牌员工"和"总裁奖"。现就职于中国移动研究院，担任资深技术专家和高级技术经理，专注于大平台建设和微架构等核心技术研发。

本书是他在大数据产品和研发技术上的十多年研究成果，总结了通用大数据平台的核心技术和前沿架构，特别适合大数据从业者，尤其适合高等院校的毕业生、在职大数据工程师和有志于从事大数据研发的工程师阅读，也可以作为工具类和实战类技术手册使用。本书是大数据全栈工程师的摇篮。

本书介绍并实践了大数据技术，包括8个通用微服务开发，是作者从事架构师工作多年的实战经验总结。本书精心总结了多种创新设计思想，全面阐述了8个通用服务引擎的开发过程，涵盖了高并发采集服务、灵活转发服务、高可扩展海量存储服务、高并发海量存储服务、高可靠海量存储服务、实时计算服务、智能分析服务和自定义迁移服务。每一位读者都可以借助这些微服务快速开发出个性化的平台和应用。

"在科学的道路上没有平坦的大道，只有不畏艰险沿着陡峭山路向上攀登的人，才有希望达到光辉的顶点。"希望作者和每一位读者都能求真务实，努力攀登数据科学的高峰，为新时代的科技发展贡献自己的智慧和力量！

北京邮电大学博士生导师　邓中亮教授

推荐序二

大数据概念最早由世界领先的咨询公司麦肯锡提出,之后短短几年在全球范围工业界、学术界、商业界等获得巨大的应用和推动。在美国硅谷大量的公司都投身其中,老牌的 IT 公司,如谷歌的 APP Engine、微软的 Azure 等;初创公司更是层出不穷。越来越多的公司意识到大数据蕴含的商业机会,信息社会的数据爆炸也不断促使了大数据技术向前飞跃发展,然而大数据对技术的高要求和数据本身的私密性决定了大数据不是人人都有资格参与,必须具备相应的技术储备和资质才可能投身其中。

随着全球数据的爆发式增长,大数据在中国也从政策层面备受关注。2014 年,大数据首次写入政府工作报告,大数据逐渐成为各级政府关注的热点,政府数据开放共享、数据流通与交易、利用大数据保障和改善民生等概念深入人心。2015 年 8 月 31 日,国务院印发《促进大数据发展的行动纲要》,成为中国发展大数据产业的战略性指导文件。作为我国推动大数据发展的战略性、指导性文件,充分体现了国家层面对大数据发展的顶层设计和统筹布局,为中国大数据应用、产业和技术的发展提供了行动指南。2016 年,《中华人民共和国国民经济和社会发展第十三个五年规划纲要》正式公布,"十三五"规划纲要对国家大数据战略的阐释,成为各级政府制订大数据发展规划和配套措施的重要指导,对我国大数据发展具有深远的意义。2016 年年底,工信部正式发布《大数据产业发展规划(2016—2020 年)》,该发展规划以大数据产业发展中的关键问题为出发点和落脚点,明确了"十三五"时期大数据产业发展的指导思想、发展目标、重点任务、重点工程及保障措施等内容,成为大数据产业发展的行动纲领。

大数据技术究竟包括哪些?大数据技术是一个交叉学科,涉及云计算、物联网、人工智能等技术,可以在各行各业得到相关的应用,如环保、医疗、人口、能源交通等各个领域,可以共性地概括为凡是涉及数据生产、采集、存储、加工、分析等行为的领域,包括数据资源建设、大数据硬软件产品的开发等工程实现的都包含在大数据技术的范畴。

围绕国家大数据战略实施要求,一些国内知名的互联网领头企业,如华为、中兴、阿里、百度、腾讯等软硬件企业陆续推出大数据相关平台和产品。以蚂蚁金服、滴滴出行、新美大、菜鸟网络等为代表的新兴独角兽企业，以及以树根互联、徐工信息为代表的工业互联网平台服务提供商也纷纷布局大数据领域。但是各公司都实际面临大数据领域的人才奇缺的状况，单纯依靠高校培养大数据人才远远不够，需要更多的社会力量加入。

本书作者是我的多年好友，他带领团队完成了国内多个国家级云平台建设，获得了诸多项目荣誉，是大数据行业的知名专家。他从大数据平台的架构设计核心技术要求出发，全面阐述了构建一个 PB 级规模数据运营能力的云平台的具体要求，零距离传授大数据平台的三十多种主流技术的架构设计、技术原理、集成方法、性能调优，全面涵盖 Spring 生态、Hadoop 离线计算、Spark 实时计算、机器学习和设计模式五大生态领域，注重理论和实际项目相结合，代码量丰富、详尽。

在当前大数据领域书籍偏重理论分析、缺少实际项目论述的背景下，这本书显得尤其实用，不仅可以成为广大希望投身大数据领域的从业者一本很好的引路书籍和工具，而且也可以作为高等院校相关专业学生学习大数据的参考资料。

国资委网络安全专家　北京科技委专家
大唐集团科研院网络安全实验室主任　张　伟

前言

大数据在近几年逐渐步入理性发展期。未来的大数据发展会渗透到各个垂直行业领域，尤其在企业数据资源化、传统行业智能化、分析引擎产品化等方向上迎来前所未有的发展机遇。与此同时，大数据的中高端人才缺乏严重阻碍大数据产业发展，尤其是掌握技术、精通管理、拥有项目经验的高级工程师成为企业炙手可热的人才。在北上广深杭等一线城市，有一年以上工作经验的大数据工程师的月薪过万，有几年工作经验的数据分析师的年薪在30万～50万元之间，而更顶尖的大数据技术人才则是年薪超百万，成为各大互联网和一流企业争夺的对象。2018年1月，208所高校、高职院校获批开设"大数据技术与应用"专业。同时，教育部设立了"数据科学与大数据技术"本科专业，35所大学已获批开设，还有200多所高校正在申报。纵观企业的大数据人才需求和高校的课程设立布局，应试者的实操能力和项目经验积累越来越受到注重。在专业技能上，重点需求体现在大数据基础平台建设和应用开发上，涵盖了高并发采集、灵活转发、海量存储、实时计算、智能分析和高效迁移等专业技能。在个人能力要求上，重点需求体现在责任心、团队合作能力和解决问题能力等方面。本书重点讲授大数据全栈专业技能，但接下来，我先谈谈大数据工程师的个人能力要求和职业发展规划。

我从事计算机领域研发16年来，主导负责了十多个国家项目，经历了单机服务到分布式服务的项目研发模式，也实践了从百万用户到上亿用户的商用产品。但是，在新的移动互联网和大数据时代，对工程师和技术研发人员的专业技能和个人能力提出了新的要求，单兵作战模式无法成就一个商业产品，需要设计驱动和团队协同作战。团队合作和协同作战一直是我倡导的软件产品商用化的管理模式，我的很多学生在BAT等知名互联网公司担任技术主管，从某种意义上来说，得益于我很早就对团队赋予了设计引领产品的创新思想。但是，在我的职业发展中也曾多次面临发展瓶颈，我是如何面对挑战和压力呢？接下来，我想分享一下在软件研发职场上的晋升技巧和有效工作方法，让大家少走弯路，多获捷径。大数据工程师的职业发展路线大致分几个关键阶段：一是上升为项目经理阶段；二是历练为技术经理阶段；三是发展为资深架构师阶段；四是成长为首席技术官（CTO）阶段。

第一个阶段是上升为项目经理，先争取在项目中担任技术骨干，并逐步主动承担和肩负更多更具挑战的研发任务。项目经理职位在一流的互联网公司至少需要奋斗3年以上，如何缩短这个非常漫长和艰苦奋斗的过程，建议从以下几个方面做起。

一是需要责任心和主动性，不仅要按时完成项目经理交付的开发任务，最好还经常帮助同事突破技术难题。建议是：一定要为成为技术专家而不懈努力，千万别奔波在做一些事务性的工作而忽略技术本身，软件工程师的核心竞争力就是拥有全面的核心技术，并具备快速解决技术难题的能力。

二是要有很强的团队合作能力，善于发现别人优点并学会适当表扬，善于总结自己的研发成果并学会主动分享，善于表达自己并学会归纳总结。团队合作能力也是需要不断提升的，多听取别人的忠告而改变自己，多帮助别人解决问题而感受快乐，多用心学习核心技术而不搬弄是非。

三是要学会成就团队，就是能创造一个环境，让每个人都能在其中发挥出更多的能力，也就是一种领导力。我在华为工作期间，主动承担了一些技术维护工作，经常为大家管理服务器并配置环境，很快被同事们赋予"大管家"称号，自己在享受称号的同时也得到了领导的认可，虽然技术维护工作本身不计算在绩效中，但是其对团队的运营能力提升是举足轻重的。学会成就团队是树立你在团队中威望的很重要品质。

四是要提升汇报和总结能力。不管是民营企业还是国有企业，汇报能力对职场人都是非常重要的，究其原因是汇报能体现一个人的综合能力，需要有严谨的逻辑思维和优秀的写作能力，让领导在短时间内掌握一个项目的开发现状、存在问题、解决方案和创新工作等。更好地规划工作、布局工作、超预期完成工作，并在适当时候提出有建设性的宝贵意见是至关重要的。汇报工作的核心是分析和解决方案，领导都是团队中最忙和承担压力最大的人，比起发现问题来说，他更关注的是问题分析和最优解决方案。工程师学会多思考问题并有针对性地提出优秀解决方案，对团队和个人发展都是至关重要的。比如我们的项目因缺乏设计而导致开发周期太长，比如我们的项目因不能定期和客户沟通需求而导致偏离实际需求，比如我们不能按期交付项目成果而导致领导不满意。

第二个阶段是历练为技术经理。这个阶段对于一般的项目经理而言就是一个项目接着一个项目交付，上升空间遥不可及，如果要突破晋升空间也是有工作方法和拓展思路，建议从以下几个方面做起。

一是把控好项目的里程碑并学会提升管理水平。项目要有合理规划，从项目工作计划到项目任务分解、从技术选型到技术验证成功、从总体设计规划到架构设计细化、从架构设计分解到概要设计说明、从概要设计到详细设计落实、从详细设计规划到核心代码编写等，都是需要不同阶段的技术评审和质量审查，都是需要分时段交付研发成果，都是需要管理和技术能够协同推进。

二是要加强团队建设，更关注人才的能力和培养。带团队就是带人心，在公司规则之内多考虑员工的合理想法，切不能顾此失彼地加压。从团队建设力度就可以看出公司的发展动向，如果以人为本重视长期发展，常常会考虑激励和培养员工，这恰恰也是小公司的努力方向，希望多重视程序员的意见并采纳实施，其实这样做之后最大受益者也是老板。

三是要多输出具备影响力的项目成果，如商用产品、项目奖项、核心专利和高水平学术论文。任何公司都是需要产品布局和发展规划的，尤其是短时间内要占领行业制高点，最好是有核心竞争力的产品或者专利来支撑公司在行业的领航地位。这个核心竞争力就是来源于项目而高于项目本身的拳头产品。我们在研发中多积累优秀设计思想、多总结提炼核心算法、多琢磨技术难题的创新解决方案、多讨论问题碰撞思想火花、多研究论文学习前辈的前瞻思想。

四是多读书，多学习优秀管理思想，领会分层管理的领导艺术和对结果负责的管理体制，更不能越权管理。技术经理往往需要掌握分级分层的管理思想。如果我们的日常工作都聚焦在具体事务上，如果不关注产品运营而拘泥于任务细节上，如果没有远大的理想和成就一番事业的抱负，如果没有带领团队打造核心产品的目标，那么在行动和执行力上就会出现小格局、小思维，最终因为延误战机而失去创造奇迹的机会，没有成功的团队就不会有成功的个人，没有成功的个人谈何脱颖而出的成功技术领导人？华为公司成功的原因之一就是层层管理者都要保证按期交付而不越权管理，高层领导负责战略和市场，中层领导负责战术落地实施并跟踪任务，基层人员负责细节实施和按期交付，一个完备的权责明确的分层管理机制一定会推动公司高效的运营。越权管理不仅会导致基础管理员失去权力而懈怠，更会导致不能细化管理而延误进度，大目标都是小里程碑积累完成的，不积跬步何以至千里。

第三个阶段是发展为资深架构师。这个阶段需要在知名企业的一线产品上历练十年以上，架构师是一个既需要掌控整体，又需要洞悉局部瓶颈的技术领袖。架构师在整个产品研发的生命周期中都起着至关重要的作用，随着开发进程的推进，其职责或关注点不断地加深。在需求分析阶段，软件架构师主要负责梳理非功能性系统需求，如软件的高可维护性、高性能、高复用性、高可靠性、有效性和可测试性等，另外，架构师还要经常分析客户不断变化的需求，确认开发团队所提出的设计；在总体设计阶段，架构师的关注点主要在开发团队的技术能力和开发模式；在软件概要和详细设计阶段，架构师负责对整个软件体系结构、关键构件、接口和开发策略的设计；在代码编写阶段，架构师则成为详细设计者和代码编写者的老师，并且要经常性地组织一些技术研讨会、技术培训班等来提升团队的技术能力；在软件测试交付阶段，架构师跟踪关注性能需求，同时开始为下一版本的产品是否应该增加新的功能模块进行决策。从架构师的工作职责上说：一是必须具有丰富的软件设计与研发经验，并验证所进行的设计是如何映射到实现中去；二是要具有领导能力与团队协作能力，架构师必须是一个团队最核心的技术领导人，能在关键时刻对技术的选择做出及时、有效的决定；三是要有不断积累新技术和新架构的技术能力，架构师需要掌握的知识是多维度和多方面的，如精通各种标准的通信协议、网络服务、面对对象数据库、关系数据库或者 NoSQL 数据库、数据处理和分析等知识，另外，架构师应与时俱进地学习新软件设计和开发思想，并不断探索更有效的新方法。开发语言、设计模式和开发平台不断地升级，架构师需要吸收这些新技术、新知识，并将它们用于软件产品和项目开发工作中。总的来说，架构师是一个技术高端职位，技术经理如何得到这样的机会，如何利用所掌握的技能进行应用的合理构架，如何不断地抽象和归纳自己的构架模式，如何深入行业成为一流公司的架构师，确实需要不断地磨炼。

第四个阶段是成长为公司 CTO。这个阶段需要有敢为人先的胆识、阅历丰富的见识、与时俱进的学识。作为 CTO，要想突破自己的领导位置，要敢于大胆提出创新思想和超前理念来带领团队脱颖而出，同一个起跑线上更需要与众不同的决心和勇气，好的机会永远是留给有胆识的技术领导人，切勿在关键时刻瞻前顾后或者犹豫不决，否则很难成就一番大业，最有说服力的例子就是 BAT 的创始人，他们用超人的胆识抓住了移动互联网高速发展契机，各自打造自己擅长的垂直领域，通过核心技术让本地化和移动设备完美结合，改变了新时代下的人的消费和社交模式。要想突破自己的技术职级，要提升自己对行业信息和外界发展的见识，不能守旧在自己的技术领域。很多技术经理习惯停留在技术舒适区，不愿意进入挑战区，而且很少参加国际或者国内的主流峰会，很难提出高瞻远瞩的创新性的解决方案。作为技术领导人，要想超越同行成为佼佼者，要静下心来沉淀和历练，只有拼出来的美丽，没有等出来的辉煌。

有了如上的职业奋斗目标，我们要脚踏实地地开始走入大数据研发工作中了，下面是本书章节安排。

- 第 1 章：大数据平台服务总体设计，从总体技术要求阐述微服务的核心需求和设计方法。
- 第 2 章：深入阐释"高并发采集微服务"的架构设计和技术实现。
- 第 3 章：着重论述"灵活转发微服务"的架构设计和技术实现。
- 第 4 章：详细介绍"高可扩展海量存储微服务"的架构设计和技术实现。
- 第 5 章：重点讲述"高并发海量存储微服务"的架构设计和技术实现。
- 第 6 章：重点说明"高可靠海量存储微服务"的架构设计和技术实现。
- 第 7 章：详细论述"实时计算微服务"的架构设计和技术实现。
- 第 8 章：主要介绍"智能分析微服务"的架构设计和技术实现。
- 第 9 章：深入论述"自定义迁移微服务"的架构设计和技术实现。

本书的创作成果是我多年的架构经验积累和核心技术提炼，从开始撰写到完成初稿大概花费了一年多时间，我钻研了上百本大数据技术专著，希望把这些前瞻技术运用于项目实战，让大数据工程师们快速学到真本领，把知识转化为科技生产力。在此，我首先感谢我的团队和同事们,一起走过了科研之路,一起孵化了商用产品,一起收获了专业技能。

感谢郑智民、周贤波、于路、马超、曹权红、郭义华、王威等技术专家多年对我工作的积极支持和热心帮助。尤其感谢曹权红,他年轻有为,勤奋好学,专注于大数据技术研究,在项目研发上已经能独当一面,是团队中的技术达人和奋斗者。马超也是团队中的佼佼者,在技术上积累了丰富的研发经验。

感谢我的导师邓中亮教授,传授给我很多探索科学的方法,是我从事科研工作的领路人,作为一名著名科学家,他获得了诸多国家科技荣誉,但一直奋斗在科研一线,非常值得我敬重。他能在百忙中抽出时间给我作序,非常感谢。

感谢我的家人,我的妻子和女儿,是她们对我的科研工作的默默支持,才让我能全力以赴地投入到技术研发上。

感谢我的父母,他们不仅给了我生命,更多的是培养我锲而不舍地去攀登科学高峰的品质。

感谢对我的书进行推荐的朋友们,是你们对我的支持才有了这本书创作的源泉。

感谢我的学生们,你们的青春都是用来奋斗的,相信大数据产业发展的未来属于你们。

特别说明:本书代码都是基于前瞻技术 Spring Cloud 分布式架构设计,读者如有研究需要,可以联系作者。

常耀斌

2018 年 10 月于北京

第 1 章 企业级大数据平台服务的总体设计

1.1 平台架构设计的总体技术要求 ·· 2
1.2 微服务引擎的可扩展性设计 ·· 6
1.3 微服务引擎的优秀解决方案 ·· 8
    1.3.1 高并发采集微服务 ·· 11
    1.3.2 灵活转发微服务 ·· 12
    1.3.3 高可扩展海量存储服务 ·· 13
    1.3.4 高并发海量存储服务 ··· 13
    1.3.5 高可靠海量存储服务 ··· 15
    1.3.6 实时计算服务 ·· 16
    1.3.7 基于机器学习的智能分析服务 ······································ 17
    1.3.8 自定义迁移服务 ·· 17
1.4 设计小结 ··· 17

第 2 章 大数据高并发采集微服务引擎

2.1 核心需求分析和优秀解决方案 ·· 20
2.2 服务引擎的技术架构设计 ··· 20
    2.2.1 Maven 与 Eclipse 集成配置 ··· 22
    2.2.2 Mina2.0 框架以及业务设计 ··· 24

  2.2.3 设备协议规范制定及数据包设计 26
  2.2.4 按照设备和数据类型进行业务树构建 30
  2.2.5 按照设备的数据包状态进行解析 32
  2.2.6 按照通用方式进行高并发入库 34
 2.3 核心技术讲解及模块化设计 35
  2.3.1 Spring Maven Web 服务构建 35
  2.3.2 Spring Boot 微服务构建 42
  2.3.3 数据包定义和实现 51
  2.3.4 业务树构建和实现 62
  2.3.5 数据包状态进行解析实现 79
  2.3.6 按照通用方式进行高并发入库实现 90
  2.3.7 客户端模拟器工具类进行高并发测试 126
 2.4 项目小结 139

第 3 章 大数据灵活转发微服务引擎

 3.1 核心需求分析和优秀解决方案 142
 3.2 服务引擎的技术架构设计 142
 3.3 核心技术讲解及模块化实现 145
  3.3.1 Spring MVC Web 服务构建 145
  3.3.2 Spring Boot 微服务构建 152
  3.3.3 灵活配置和通用工具类构建 156
  3.3.4 创建发送数据主题，注册观察者对象 159
  3.3.5 启动多线程进行数据发送 165
  3.3.6 采用 Post 策略模式进行数据发送 168
  3.3.7 采用 ActiveMQ 策略模式进行数据发送 169
 3.4 项目小结 173

第 4 章 大数据高可扩展海量存储微服务引擎

 4.1 核心需求分析和优秀解决方案 176

4.2 服务引擎的技术架构设计	177
4.3 核心技术讲解及模块化实现	179
4.3.1 Spring MVC 的工作原理及执行流程	179
4.3.2 Spring MVC Web 服务构建	180
4.3.3 Spring Boot Web 微服务构建	187
4.3.4 统一对外数据接收接口及通用类	191
4.3.5 MySQL 对智能终端运动数据的分状态和分策略处理	203
4.3.6 MySQL 对智能终端运动数据的分职责处理	210
4.3.7 MySQL 对智能终端运动数据的统一入库处理	214
4.4 项目小结	228

第 5 章　大数据高并发海量存储微服务引擎

5.1 核心需求分析和优秀解决方案	230
5.2 服务引擎的技术架构设计	230
5.3 核心技术讲解及模块化实现	231
5.3.1 Spring MVC 和 Spring Boot 集成 MongoDB	232
5.3.2 MongoTemplate 核心类实现 Dao 层接口	234
5.3.3 基于 MongoDB 处理智能终端运动数据	236
5.3.4 基于 MongoDB 管道技术处理体检数据	243
5.3.5 基于 AngularJS 架构可视化体检数据	264
5.4 项目小结	270

第 6 章　大数据高可靠海量存储微服务引擎

6.1 核心需求分析和优秀解决方案	272
6.2 服务引擎的技术架构设计	272
6.3 核心技术讲解及模块化实现	273
6.3.1 Hadoop 完全分布式集群构建	274
6.3.2 Spring MVC 和 Spring Boot 集成 Hbase	279
6.3.3 HbaseTemplate 核心类实现 Dao 层接口	280

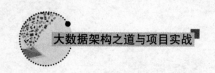

	6.3.4 Hbase 集群的智能终端运动数据 Controller 接口	286
	6.3.5 Hbase 集群的智能终端运动数据 Service 接口	288
	6.3.6 Hbase 集群的智能终端运动数据 Dao 接口	292
6.4	项目小结	300

第 7 章 大数据实时计算微服务引擎

7.1	核心需求分析和优秀解决方案	302
7.2	服务引擎的技术架构设计	302
7.3	核心技术讲解及模块化实现	303
	7.3.1 分布式采集服务 Flume 部署及数据采集	303
	7.3.2 分布式消息服务 Kafka 部署及数据发送	305
	7.3.3 创建 HBase 数据库和 Spark 环境	307
	7.3.4 分布式实时处理引擎 Spark Streaming 原理及数据处理	308
	7.3.5 构建 BD_RTPServer_DP 工程实现数据处理	309
	7.3.6 构建 BD_RTPServer_Boot 服务实现可视化	318
7.4	项目小结	327

第 8 章 大数据智能分析微服务引擎

8.1	核心需求分析和优秀解决方案	330
8.2	服务引擎的技术架构设计	330
8.3	核心机器学习算法讲解和应用	331
	8.3.1 逻辑回归的原理分析	331
	8.3.2 支持向量机原理分析	333
	8.3.3 决策树原理分析	334
	8.3.4 聚类算法原理分析	336
	8.3.5 关联规则算法原理分析	336
	8.3.6 协同过滤原理分析	336
8.4	Spark 架构原理与数据预测	337
	8.4.1 YARN 运行架构工作原理	339

8.4.2	Spark Mlib 核心技术	341
8.4.3	Spring Maven 工程构建	342
8.4.4	决策树预测体检费用	345
8.4.5	逻辑回归预测体检费用	347
8.4.6	随机森林预测体检费用	349
8.4.7	支持向量机预测疾病概率	350
8.4.8	协同过滤推荐药品	352
8.5	项目小结	353

第 9 章 大数据自定义迁移微服务引擎

9.1	核心需求分析和优秀解决方案	356
9.2	服务引擎的技术架构设计	356
9.3	核心技术讲解及模块化实现	357
9.3.1	Hadoop 生态的核心组件	357
9.3.2	HBase 工作原理	358
9.3.3	Sqoop 工作原理	360
9.3.4	MapReduce 工作原理	360
9.3.5	Sqoop 抽取历史数据到 HDFS	361
9.3.6	构建工程 BD_CustomTransfer_Maven	364
9.3.7	智能终端运动数据从 MySQL 数据迁移到 Hive	374
9.4	项目小结	377

第 1 章

企业级大数据平台服务的总体设计

架构之道分享之一：孙子兵法的《计篇》论述了战争之前对战争进行战略顶层设计问题。孙子认为，战争是国家大事，一定要认真研究、周密筹划、慎重决策。《计篇》探讨了两个重点问题：一是战争的决定因素；二是战争的根本原则。这些指导思想也适用于项目实战，决定项目成功的两个重要因素：一是项目的总体规划和目标分解；二是项目的架构设计和完备的体系建设。

本章学习目标

★ 掌握构建大数据平台的总体技术要求
★ 掌握构建大数据平台的微服务构建方法
★ 掌握8个微服务架构的设计思想和解决方案
★ 掌握构建微服务的非功能需求设计

1.1 平台架构设计的总体技术要求

本章是项目实战前的总体战略设计篇，将深入探讨项目的总体架构设计思想。我们从构建一个商用的企业级大数据平台服务的总体技术要求出发，详细阐述从核心需求到设计方案的运筹过程。

2016年是大数据时代的元年，2017年是人工智能时代的元年，2018全球人工智能产品应用博览会在苏州举办，更推进了大数据和人工智能在智能制造、软硬件终端和服务业等领域的广泛应用。大数据与人工智能、云计算、物联网、区块链等技术日益融合，成为全球最热的战略性技术，给大数据从业者带来了前所未有的发展机遇，同时也对大数据工程师提出了高标准的技能要求。大数据具有海量性、多样性、高速性和易变性等特点，映射到大数据平台建设要求，不仅要具备海量数据采集、并行存储、灵活转发、高效调用和智能分析的通用Paas服务能力，而且能快速孵化出各种新型的Saas应用的能力。要实现这个目标，架构设计至少要满足3个总体技术要求：一是把分布式大数据平台的基础数据服务能力建设摆在首位，规划出支撑PB级规模数据运营能力的云平台架构，运用经典设计原则和设计模式的架构之美，吸纳业内主流分布式技术的思想精髓，深耕主流平台服务模式到现代微架构的演变内涵；二是用系统架构设计和微服务建设思想武装团队，持续撰写多维度的架构蓝图，推动团队协同作战；三是围绕大数据全栈技术体系解决项目实战中的各类难题，制定主流技术规范和设计标准，通过平台核心组件方式快速迭代出新型业务。从设计要求来讲，大数据平台服务的整体设计要具备全面、全局、权衡的关键技术要求，不仅能全面提炼国内外优秀架构和解决方案的精华，而且要理解分布式技术的底层设计思想；不仅能全局了解上下游技术生态和业务结合的设计过程，而且要游刃有余地处理系统功能和性能问题；不仅能权衡新技术引入和改造旧系统的成本估算，而且要推动作战团队轻松驾驭新技术。

1. 第一个总体技术要求

把分布式大数据平台的基础数据服务能力建设摆在首位。规划出支撑 PB 级规模数据运营能力的创新云平台架构，运用经典设计原则和设计模式的架构之美，吸纳业内主流分布式技术的思想精髓，深耕主流平台服务模式到现代微架构的演变内涵。下面分别从应用场景、架构设计、架构演变等 5 个方面来详细阐述。

（1）从应用场景角度来考虑，实现垂直行业数据运营方优良资源聚合，促进移动终端和应用服务融合，加强线上和线下服务无缝对接，为用户提供全生命周期的衣食住行等服务，并在大数据产业中孵化有价值的新型应用，促进服务模式的创新。

（2）从应用架构设计来考虑，大系统拆分为多个微服务后，每一个微服务要围绕耦合度较高的业务单元进行构建，并建设自己的数据存储、业务开发、自动化测试以及独立部署机制。在微服务之间实现接口互联互通时，能灵活抽取微服务内部的功能组件，实现分布式事务等协作问题。

（3）从数据架构设计来考虑，支持 PB 级规模数据运营能力的数据架构，包括微服务引擎的数据库建设、分布式数据库和消息服务中间件的建设等，同时考虑可扩展的弹性设计来支撑后续系统升级迭代。

（4）从架构模式选型来考虑，设计模式是具有大智慧的软件设计经验的总结，是软件行业的《孙子兵法》。设计模式总结了面向对象设计中最有价值的经验，并且从可复用和可扩展角度描述了代码架构的思想精髓。下面从基本原理和应用场景角度两个方面，讲一下六大原则和三类设计模式。

① 一是开闭原则（Open Close Principle），强调对扩展开放和对修改关闭。应用场景：当代码架构在迭代演进时，不能去修改原有的代码，而是抽象出父类接口，修改子类即可。

② 二是里氏代换原则（Liskov Substitution Principle），强调父类和子类的关系。应用场景：在定义时使用父类对象，而在运行时再关联子类类型。

③ 三是依赖倒转原则（Dependence Inversion Principle），强调接口的重要性，接口就是把一些公用的方法和属性声明，然后具体的业务逻辑是可以在实现接口的具体类中实现的。当依赖对象是接口时，就可以屏蔽实现这个接口的具体类改变。应用场景：通过抽象（接口或抽象类）使各个类或模块的实现彼此独立，不互相影响，实现模块间的松耦合。

④ 四是接口隔离原则（Interface Segregation Principles），强调接口的职责要明确，根据职责定义"较小"的接口，不要定义"高大全"的接口。也就是说，接口要尽可能的职责单一，暴露给客户端的方法更具有"针对性"，往往使用多个隔离的接口比使用

单个接口要好。应用场景：在使用接口时要注意控制接口的粒度，接口定义的粒度不能太细，也不能太粗。

⑤ 五是单一职责原则（Single Responsibility Principle），强调一个类只负责一个功能领域中的相应职责，应用场景：一个类是和一组相关性很高的函数、数据的封装，如单例模式可以降低内存的开销。

⑥ 六是迪米特法则（Law of Demeter），强调应该尽量减少对象之间的交互，如果其中的一个对象需要调用另一个对象的某一个方法，可以通过第三者转发这个调用。应用场景：通过引入一个合理的第三者来降低现有对象之间的耦合度。

在遵循六大设计原则的战略思维之后，又演变出了23种设计模式。这23种设计模式从战术角度阐述了如何协同作战部署和兵力调动的问题，作战前的兵力部署和组合分别是创建型模式和结构型模式，作战时的兵力调动就是行为型模式。

① 创建型模式就是作战部署，侧重类的构建方法。主要包含5种设计模式：工厂方法模式（Factory Method Pattern）、抽象工厂模式（Abstract Factory Pattern）、建造者模式（Builder Pattern）、原型模式（Prototype Pattern）和单例模式（Singleton Pattern）。其中最常用的就是工厂方法模式、抽象工厂模式和单例模式。

② 结构型模式就是作战前的兵力组合，用来处理类或者对象的组合。主要包含7种设计模式：适配器模式（Adapter Pattern）、桥接模式（Bridge Pattern）、组合模式（Composite Pattern）、装饰者模式（Decorator Pattern）、外观模式（Facade Pattern）、享元模式（Flyweight Pattern）和代理模式（Proxy Pattern）。其中最常用的就是适配器模式、桥接模式和代理模式。

③ 行为型模式就是兵力调动，用来描述类或对象之间如何交互和担当职责，主要包含11种设计模式：责任链模式（Chain of Responsibility Pattern）、命令模式（Command Pattern）、解释器模式（Interpreter Pattern）、迭代器模式（Iterator Pattern）、中介者模式（Mediator Pattern）、备忘录模式（Memento Pattern）、观察者模式（Observer Pattern）、状态模式（State Pattern）、策略模式（Strategy Pattern）、模板方法模式（Template Method Pattern）和访问者模式（Visitor Pattern）。其中最常用的就是责任链模式、迭代器模式、观察者模式、状态模式和策略模式。

（5）从主流微服务发展来考虑，Spring生态是上述设计模式的最好实践，全部底层源码是采用设计模式实现的，是一枝独秀的企业级框架，可扩展性和可维护性非常好。Spring框架不仅是高度可配置的，而且可以集成多种中间件和视图。Spring框架贯穿了整个中间层，将Web层、Service层、DAO层及PO无缝整合，数据服务层用来存放数据。第一代以SSH（Struts+Spring+Hibernate）框架为代表，在PC Web应用互联网时代非常受追捧，但在敏捷开发模式下协同作战效率很低，而且用户和数据规模巨大情况下已经无法满足运营需求。第二代架构是以Spring MVC为代表的单体架构，在移动互联网时代非常主流，Spring MVC 分离了控制器、模型对象、分派器以及处理程序对象的角色，这种分离让它们更容易进行定制。第三代架构是 Spring Cloud 和 Spring Boot 微架构，强调 Spring Cloud 分布式的五大能力，包括服务器的注册与发现（Netflix Eureka）、客户端负载均衡（Netflix Ribbon）、断路器（Netflix Hystrix）、服务网关（Netflix Zuui）和分布式配置（Spring Cloud Config）在后面要详细阐述。

2. 第二个总体技术要求

用系统架构设计和微服务建设思想武装团队，持续撰写多维度的架构蓝图，推动团队协同作战。架构师不仅要具备大型云平台架构的实战经验，更要有大智慧和战略思维，通过蓝图来推动和管理好每一个产品的全生命周期。我从事架构师十多年以来，带领团队实践了二十多个大型项目，至今还在全力以赴地奋斗在开发一线，从未脱离核心代码实战，因为真理源于实践。我归纳总结了一些心得，分享给即将肩负重任的工程师们，相信会让大家受益匪浅。一般来说，一个项目成功的决定因素有两点：一是制订了具有前瞻性的、分布式架构设计蓝图，着重考虑系统具备高可扩展和高可靠能力，并保证项目在预算范围内按期交付；二是架构师严格按照架构蓝图来推动协同作战，通过蓝图来量化工作内容，保证工程师能在关键的交付点完成里程碑任务。每一个优秀产品在上线之前都需要架构师全力以赴地运筹帷幄，灵活运用大智慧和架构蓝图来协作管理。《孙子兵法》是中国古代军事思想史上现存最重要的兵学著作，其中很多优秀作战思想都可以运用在军事、商业、职场、项目等诸多领域，是每一位资深架构师的必修课。下面借鉴《孙子兵法》来分析项目成功的关键因素。

（1）项目的架构蓝图。这是对架构师技术能力和作战经验的考量。按照我的项目经验，技术架构蓝图主要包括开发架构图、逻辑架构图、运行架构图、部署架构图、数据架构图等，架构师着重关注架构设计、业务分解、组件设计、组件组合、系统交互设计、性能调优和技术攻关等，产品经理负责系统交互设计，研发经理负责项目任务分解，高级工程师负责组件实现，团队成员按照架构蓝图来分工协作，从而形成有效的管理体制。《孙子兵法》在《势篇》论述了如何灵活地使用自己团队的军事力量，就是要建立有效的军队管理组织结构，建立有效的军队命令传达体系。

（2）推动协同作战。《孙子兵法》的《作战篇》的"作战"就是始战和战争准备，讲的是战争要有物质基础与后勤保障，战争会带来兵资巨耗，尤其是旷日持久战会造成兵久国疲，产生内忧外患，危害国家的安全，因此战争应该"速胜"。完成一个项目就像打一场战争，对项目要有整体规划和里程碑管控，不断地输出阶段性项目成果，不断地发布产品小版本，不断地让客户体验产品并反馈问题，不断地让领导看到项目进展，不断地让工程师们产生荣誉感和认可感。打持久战势必会影响团队士气，甚至会让团队、领导、客户等项目干系人失去信心。

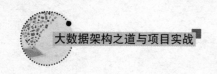

3. 第三个总体技术要求

围绕大数据全栈技术体系解决项目实战中的各类难题，制定主流技术规范和设计标准，通过平台核心组件方式快速迭代出新型业务。对于设计规范的重要性，我们不妨用《孙子兵法》的大智慧来分析一下。《孙子兵法》在《谋攻篇》提出"不战而屈人之兵"的军事思想，首先说"战争是政治的延续"的思想，然后提出"上兵伐谋，其次伐交，其次伐兵，其次攻城"的思想，从研发角度理解，就是借助商用中间件的核心思想和接口能力，来提升平台的技术性能指标，其次再借助设计模式来强化自己的业务分层设计，最后自主研发满足需求的业务系统。大数据发展正当时，实现各垂直行业数据运营方优良资源聚合，移动终端和应用服务融合，线上和线下服务无缝对接，促进服务模式的创新，对研发人员来说是很大的挑战和壁垒。从项目交付成果来看，这不仅要交付一个对来自多源异构（时间序列）数据进行采集、存储、转发、计算、迁移、分析等提供各种公共能力的系统群，也要为用户和各机构提供安全可靠交互的控制中心，还要包括大数据平台的开放性、模块化、灵活性和可扩展性等非功能需求。

接下来，针对上述 3 个总体技术要求，我们继承设计模式的架构之美，吸纳分布式技术精髓，实战主流微架构设计方案。

1.2 微服务引擎的可扩展性设计

微服务是系统架构的一种设计风格，它是把一个独立的大系统拆分成多个小服务，让这些小服务都在各自的进程中运行，服务之间通过安全的 Http Restful 接口进行协同通信。微服务的产生是为解决一个单体应用在庞大业务发展后导致的不可维护性，当开发团队在敏捷开发和部署中举步维艰时，最主要问题就是这个应用太复杂，以至于任何单个开发者都不可能独自承担。总结单体应用的主要存在问题：一是在不同模块发生资源冲突时，扩展非常困难；二是系统可靠性问题，因为所有模块都运行在一个进程中，任何一个模块发生类似内存泄漏的问题，将会有可能弄垮整个进程；三是系统升级时依赖包的版本冲突问题，不同中间件直接引用的依赖包都是需要版本支撑的，很容易导致相互冲突而不可控。

目前知名互联网公司都是通过采用微服务架构解决了上述问题。其思路不是开发一个巨大的单体式应用，而是将应用分解为小的微服务。一个微服务一般完成某个指定的任务或者功能，每一个微服务都有自己的业务逻辑和适配器。一些微服务还会发布 API 给其他微服务和应用客户端使用。每一个应用功能区都使用微服务完成，和以往的多个服务共享一个数据库不一样，微服务架构要求每个服务都有自己的数据库。总结一下，微服务发展多年以来，它的架构模式有诸多好处。具体说明如下：

（1）分解庞大单体应用来解决多个服务之间的相互依赖。在功能不变的情况下，应用被分解为多个可管理的分支或服务。每个服务都通过消息通信机制来发生交互。微服务架构模式提供了模块化和产品化的平台级解决方案。

（2）更适合敏捷开发和小团队协同作战。开发者可以自由选择开发技术并提供 API 服务。开发者不需要被迫使用某项目指定好的技术工具。

（3）实现了独立的部署。开发者不再需要协调其他服务部署对本服务的影响。这种改变可以加快部署速度，微服务架构模式使得持续化部署成为可能。

（4）让每个服务实现高可扩展。开发者根据用户规模来部署集群服务。

针对上述多个好处，我们从系统整体技术能力出发，提出物联网大数据平台的 8 个通用微服务的技术要求，包括大数据的高并发采集服务、灵活转发服务、高可扩展海量存储服务、高并发海量存储服务、高可靠海量存储服务、自定义迁移服务、基于机器学习的智能分析服务和基于 Spark 生态的实时计算服务，具体如下，如图 1-1 所示。

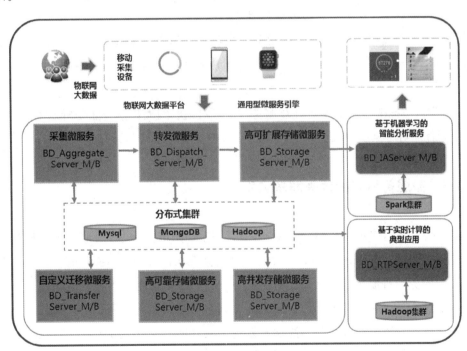

图 1-1　大数据实时计算服务引擎的模块化设计

（1）高并发采集服务：支持多种移动终端和物联网数据的可扩展接入，并具备大规模接入并发处理能力。能够兼容主流行业通用的可扩展协议和规范，并采用高可靠的集群或者负载均衡技术框架来解决。如引入 Mina 或者 Netty 技术框架后适配多种移动终端接入。标准化接入包括常用的字节流、文件、JSON 等数据格式属于主流的数据交换格式。

（2）灵活转发服务：按照分析应用需求，转发不同的数据类型和数据格式，交互方式之一是主流的消息中间件 MQ 或者 Kafka，保证高效的转发并转换数据给数据服务运营方。交互的方式之二是 Restful 方式，保证数据可以按照协议规范进行安全可靠的数据转发和传输。

（3）高可扩展海量存储服务：支持数据类型和数据表可扩展，对物联网大数据进行海量存储和计算，尤其适用于初创公司在百万级用户之内的大数据平台。

（4）高并发海量存储服务：支持数据类型和数据量的高速增长，对物联网大数据进行批处理，适合构建 PB 级数据量和千万级用户量的云平台。

（5）高可靠海量存储服务：支持物联网多源异构数据的统一高效和海量存储，并提供易于扩展的行业数据的离线计算和批处理架构，适合构建 ZB 级数据量和亿级用户量的分布式大平台。

（6）自定义迁移服务：支持对物联网大数据的整体迁移和同步，通过数据转换和数据迁移工具对不同数据类型和数据格式进行整体迁移，实现数据集的自定义生成。

（7）基于机器学习的智能分析服务：支持安全高效的机器学习算法，通过支持分布式分类、聚类、关联规则等算法，为用户和物联网机构提供个性化的智能分析服务。

（8）基于 Spark 生态的实时计算服务：支持对物联网大数据智能分析能力，通过企业级中间件服务框架提供安全可靠接口，实现数据实时统计和计算。

1.3 微服务引擎的优秀解决方案

接下来，我们构建一个微服务中心，可以把 8 个微服务进行互联互通，构建一个大数据服务平台。Spring Cloud 是我们主流的微架构，其中 Eureka 是 Spring Cloud Netflix 微服务一种服务注册套件，通常与 Spring Boot 搭建的微服务搭配使用，主要是作为微服务之间的消息中间件。Eureka 分为客户端与服务端两部分。服务端就是服务注册中心，主要用于微服务的注册和发现，其功能和 zookeeper 类似。Eureka 可以集群的方式运行，这样就充分保证了服务的高可用性。对于客户端而言，主要分为服务生产者和消费者。服务的生产者主要是向 Eureka 注册服务。服务的消费者主要是周期性查询注册中心，发现新服务并进行消费。所以，在启动微服务时，必须先启动 Eureka。在本书的第 2 章和第 3 章中，两个微服务之间的通信就采用了 Eureka 作为服务通信的桥梁。

（1）构建 Spring Cloud 微服务中心的代码工程架构，如图 1-2 所示。

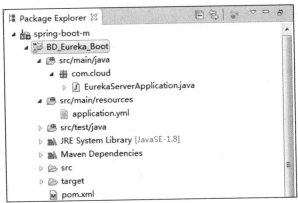

图 1-2　大数据微服务中心的代码架构设计

（2）添加工程所需要用到的依赖，在 pom.xml 文件中添加如下代码：

```xml
<?xml version="1.0" encoding="UTF-8"?>
<project xmlns="http://maven.apache.org/POM/4.0.0" xmlns:xsi="http://
    www.w3.org/2001/XMLSchema-instance"
xsi:schemaLocation="http://maven.apache.org/POM/4.0.0 http://maven.apache.
    org/xsd/maven-4.0.0.xsd">
<modelVersion>4.0.0</modelVersion>
<groupId>com.forezp</groupId>
<artifactId>eureka-server</artifactId>
<version>0.0.1-SNAPSHOT</version>
<packaging>jar</packaging>
<name>eureka-server</name>
<description>Demo project for Spring Boot</description>
<parent>
    <groupId>org.springframework.boot</groupId>
    <artifactId>spring-boot-starter-parent</artifactId>
    <version>1.5.2.RELEASE</version>
    <relativePath/><!-- lookup parent from repository -->
</parent>
<properties>
    <project.build.sourceEncoding>UTF-8</project.build.sourceEncoding>
    <project.reporting.outputEncoding>UTF-8</project.reporting.
        outputEncoding>
    <java.version>1.8</java.version>
</properties>
<dependencies>
    <dependency>
        <groupId>org.springframework.cloud</groupId>
        <artifactId>spring-cloud-starter-eureka-server</artifactId>
    </dependency>
```

```xml
        <dependency>
            <groupId>org.springframework.boot</groupId>
            <artifactId>spring-boot-starter-test</artifactId>
            <scope>test</scope>
        </dependency>
    </dependencies>
    <dependencyManagement>
        <dependencies>
            <dependency>
                <groupId>org.springframework.cloud</groupId>
                <artifactId>spring-cloud-dependencies</artifactId>
                <version>Dalston.RC1</version>
                <type>pom</type>
                <scope>import</scope>
            </dependency>
        </dependencies>
    </dependencyManagement>
    <build>
        <plugins>
            <plugin>
                <groupId>org.springframework.boot</groupId>
                <artifactId>spring-boot-maven-plugin</artifactId>
            </plugin>
        </plugins>
    </build>
    <repositories>
        <repository>
            <id>spring-milestones</id>
            <name>Spring Milestones</name>
            <url>https://repo.spring.io/milestone</url>
            <snapshots>
                <enabled>false</enabled>
            </snapshots>
        </repository>
    </repositories>
</project>
```

（3）编写服务核心配置文件 application.yml，主要用于指定该服务的路径、一些常用配置参数等。相关配置如下：

```yaml
server:
  port:8761
eureka:
  instance:
    hostname:localhost
  client:
```

```
#实例是否在eureka服务器上注册自己的信息以供其他服务发现,默认为true
    registerWithEureka:false
#此客户端是否获取eureka服务器注册表上的注册信息,默认为true
    fetchRegistry:false
    serviceUrl:
        defaultZone:http://${eureka.instance.hostname}:${server.port}/
        eureka/
```

（4）编写微服务启动类EurekaServerApplication,主要通过注解@EnableEurekaServer声明该服务是注册中心。代码实现如下：

```
package com.cloud;

import org.springframework.boot.SpringApplication;
import org.springframework.boot.autoconfigure.SpringBootApplication;
import org.springframework.cloud.netflix.eureka.server.EnableEurekaServer;

@EnableEurekaServer
@SpringBootApplication
public class EurekaServerApplication {
 public static void main(String[] args){
    SpringApplication.run(EurekaServerApplication.class,args);
 }
}
```

（5）启动服务，在启动类EurekaServerApplication通过main方法运行，在浏览器中输入http://${eureka.instance.hostname}:${server.port}/eureka/会看到注册中心基本信息以及服务信息，结果如图1-3所示。

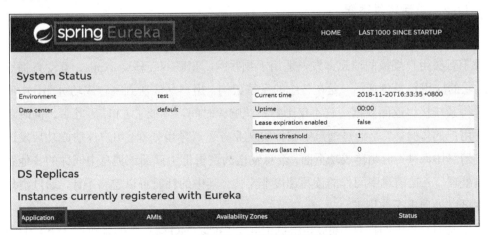

图1-3　微服务中心的服务注册信息

1.3.1　高并发采集微服务

面对用户量上千倍和数据量上万倍的增长速度，如何保证物联网大数据在比较快的时间内进入平台？应对用户量的增长，如何在规定的时间内完成采集？在硬件设备处理能力之外，让数据更快地汇聚到平台是核心需求。具体考虑如下。

（1）满足采集来自不同的厂家、移动设备类型、传输协议的行业数据的需求。我们在接口设计中完全可以针对不同设备和传输协议来设计，就是借用"分而治之"的用兵之道，"分而治之"就是把一个复杂的算法问题按一定的"分解"方法分为等价的规模较小的若干部分，然后逐个解决，分别找出各部分的解，把各部分的解组成整个问题的解，这种朴素的思想也完全适合于技术设计，软件的体系结构设计、模块化设计都是分而治之的具体表现。其中策略模式就是这个思想的集中体现。策略模式定义了一个公共接口，各种不同的算法以不同的方式实现这个接口。

（2）满足高并发需求。需要借助消息队列、缓存、分布式处理、集群、负载均衡等核心技术，实现数据的高可靠、高并发处理，有效降低端到端的数据传输时延，提升用户体验。借用"因粮于敌"的思想。"因粮于敌"的精髓是取之于敌，胜之于敌，以战养战，动态共存。我们常说的借用对手优势发展自己并整合资源就是这个思想的集中体现。正式商用的系统需要借助高性能中间件来并行处理数据，达到不丢包下的低延迟。我们采用商用的 Mina 负载均衡技术框架，可以支持多种设备和传输协议（HTTP、TCP、UDP）的数据接入，可以满足每秒上万并发数的数据接入需求。针对以上的核心需求分析和技术定位，我们可以借助第三方中间件和采用设计模式实现个性化业务，来解决接口的集中化、可扩展性、灵活性等问题，借助 Mina 的 Socket NIO 技术魅力，适配高并发的数据接口 IOFilterAdapter 进行反序列化编码，适配高并发的数据接口 IOHandlerAdapter 进行业务处理。

1.3.2 灵活转发微服务

灵活转发能力的总体设计中要考虑接口和消息中间件两种方式，其中消息中间件可支撑千万级用户规模的消息并发，适用于物联网、车联网、移动 Apps、互动直播等领域。它的应用场景包括：一是在传统的系统架构，用户从注册到跳转成功页面，中间需要等待系统接口返回数据。这不仅影响系统响应时间，降低了 CPU 吞吐量，同时还影响了用户的体验。二是通过消息中间件实现业务逻辑异步处理，用户注册成功后发送数据到消息中间件，再跳转成功页面，消息发送的逻辑再由订阅该消息中间件的其他系统负责处理。三是消息中间件的读写速度非常快，其中的耗时可以忽略不计。通过消息中间件可以处理更多的请求。

主流的消息中间件有 Kafka、RabbitMQ、RocketMQ 等，下面来对比一下它们的性能。Kafka 是开源的分布式发布-订阅消息系统，归属于 Apache 顶级项目，主要特点是基于 Pull 模式来处理消息消费，追求高吞吐量，主要用于日志收集和传输。自从 0.8 版本开始支持复制，不支持事务，对消息的重复、丢失、错误没有严格要求，适合产生大量数据的互联网服务的数据收集业务；RabbitMQ 是 Erlang 语言开发的开源消息队列系统，基于 AMQP 协议来实现。AMQP 的主要特征是面向消息、队列、路由（包括点对点和发布/订阅）、可靠性、安全。AMQP 协议用在企业系统内，对数据一致性、稳定性和可靠性要求很高的场景，对性能和吞吐量的要求还在其次。RocketMQ 是阿里开源的消息中间件，由 Java 语言开发，具有高吞吐量、高可用性、适合大规模分布式系统应用的特点。RocketMQ 的设计思想源于 Kafka，但并不是 Kafka 的一个 Copy，它对消息的可靠传输及事务性做了优化，目前在阿里集团被广泛应用于交易、充值、流计算、消息推送、日志流式处理、Binglog 分发等场景。结合上述服务优势对比，在第 3 章我们会使用最主流的 ActiveMQ 消息中间件来处理数据转发，在第 7 章我们采用分布式的 Kafka 实现数据转发。

1.3.3 高可扩展海量存储服务

高可扩展是大数据处理的核心需求之一。实际工作中，当用户量在 100 万以内，而且数据量在 TB 级别以内，常常可以选择用 MySQL 数据库，灵活、成熟和开源的 MySQL 数据库是初创公司的首选。我们考虑使用纵表实现系统灵活可扩展，让经常使用的数据放在一个数据表中，让灵活变化的字段实现字典表模式，让内容常发生变化的数据对象尽量采用 JSON 格式。著名的 OpenMRS 系统在 MySQL 数据库中实现了自定义表格，让医生可以实现灵活自定义表格，收集自己的临床试验数据，让用户每天可以记录自己的饮食信息。这样的设计就实现了应用场景的普适性。我们借鉴 OpenMRS 的核心思想来构建一个基于 MySQL 的小规模的物联网大数据模型。应用场景就是：一个患者到多个医院进行体检并记录了各个生理指标。我们根据应用场景来建立数据模型。患者表构建为 Patient 表，医院表构建为 Location 表，体检构建为 Encounter 表，测量构建为 Observation 表，体检类型描述构建为 Concept 表，采用 5 张表的多表关联实现了普适的可扩展数据模型，在第 3 章会详细阐述。

高可扩展的另外一个接口实现就是 Restful 架构。Restful 接口是安全开放平台的主流接口风格。一般的应用系统使用 Session 进行登录用户信息的存储和验证，而大数据平台的开放接口服务的资源请求则使用 Token 进行登录用户信息的验证。Session 主要用于保持会话信息，会在客户端保存一份 Cookie 来保持用户会话有效性，而 Token 则只用于登录用户的身份鉴权。所以在移动端使用 Token 会比使用 Session 更加简易并且有更高的安全性。Restful 架构遵循统一接口原则，统一接口包含了一组受限的预定义的操作，不论什么样的资源，都是通过使用相同的接口进行资源的访问。接口应该使用预先定义好的主流的、标准的 Get/Put/Delete/Post 操作等。

1.3.4 高并发海量存储服务

MongoDB 是适用于垂直行业应用的开源数据库，是我们高并发存储和查询的首选数据库。MongoDB 能够使企业业务更加具有扩展性，通过使用 MongoDB 来创建新的应用，能使团队提升开发效率。

我们具体分析一下关系模型和文档模型的区别。关系模型是按照数据对象存到各个相应的表里，使用时按照需求进行调取。举例来说，针对一个体检数据模型设计，在用户管理信息中包括用户名字、地址、联系方式等。按照第三范式，我们会把联系方式用单独的一个表来存储，并在显示用户信息时通过关联方式把需要的信息取回来。但是 MongoDB 的文档模式，存储单位是一个文档，可以支持数组和嵌套文档，这个文档就可以涵盖这个用户相关的所有个人信息，包括联系方式。关系型数据库的关联功能恰恰是它的发展瓶颈，尤其是用户数据达到 PB 级之后，性能和效率会急速下降。

我们采用 MongoDB 设计一个高效的文档数据存储模式。首先考虑内嵌，把同类型的数据放在一个内嵌文档中。内嵌文档和对象可以产生一一映射关系，如 Map<String, String>可以实现存储一个内嵌文档。如果是多表关联，可以在主表里存储一个 id 值，指向另一个表中的 id 值，通过把数据存放到两个集合里实现多表关联，目前在 MongoDB 4.0 之后版本开始支持多文档的事务处理。

我们采用 AngularJS 框架设计一个高并发调用系统。一提到数据调用就想到了 JQuery 框架，JQuery 框架的设计思想是在静态页面基础上进行 DOM 元素操作。目前最成熟的数据调用的主流框架之一是 AngularJS 框架，AngularJS 特别适合基于 CRUD 的 Web 应用系统。它简化了对 Web 开发者的经验要求，同时让 Web 本身变得功能更强。AngularJS 对 DOM 元素操作都是在 Directive 中实现的，而且一般情况下很少自己直接去写 DOM 操作代码，只要监听 Model，Model 发生变化后 View 也会发生变化。AngularJS 框架强调 UI 应该是用 Html 声明式的方式构建，数据和逻辑由框架提供的机制自动匹配绑定。AngularJS 有着诸多优势的设计思想，最为核心的是：数据理由、依赖注入、自动化双向数据绑定、语义化标签等。依赖注入思想实现了分层解耦，包括前后端分离和合理的模块化组织项目结构，让开发者更关注于每一个具体的逻辑本身，从而加快了开发速度，提升了系统的质量。双向绑定是它的精华所在，就是界面的操作能实时反映到数据，数据的变更能实时展现到界面，数据模型 Model 和视图 View 都是绑定在了内存映射 $Scope 上。

下面是我设计的 AngularJS 的项目框架，可以应用于所有业务系统，在第 4 章的体检报告可视化展示中会详细阐述。建立 MVC 的三层框架，先建立一个单页视图层 Main.html，然后创建一个模型层 Service.js，最后创建一个控制层 App.js，App.js 中包括多个模块的 JS 和 Html 文件，这样就构建了一个完整的 AngularJS MVC 框架，如图 1-4 所示。

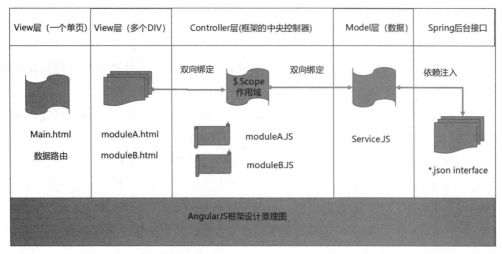

图 1-4　AngularJS MVC 框架设计

1.3.5　高可靠海量存储服务

高可靠海量存储是大数据处理的核心需求之一。实际工作中，常常需要实现多模态、不同时间颗粒度的行业数据的统一高效和海量存储，并提供易于扩展的离线计算和批处理架构。例如，引入 Hadoop 和 Spark 的大数据存储与计算方案。高可靠数据海量存储的总体设计中要吸纳主流的 Hadoop 架构，Hadoop 集群是一个能够让用户轻松架构和使用的分布式计算平台，用户可以在 Hadoop 上开发和运行处理海量数据的应用程序，如图 1-5 所示。

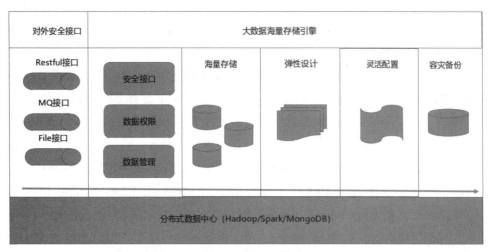

图 1-5　高可靠海量存储服务框架设计

高可靠海量存储服务主要有以下几个优点。

（1）高可靠性。Hadoop 按列存储和处理数据的能力值得信任。Hadoop 能够在节点之间动态地移动数据，并保证各个节点的动态平衡，因此处理速度非常快。

（2）高扩展性。Hadoop 是在可用的列簇中分配数据并完成计算任务的，这些集簇可以方便地扩展到数以千计的节点中。

（3）高容错性。Hadoop 能够自动保存数据的多个副本，并且能够自动将失败的任务重新分配。

数据海量存储的弹性设计中要吸纳主流的 HBase 架构。它是一个高可靠性、高性能、面向列、可伸缩的分布式存储系统，适用于结构化的存储，底层依赖于 Hadoop 的 HDFS，利用 HBase 技术可在廉价 PCServer 上搭建起大规模结构化存储集群。因此 HBase 被广泛使用在大数据存储的解决方案中。从应用场景分析，因为 HBase 存储的是松散的数据，如果应用程序中的数据表每一行的结构是有差别的，使用 HBase 最好，因为 HBase 的列可以动态增加，并且列为空就不存储数据，所以如果你需要经常追加字段，且大部分字段是 NULL 值的，可以考虑 HBase。因为 HBase 可以根据 Rowkey 提供高效的查询，所以你的数据都有着同一个主键 Rowkey。具体实现见第 6 章。

1.3.6 实时计算服务

实时计算的总体设计中要考虑 Spark 生态技术框架。Spark 使用 Scala 语言进行实现，Scala 语言是一种面向对象、函数式编程语言，能够像操作本地集合对象一样轻松地操作分布式数据集（Scala 提供一个称为 Actor 的并行模型）。Spark 具有运行速度快、易用性好、通用性强等特点，是在借鉴了 MapReduce 思想之上发展而来的，继承了其分布式并行计算的优点并改进了 MapReduce 明显的缺陷，具体优势分析如下：

（1）Spark 把中间数据放到内存中，迭代运算效率高。MapReduce 中计算结果需要落地，保存到磁盘上，这样势必会影响整体速度，而 Spark 支持 DAG 图的分布式并行计算的编程框架，减少了迭代过程中数据的落地，提高了处理效率。

（2）Spark 容错性高。Spark 引进了弹性分布式数据集 RDD（Resilient Distributed Dataset）的抽象，它是分布在一组节点中的只读对象集合，这些集合是弹性的，如果数据集一部分丢失，则可以根据"血统"对它们进行重建。另外，在 RDD 计算时可以通过 CheckPoint 来实现容错。

（3）Spark 具备通用性。在 Hadoop 提供了 Map 和 Reduce 两种操作基础上，Spark 又提供了很多数据集操作类型，大致分为 Transformations 和 Actions 两大类。Transformations 包括 Map、Filter、FlatMap、Sample、GroupByKey、ReduceByKey、Union、oin、Cogroup、MapValues、Sort 和 PartionBy 等多种操作类型，同时还提供 Count 和 Actions（包括 Collect、Reduce、Lookup 和 Save）等操作。

（4）强大的 Spark MLlib 机器学习库，旨在简化机器学习的工程实践工作，并方便扩展到更大规模。MLlib 由一些通用的学习算法和工具组成，包括分类、回归、聚类、协同过滤、降维等，同时还包括底层的优化原语和高层的管道 API。

1.3.7 基于机器学习的智能分析服务

智能分析服务的总体设计中要考虑 Spark MLlib 工具。当今主流的建模语言包括 R 语言、Weka、Mahout 和 Spark 等，下面来分析它们的基因和应用场景。

R 是一种数学语言，里面封装了大量的机器学习算法，但是它是单机的，不能够很好地处理海量的数据。Weka 和 R 语言类似，里面包含大量经过良好优化的机器学习和数据分析算法，可以处理与格式化、转换相关的各种任务，唯一的不足就是它对高内存要求的大数据处理遇到瓶颈。

Mahout 是 Hadoop 的一个机器学习库，有海量数据的并发处理能力，主要的编程模型是 MapReduce。而基于 MapReduce 的机器学习在反复迭代的过程中会产生大量的磁盘 I/O，即本次计算的结果要作为下一次迭代的输入，这个过程中只能把中间结果存储于磁盘，然后在下一次计算时重新读取，这对于迭代频发的算法显然是致命的性能瓶颈，所以计算效率很低。现在 Mahout 已经停止更新 MapReduce 算法，向 Spark 迁移。另外，Mahout 和 Spark MLlib 并不是竞争关系，Mahout 是 MLlib 的补充。

MLlib 是 Spark 对常用的机器学习算法的实现库，同时包括相关的测试和数据生成器。Spark 的设计就是为了支持一些迭代的工作，这正好符合很多机器学习算法的特点。在逻辑回归的运算场景下，Spark 比 Hadoop 快了 100 倍以上。Spark MLlib 立足于内存计算，适应于迭代式计算。而且 Spark 提供了一个基于海量数据的机器学习库，它提供了常用机器学习算法的分布式实现，工程师只需要有 Spark 基础并且了解机器学习算法的原理，以及方法相关参数的含义，就可以轻松地通过调用相应的 API 来实现基于海量数据的机器学习过程。具体实现见第 8 章。

1.3.8 自定义迁移服务

数据迁移能力的总体设计中要考虑 Sqoop 框架。Sqoop 是目前 Hadoop 和关系型数据库中的一个数据相互转移的主流工具，可以将一个关系型数据库（如 MySQL、Oracle、Postgres 等）中的数据导入 Hadoop 的 HDFS 中，也可以将 HDFS 的数据导入关系型数据库中。作为 ETL 工具，使用元数据模型来判断数据类型并在数据从数据源转移到 Hadoop 时确保类型安全的数据处理。Sqoop 框架可以进行大数据批量传输设计，能够分割数据集并创建 Hadoop 任务来处理每个区块。具体实现见第 9 章。

1.4 设计小结

本章是总体设计篇，由总体技术要求推导出微服务的核心需求和设计方法，最后针对 8 个微服务在各章节中提出了架构设计和实现方法。具体如下。

- 第2章：深入阐释"高并发采集微服务"的架构设计和技术实现。
- 第3章：着重论述"灵活转发微服务"的架构设计和技术实现。
- 第4章：详细介绍"高可扩展海量存储微服务"的架构设计和技术实现。
- 第5章：重点讲述"高并发海量存储微服务"的架构设计和技术实现。
- 第6章：重点说明"高可靠海量存储微服务"的架构设计和技术实现。
- 第7章：详细论述"实时计算微服务"的架构设计和技术实现。
- 第8章：主要介绍"智能分析微服务"的架构设计和技术实现。
- 第9章：深入论述"自定义迁移微服务"的架构设计和技术实现。

第 2 章

大数据高并发采集微服务引擎

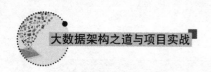

> 架构之道分享之二：孙子兵法的《作战篇》提出了降低战争消耗的两个主张：一是缩短战争持续时间，迅速赢得战争；二是就地解决粮草问题，减少长途运输耗费。映射到项目设计上，我们提升项目开发效率有两个方法：一是采用微服务设计和敏捷开发，迅速迭代和演进；二是采用国内主流成熟框架，避免重复开发带来的资源浪费。

本章学习目标

- ★ 掌握基于 Spring Boot 和 Spring MVC 高并发采集微服务的构建
- ★ 掌握 Mina 框架的工作原理和实战技巧
- ★ 掌握组合、迭代、策略、状态模式的工作原理和实战技巧
- ★ 掌握 C3P0 技术原理和高可靠入库方法

2.1 核心需求分析和优秀解决方案

为了满足物联网数据高并发接入需求的同时，要考虑各种设备接口规范性和可扩展性设计，还要保证后续新设备接入和旧设备的升级改造。性能方面要保证单机达到 3 000TPS 以上的吞吐量。为了达到如上的功能和性能需求，在框架选型方面，我们借助了负载均衡技术框架 Mina2.0 的高性能采集能力。在架构设计方面，自主设计业务树来加载不同设备和不同厂家的物联网数据，保证设备和数据的可扩展性接入能力。业务的可扩展性一般采用树或者图的数据结构来设计，满足可扩展性和高效遍历的需求。针对物联网数据的不同设备、不同数据类型、不同数据格式的业务特点，采用树的数据结构设计是最合理的。业务树构建要考虑不同类型设备的灵活接入，可以采用策略设计模式来解决扩展性。业务树构建要考虑不同类型数据包的灵活接入，可以采用状态设计模式来解决扩展性。业务树构建要考虑不同数据格式的解析，可以采用组合和迭代模式来遍历业务树。

2.2 服务引擎的技术架构设计

（1）大数据高并发采集服务包括 5 个核心模块，如图 2-1 所示，每一个模块要考虑可扩展性和高性能两个关键因素，具体说明如下。

① 核心模块一：通过 Spring MVC 和 Spring Boot 微服务构建采集服务框架，借助 Mina2.0 的负载均衡框架实现高并发处理能力。

② 核心模块二：阐述数据协议规范制定及数据包设计，采用业务树进行数据包设计。

③ 核心模块三：按照设备和数据类型进行业务树构建，采用组合和迭代模式设计业务树。

④ 核心模块四：按照设备的数据包状态进行解析数据包，采用策略和状态模式设计业务树。

⑤ 核心模块五：为了高效率入库，采用 C3P0 数据库连接池技术，保证数据高并发入库。

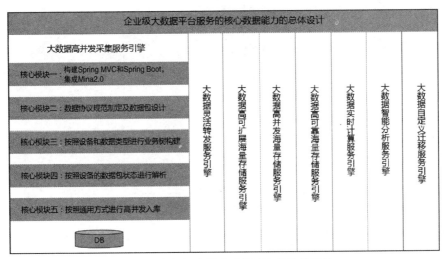

图 2-1　大数据高并发采集服务模块化设计

（2）构建 Spring MVC 版本的 BD_AggregateServer_Maven 服务工程框架，如图 2-2 所示。

图 2-2　大数据采集服务 Spring MVC 版本工程

（3）构建 Spring Boot 版本的 BD_AggregateSever_Boot 服务工程框架，如图 2-3 所示。

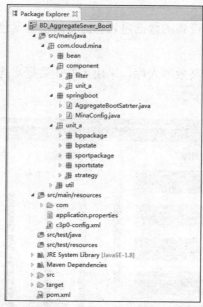

图 2-3　大数据采集服务 Spring Boot 版本工程

2.2.1　Maven 与 Eclipse 集成配置

Maven 项目作为时下最火的项目管理工具，广泛地应用在大部分的项目开发中。利用它可以快速方便地完成项目的构建、测试、打包、发布等功能。其核心配置文件 pom.xml 主要描述了开发者需要遵循的规则、组织和 licenses、项目的 url、项目的依赖性以及其他所有的项目相关因素。Maven 项目之间存在着 3 种关系：继承、依赖、聚合。相关概念总结如表 2-1 所示。

表 2-1　Maven 常用概念简介表

名　称	含　义　说　明
modelVersion	Maven 模块版本
groupId	组织名以及项目名称
artifactId	子模块名称
packaging	打包类型，可取值：jar、war、pom 等，这个配置用于 package 的 phase，具体可以参见 package 运行时启动的 plugin
scope	依赖项的适用范围： ● compile：默认值，适用于所有阶段，会随着项目一起发布 ● provided：类似 compile，期望 JDK、容器或使用者会提供这个依赖。如 servlet.jar ● runtime：只在运行时使用，如 JDBC 驱动、适用运行和测试阶段 ● test：只在测试时使用，用于编译和运行测试代码。不会随项目发布 ● system：类似 provided，需要显式提供包含依赖的 jar，Maven 不会在 Repository 中查找它
exclusions	排除项目中的依赖冲突时使用

Maven 也能与常见的开发工具 eclipse、idea 集成，下面是与 eclipse 集成相关配置。

（1）下载 Maven 插件，下载地址是 https://Maven.apache.org/，版本是 apache-Maven-3.5.2.zip。

（2）打开 eclipse 菜单栏 preperences 选项，在搜索框中输入 maven，选择 Installations 选项，单击 Add 按钮，选择已解压的 Maven 安装文件路径，单击 Apply 按钮，如图 2-4 和图 2-5 所示。

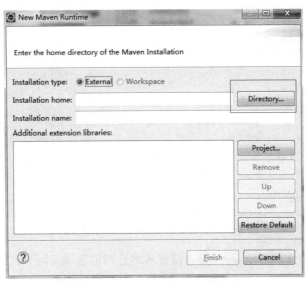

图 2-4　eclipse 安装 Maven 路径设置

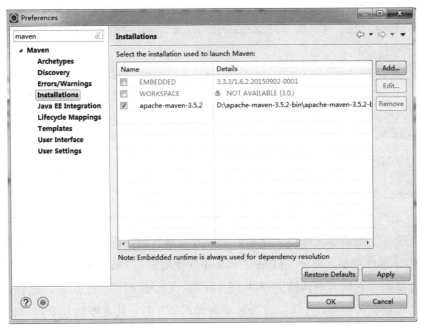

图 2-5　Maven 默认版本设置

（3）单击左栏的 User Settings 选项，选择 Maven 的核心配置文件 setting.xml 的路径，如图 2-6 所示。

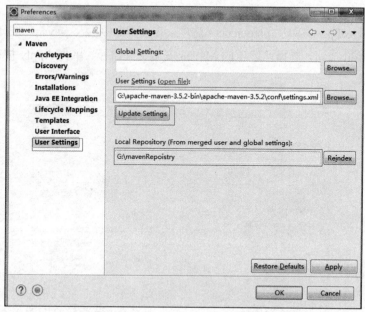

图 2-6　Maven 配置文件 setting.xml 设置

（4）修改配置文件 settings.xml，设置本地仓库位置以及阿里云镜像（可选配置，使用阿里云镜像相对比中央仓库镜像下载 jar 包依赖速度快），并单击 Update Settings 按钮。配置实现如下：

```
<localRepository>C:\Users\changyaobin\.m2\repository</localRepository>
<mirror>
    <id>nexus-aliyun</id>
    <mirrorOf>central</mirrorOf>
    <name>Nexus aliyun</name>
    <url>http://Maven.aliyun.com/nexus/content/groups/public</url>
</mirror>
```

2.2.2　Mina2.0 框架以及业务设计

1. Mina2.0 核心技术讲解

Mina 是一个主要对基于 TCP/IP、UDP/IP 协议栈的主流网络通信应用框架，被支付宝等商用产品使用多年，可以帮助我们快速开发高性能、高扩展性的网络通信应用。Mina 提供了事件驱动、异步（Mina 的异步 I/O 使用的是 JAVA NIO 作为底层支持）操作的编程模型。它同时对网络通信的 Server 端、Client 端进行了封装，这样，开发者只需要关心数据的接受、发送以及业务处理即可。下面是其重要的几个接口，如图 2-7 所示。

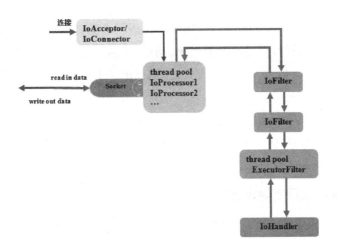

图 2-7 Mina 框架流程

（1）IoService：这个接口在一个线程上负责套接字的建立，拥有自己的 Selector，监听是否有连接被建立。这个接口是服务端 IoAcceptor、客户端 IoConnector 的抽象，提供 I/O 服务和管理 IoSession 的功能。IoAcceptor 进程用于监听客户端的连接，每监听一个端口建立一个线程，可以同时监听多个端口。IoConnector 进程用于与服务端建立连接，每连接一个服务端就建立一个线程。这两种线程都是通过线程池建立的，我们可以在构建对象时就指定线程池类型，默认的线程池类型为 newCachedThreadPool。

（2）IoProcessor：这个接口在另一个线程上，负责检查是否有数据在通道上读写和 IO 的处理，也就是说它也拥有自己的 Selector，这是与我们使用 JAVA NIO 编码时的一个不同之处。通常在 JAVA NIO 编码中，我们都是使用一个 Selector，也就是不区分 IoService 与 IoProcessor 两个功能接口。另外，IoProcessor 负责调用注册在 IoService 上的过滤器，并在过滤器链之后调用 IoHandler。对于一个 IoAcceptor 或 IoConnector 线程对应一个 IoProcessor 线程，这个 IoProcessor 线程从 IoProcessor 线程池中取出，IoProcessor 线程池的大小默认为机器的 CPU 核数+1。

（3）IoFilter：这个接口通常都是成组存在的，这些过滤器按照一定的顺序组成一条过滤器链，主要用于日志输出、黑名单过滤、数据的编码与解码等。其中数据的 encode 与 decode 是最为重要的，也是在使用 Mina 时最主要关注的地方。

（4）IoHandler：这个接口负责编写业务逻辑，也就是接收、发送数据的处理中心。

（5）服务端流程：

- 通过 SocketAcceptor 同客户端建立连接。
- 连接建立之后 I/O 的读写交给了 I/O Processor 线程，I/O Processor 是多线程的。
- 通过 I/O Processor 读取的数据按照先后顺序经过 IoFilterChain 里所有配置的 IoFilter，IoFilter 进行消息的过滤，格式的转换。

- IoFilter 链处理完毕后将数据交给 Handler 进行业务处理，完成了数据读取处理。
- 写入过程也是类似，只是刚好倒过来，通过 IoSession.write 写出数据，然后 Handler 进行写入的业务处理，处理完成后交给 IoFilterChain，进行消息过滤和协议的转换，最后通过 I/O Processor 将数据写出到 socket 通道。

2. Mina 的业务实现流程

首先架构业务解码器树，然后将创建自定义解码器类进行解码。解码完毕后，调用自定义 IoHandler 类完成业务数据处理。具体实现过程可以参照 2.3.4 节。

2.2.3 设备协议规范制定及数据包设计

数据包设计首先采用字节流方式，可以节省设备的存储成本。考虑到数据完整性和交互可靠性，采用 TCP 协议。通过消息请求和消息确认机制，实现设备和服务之间的信息交互。设计如下：

1. 设备发送到平台的登录包

（1）消息头 Heaher，占用 4 字节，例如 0xa8 0x4 0x00 0x01。

（2）消息长度 Length，占用 4 字节，表明此次数据包的长度，例如 0xa8 0x6 0x00 0x09。

（3）数据包类型 Type，占用 2 字节，声明此数据包的类型，例如 0x01 0x80。

（4）设备 id，占用 16 字节，声明此数据包设备来源 id，例如 0101100110010101。

（5）密码 password，占用 16 字节，声明用户的密码，例如 1234567821321242。

2. 平台回复设备登录数据包的 ACK

（1）消息头 Heaher，占用 4 字节，例如 0xa7 0xb8 0x00 0x01。

（2）消息长度 Length，占用 4 字节，例如 0xa8 0x6 0x00 0x09。

（3）类型 Type，占用 2 字节，例如 0x01 0x01。

（4）年份 Year，占用 2 字节。

（5）月份 Month，占用 1 字节。

（6）日 Day，占用 1 字节。

（7）小时 Hour，占用 1 字节。

（8）分钟 Minute，占用 1 字节。

（9）秒 Second，占用 1 字节。

3. 设备发送到平台的 1 号数据包

（1）消息头 Heaher，占用 4 字节，例如 0xa7 0xb8 0x00 0x01。

（2）消息长度 Length，占用 4 字节，例如 0xa8 0x6 0x00 0x09。

（3）类型 Type，占用 2 字节，例如 0x07 0x02。

（4）设备唯一标识 DeviceId，占用 21 字节，使用 devid 占用 5 字节作为前缀，后 16 字节作为设备 id。

（5）用户数据 UserData，类型为数组结构，其详细信息如下：

① 数据包类型，占用 1 字节，主要声明数据是设备自动上传还是用户手动上传，例如，1 代表自动上传，2 代表手动上传。

② 数据包生成时间，占用 3 个字节，按照字节顺序分别代表年、月、日。

③ 总步数，占用 4 字节。

④ 设备电池电量，表明设备剩余的电池电量，占用 1 字节。

⑤ 体重，表明用户的体重信息，默认单位是 kg，占用 1 字节。

⑥ 步幅，表明用户的步幅，默认单位是 cm，占用 1 字节。

⑦ 卡路里，表明用户智能终端运动所消耗的总卡路里，默认单位是 kcal，占用 4 字节。

⑧ 总步数，表明用户行走的总步数，占用 4 字节。

⑨ 智能终端运动总距离，默认单位是 m，占用 4 字节。

⑩ 智能终端运动等级 1，表明用户在该等级的时间，默认单位是 s，占用 2 字节。

⑪ 智能终端运动等级 2，表明用户在该等级的时间，默认单位是 s，占用 2 字节。

⑫ 智能终端运动等级 3，表明用户在该等级的时间，默认单位是 s，占用 2 字节。

⑬ 智能终端运动等级 4，表明用户在该等级的时间，默认单位是 s，占用 2 字节。

4. 平台回复设备 1 号数据包的 ACK

（1）消息头 Heaher，占用 4 字节，例如 0xa7 0xb8 0x00 0x01。

（2）消息长度 Length，占用 4 字节，例如 0xa8 0x6 0x00 0x09。

（3）类型 Type，占用 2 字节，例如 0x07 0x02。

（4）操作状态响应码 ACK，占用 1 字节，例如，0x0E 代表成功，0x0F 代表失败。

5. 设备发送到平台的 2 号数据包

（1）消息头 Heaher，占用 4 字节，例如 0xa7 0xb8 0x00 0x01。

（2）消息长度 Length，占用 4 字节，例如 0xa8 0x6 0x00 0x09。

（3）类型 Type，占用 2 字节，例如 0x07 0x02。

（4）设备唯一标识 DeviceId，占用 21 字节，使用 devid 占用 5 字节作为前缀，后 16 字节作为设备 id。

（5）用户智能终端运动数据 USRDATA，占用 114 字节，表示用户单个小时的数据，数据结构为数组形式，其详情如下：

① 年 Year，占用 2 字节。

② 保留字段 Reversed，占用 1 字节。

③ 月 Month，占用 1 字节。

④ 日 Day，占用 1 字节。

⑤ 小时 Hour，占用 1 字节。

⑥ 每 5 分钟的步数，共有 12 组数据，每组数据占用 2 字节。

⑦ 每 5 分钟的消耗的卡路里，共有 12 组数据，每组数据占用 2 字节。

⑧ 每 5 分钟的智能终端运动强度等级 2，共有 12 组数据，每组数据占用 1 字节。例如，0 代表不是该强度，1 代表是该强度。

⑨ 每 5 分钟的智能终端运动强度等级 3，共有 12 组数据，每组数据占用 1 字节。例如，0 代表不是该强度，1 代表是该强度。

⑩ 每 5 分钟的智能终端运动强度等级 4，共有 12 组数据，每组数据占用 1 字节。例如，0 代表不是该强度，1 代表是该强度。

⑪ 每 5 分钟步数、卡路里、智能终端运动等级的平方和，共有 12 组数据，每组数据占用 2 字节。

6. 平台回复设备 2 号数据包的 ACK

（1）消息头 Heaher，占用 4 字节，例如 0xa7 0xb8 0x00 0x01。

（2）消息长度 Length，占用 4 字节，例如 0xa8 0x6 0x00 0x09。

（3）类型 Type，占用 2 字节，例如 0x08 02。

（4）操作状态响应码 ACK，占用 1 字节，例如，0x0E 代表成功，0x0F 代表失败。

7. 设备发送到平台的 3 号数据包

（1）消息头 Heaher，占用 4 字节，例如 0xa7 0xb8 0x00 0x01。

（2）消息长度 Length，占用 4 字节，例如 0xa8 0x6 0x00 0x09。

（3）类型 Type，占用 2 字节，例如 0x07 0x02。

（4）设备唯一标识 DeviceId，占用 21 字节，使用 devid 占用 5 字节作为前缀，后 16 字节作为设备 id。

（5）用户智能终端运动数据 USRDATA，占用 114 字节，表示用户单个小时的数据，数据结构为数组形式，其与 1 号包唯一不同的是，1 号包传来的是总步数，而该数据包只传来的是有效的步数。

8. 平台回复设备 3 号数据包的 ACK

（1）消息头 Heaher，占用 4 字节，例如 0xa7 0xb8 0x00 0x01。

（2）消息长度 Length，占用 4 字节，例如 0xa8 0x6 0x00 0x09。

（3）类型 Type，占用 2 字节，例如 0x08 03。

（4）操作状态响应码 ACK，占用 1 字节，例如，0x0E 代表成功，0x0F 代表失败。

9. 智能终端运动数据包设计

针对如上协议规范，我们采用接口或者抽象类进行设计。对于面向对象编程来说，抽象是它的重要特征之一。Java 中可以通过接口和抽象类两种形式来体现 OOP 的抽象。这两者有太多相似的地方，又有太多不同的地方。

（1）含有抽象方法的类称为抽象类。抽象方法是一种特殊的方法：它只有声明，而没有具体的实现。抽象方法必须用 abstract 关键字进行修饰，抽象类必须在类前用 abstract 关键字修饰。因为抽象类中含有无具体实现的方法，所以不能用抽象类创建对象。抽象类包含抽象方法，但并不意味着抽象类中只能有抽象方法，它和普通类一样，同样可以拥有成员变量和普通的成员方法。注意，抽象类和普通类主要有如下三点区别。

① 抽象方法必须为 public 或者 protected（因为如果为 private，则不能被子类继承，子类便无法实现该方法），默认情况下为 public。

② 抽象类不能用来创建对象。

③ 如果一个类继承于一个抽象类，则子类必须实现父类的抽象方法。如果子类没有实现父类的抽象方法，则必须将子类也定义为 abstract 类。

（2）接口在软件工程中泛指供别人调用的方法或者函数，是对行为的抽象。在 Java 中，继承接口的形式如下：class ClassName implements Interface1, Interface2, [....]{.....}。如果一个非抽象类遵循了某个接口，就必须实现该接口中的所有方法，允许一个类遵循多个特定的接口。对于遵循某个接口的抽象类，可以不实现该接口中的抽象方法。具体区别如下：

① 抽象类可以提供成员方法的实现细节，但是接口中只能存在 public abstract 方法。

② 抽象类中的成员变量可以是各种类型的，而接口中的成员变量只能是 public static final 类型的。

③ 接口中不能含有静态代码块以及静态方法，而抽象类可以有静态代码块和静态方法。

④ 一个类只能继承一个抽象类，而一个类却可以实现多个接口。

⑤ 对于抽象类，如果需要添加新的方法，可以直接在抽象类中添加具体的实现，子类可以不进行变更；而对于接口则不行，如果接口进行了变更，则所有实现这个接口的类都必须进行相应的改动。

（3）数据包的抽象类和具体类的 UML 设计图如图 2-8 所示。

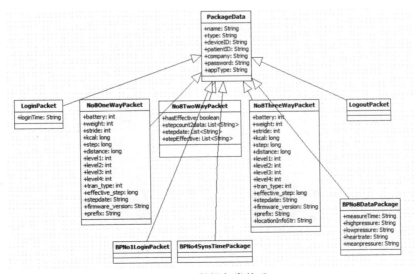

图 2-8　数据包类关系

2.2.4 按照设备和数据类型进行业务树构建

（1）迭代器（Iterator）模式，是提供一种方法访问一个容器（container）对象中各个元素，而又不需暴露该对象的内部细节。从定义可见，迭代器模式是为容器而生。因此，对容器对象的访问必然涉及遍历算法。可以一次性地将遍历方法塞到容器对象中去，或者根本不去提供什么遍历算法，让使用容器的人自己去实现。这两种情况好像都能够解决问题。然而在前一种情况，容器承受了过多的功能，它不仅要负责自己"容器"内的元素维护（添加、删除等），还要提供遍历自身的接口，而且由于遍历状态保存的问题，不能对同一个容器对象同时进行多次遍历。第二种方式虽然省事，却又将容器的内部细节暴露无遗。而迭代器模式的出现，很好地解决了上面两种情况的弊端。先来看下迭代器模式的角色定义。

① 迭代器角色（Iterator）：迭代器角色负责定义访问和遍历元素的接口。

② 具体迭代器角色（Concrete Iterator）：具体迭代器角色要实现迭代器接口，并要记录遍历中的当前位置。

③ 容器角色（Container）：容器角色负责提供创建具体迭代器角色的接口。

④ 具体容器角色（Concrete Container）：具体容器角色实现创建具体迭代器角色。

⑤ 迭代器模式的类图如图 2-9 所示。

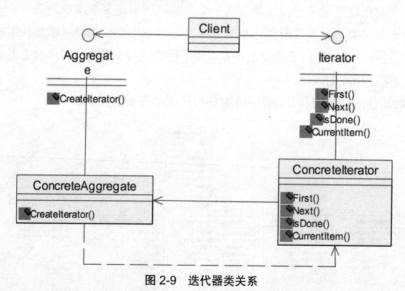

图 2-9　迭代器类关系

⑥ 迭代器的应用场景如下：

- 访问一个容器对象的内容而无须暴露它的内部表示。
- 支持对容器对象的多种遍历。
- 为遍历不同的容器结构提供一个统一的接口（多态迭代）。

（2）组合模式，是将对象以树形结构组织起来，以达成"部分和整体"的层次结构，使得客户端对单个对象和组合对象的使用具有一致性。从定义中可以得到使用组合模式的环境为：在设计中想表示对象的"部分和整体"层次结构；希望用户忽略组合对象与单个对象的不同，统一地使用组合结构中的所有对象。以下是组合模式的组成。

① 抽象构件角色（Component）：它为组合中的对象声明接口，也可以为共有接口实现默认行为。

② 树叶构件角色（Leaf）：在组合中表示叶节点对象——没有子节点，实现抽象构件角色声明的接口。

③ 树枝构件角色（Composite）：在组合中表示分支节点对象——有子节点，实现抽象构件角色声明的接口，并存储子部件。组合模式的类图表示如图 2-10 所示。

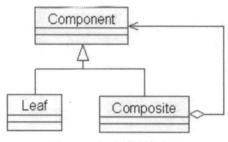

图 2-10　组合模式类关系

④ 组合模式的应用场景：

- 使客户端调用更简单，客户端可以一致地使用组合结构或其中单个对象，用户就不必关心自己处理的是单个对象还是整个组合结构，这就简化了客户端代码。
- 在组合体内加入对象部件时，客户端不必因为加入了新的对象部件而更改代码。这一点符合开闭原则的要求，对系统的二次开发和功能扩展很有利。
- 组合模式不容易限制组合中的构件。

⑤ 本章组合模式应用：建立业务树的根节点 MHRootComponent 作为所有物联网设备的根节点，根节点下面可以按照厂家和设备来进行设计业务树，图 2-11 是按照 Unit_A 厂家的智能终端运动设备 Sport 和血压设备 BP 来设计的，其中 Unit_A_SportComponent 设备可以包括 5 个数据包，Unit_A_BPComponent 设备包括 3 个数据包，如果有新的厂家或者设备接入时，可以在业务树上进行灵活添加，而不影响已经有的厂家或者设备节点，实现了真正的灵活可扩展性设计。基于组合模式的 UML 设计图如图 2-11 所示。

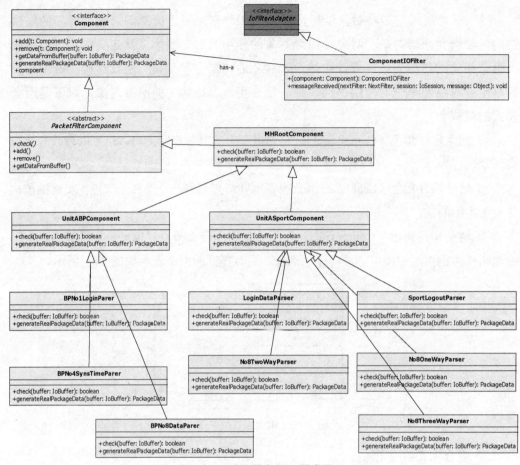

图 2-11　采集服务解码器类关系

2.2.5　按照设备的数据包状态进行解析

策略模式和状态模式都是很重要的数据业务处理思想，更是主流的分而治之的思想。

（1）策略模式（Strategy）：属于对象行为型设计模式，主要是定义一系列的算法或者业务处理方法，把这些或者业务处理方法一个个封装成拥有共同接口的单独的类，并且使它们之间可以互换。策略模式使这些算法在客户端调用它们时能够互不影响。这种模式会带来什么样的好处呢？ 它将算法的使用和算法本身分离，即将变化的具体算法封装了起来，降低了代码的耦合度，系统业务策略的更变仅需少量修改。策略模式由以下 3 个角色组成，如图 2-12 所示。

① 算法使用环境（Context）角色：算法被引用到这里和一些其他的与环境有关的操作一起来完成任务。

② 抽象策略（Strategy）角色：定义了所有具体策略角色通用接口。在 Java 中它通常由接口或者抽象类来实现。

③ 具体策略（Concrete Strategy）角色：实现了抽象策略角色定义的接口。

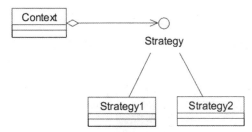

图 2-12　策略模式类关系

④ 策略模式的主要应用场景如下：
- 系统需要能够在几种算法中快速地切换。
- 系统中有一些类它们仅行为不同时，可以考虑采用策略模式来进行重构。
- 系统中存在多重条件选择语句时，可以考虑采用策略模式来重构。

（2）状态模式：允许一个对象在其内部状态改变时改变它的行为，这个对象看起来似乎修改了它的类。为了能够让程序根据不同的外部情况来做出不同的响应，最直接的方法就是在程序中将这些可能发生的外部情况全部考虑到，使用 if else 语句来进行代码响应选择。但是这种方法对于复杂一点的状态判断，就会显得杂乱无章，容易产生错误；而且增加一个新的状态将会带来大量的修改。这个时候"能够修改自身"的状态模式的引入也许是个不错的主意。状态模式可以有效地替换程序中大量的 if else 语句。主要实现过程是将不同条件下的行为封装在一个类里面，再给这些类配置一个统一的父类来约束它们。以下为状态模式的角色组成，如图 2-13 所示。

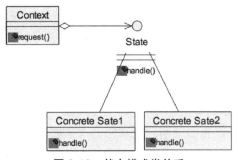

图 2-13　状态模式类关系

① 使用环境（Context）角色：客户程序是通过它来满足自己的需求。它定义了客户程序需要的接口；并且维护一个具体状态角色的实例，这个实例来决定当前的状态。

② 状态（State）角色：定义一个接口以封装与使用环境角色的一个特定状态相关的行为。

③ 具体状态（Concrete State）角色：实现状态角色定义的接口。

④ 状态模式的应用场景如下：
- 一个对象的行为取决于它的状态，并且它必须在运行时刻根据状态改变它的行为。

- 一个操作中含有庞大的多分支的条件语句，且这些分支依赖于该对象的状态。
- 状态模式和策略模式的最大区别是：各个具体的状态是有前后发生关系的，可能需要依次执行。

（3）本章的数据处理中，采用策略和状态模式，UML 设计图如图 2-14 所示。

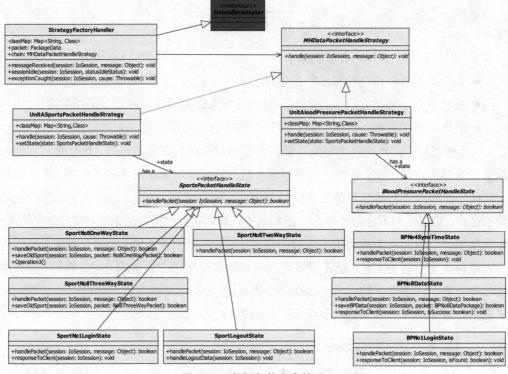

图 2-14　数据包状态类关系

2.2.6　按照通用方式进行高并发入库

JDBC 数据库连接使用 DriverManager 来获取，每次向数据库建立连接时都要将 Connection 加载到内存中，再验证用户名和密码（需要花费 0.05～1s 的时间），这样的方式将会消耗大量的资源和时间。究其原因，数据库的连接资源并没有得到很好的重复利用。若系统在线用户达到上千个，频繁地进行数据库连接操作将占用很多的系统资源，严重的甚至会造成服务器的崩溃。同时如果程序出现异常而未能关闭，将会导致数据库系统中的内存泄漏，最终将导致重启数据库。

为解决以上问题，可以采用数据库连接池技术。数据库连接池的基本思想：为数据库连接建立一个"缓冲池"。预先在缓冲池中放入一定数量的连接，当需要建立数据库连接时，只需从"缓冲池"中取出一个，使用完毕之后再放回去。数据库连接池负责分配、管理和释放数据库连接，它允许应用程序重复使用一个现有的数据库连接，而不是重新建立一个，连接池可以设置初始化连接数和最大连接数。数据库连接池技术有如下几个优点。

（1）资源可重用：由于数据库连接得以重用，避免了频繁创建，释放连接引起的大量性能开销。在减少系统消耗的基础上，但是也增加了系统运行环境的平稳性。

（2）系统可加速：数据库连接池在初始化过程中，往往已经创建了若干数据库连接置于连接池中备用。此时连接的初始化工作均已完成。对于业务请求处理而言，直接利用现有可用连接避免了数据库连接初始化和释放过程的时间开销，从而减少了系统的响应时间。

（3）连接数可共享：新的资源分配手段对于多应用共享同一数据库的系统而言，可在应用层通过数据库连接池的配置实现某一应用最大可用数据库连接数的限制，避免某一应用独占所有的数据库资源。

（4）连接可管理：避免数据库连接泄露在较为完善的数据库连接池实现中，可根据预先的占用超时设定，强制回收被占用连接，从而避免了常规数据库连接操作中可能出现的资源泄露。

本章所采用的是 C3P0 连接池。C3P0 是目前最成熟的数据库连接池技术之一，具体内容将在后面进行介绍。

2.3 核心技术讲解及模块化设计

2.3.1 Spring Maven Web 服务构建

（1）Web 服务的入口 web.xml 文件是中央控制器，主要实现 spring 和日志的配置。配置实现如下：

```xml
<?xml version="1.0" encoding="UTF-8"?>
<web-app xmlns:xsi="http://www.w3.org/2001/XMLSchema-instance" xmlns=
    "http://java.sun.com/xml/ns/javaee" xsi:schemaLocation="http://java.sun.
    com/xml/ns/javaee http://java.sun.com/xml/ns/javaee/web-app_2_5.xsd" id=
    "WebApp_ID" version="2.5">
<display-name>AggregateServer</display-name>
<welcome-file-list>
    <welcome-file>index.html</welcome-file>
    <welcome-file>index.htm</welcome-file>
    <welcome-file>index.jsp</welcome-file>
    <welcome-file>default.html</welcome-file>
    <welcome-file>default.htm</welcome-file>
    <welcome-file>default.jsp</welcome-file>
</welcome-file-list>
<context-param>
    <param-name>log4jConfigLocation</param-name>
```

```xml
    <param-value>classpath:log4j.properties</param-value>
</context-param>
<context-param>
    <param-name>log4jRefreshInterval</param-name>
    <param-value>60000</param-value>
</context-param>
<context-param>
    <param-name>contextConfigLocation</param-name>
    <param-value>WEB-INF/applicationContext.xml</param-value>
</context-param>
<filter>
    <filter-name>CharacterEncodingFilter</filter-name>
    <filter-class>org.springframework.web.filter.CharacterEncodingFilter
        </filter-class>
    <init-param>
      <param-name>encoding</param-name>
      <param-value>UTF-8</param-value>
    </init-param>
    <init-param>
      <param-name>forceEncoding</param-name>
      <param-value>true</param-value>
    </init-param>
</filter>
<filter-mapping>
    <filter-name>CharacterEncodingFilter</filter-name>
    <url-pattern>/*</url-pattern>
</filter-mapping>
<listener>
<listener-class>org.springframework.web.context.ContextLoaderListener
    </listener-class>
</listener>
<listener>
    <listener-class>org.springframework.web.util.Log4jConfigListener
        </listener-class>
</listener>
<welcome-file-list>
    <welcome-file>main.html</welcome-file>
</welcome-file-list>
<session-config>
    <session-timeout>720</session-timeout>
 </session-config>
</web-app>
```

（2）Spring 核心配置 application.xml 文件，主要用于数据库配置文件的读取、Spring 集成 mina 配置文件的加载。配置实现如下：

```xml
<?xml version="1.0" encoding="UTF-8"?>
<beans xmlns="http://www.springframework.org/schema/beans"
   xmlns:xsi="http://www.w3.org/2001/XMLSchema-instance" xmlns:aop= "http://
      www.springframework.org/schema/aop"
   xmlns:tx="http://www.springframework.org/schema/tx" xmlns:jms="http://
      www.springframework.org/schema/jms"
   xmlns:context="http://www.springframework.org/schema/context"
   xsi:schemaLocation="http://www.springframework.org/schema/beans http://
      www.springframework.org/schema/beans/spring-beans-3.2.xsd
         http://www.springframework.org/schema/context http://www.
            spring framework.org/schema/context/spring-context-3.2.xsd
         http://www.springframework.org/schema/tx
         http://www.springframework.org/schema/tx/spring-tx-3.2.xsd
         http://www.springframework.org/schema/aop
         http://www.springframework.org/schema/aop/spring-aop-3.2.xsd
         http://www.springframework.org/schema/jms
         http://www.springframework.org/schema/jms/spring-jms-3.2.xsd">
   <!-- 注解扫描器 -->
   <!-- 数据库配置文件读取 -->
   <bean
      class="org.springframework.beans.factory.config.
         PropertyPlaceholderConfigurer">
      <property name="locations">
         <list>
            <value>classpath:com/Config/SysConf.properties</value>
         </list>
      </property>
   </bean>
    <import resource="minaContext.xml"/>
</beans>
```

（3）配置 Spring 集成 Mina 配置文件 minaContext.xml，主要实现 mina 的 ioAcceptor、IoSession、IoFilter、IoHnadler 以及自定义解码器组装等相关配置。配置实现如下：

```xml
<?xml version="1.0" encoding="UTF-8"?>
<beans xmlns="http://www.springframework.org/schema/beans"
   xmlns:xsi="http://www.w3.org/2001/XMLSchema-instance" xmlns:context=
      "http://www.springframework.org/schema/context"
   xsi:schemaLocation="http://www.springframework.org/schema/beans http://
      www.springframework.org/schema/beans/spring-beans.xsd
         http://www.springframework.org/schema/context http://www.
            spring framework.org/schema/context/spring-context.xsd">
   <!-- 处理逻辑 -->
   <bean id="handler" class="com.cloud.mina.unit_a.strategy.
      StrategyFactroy Handler" />
   <bean id="unitASportsComponent"
```

```xml
        class="com.cloud.mina.component.filter.UnitASportComponent">
        <property name="list">
            <list>
                <bean class="com.cloud.mina.component.unit_a.sport.
                    SportLo ginParser" />
                <bean class="com.cloud.mina.component.unit_a.sport.
                    No8One WayParser" />
                <bean class="com.cloud.mina.component.unit_a.sport.
                    No8Two WayParser" />
                <bean class="com.cloud.mina.component.unit_a.sport.
                    No8Three WayParser" />
                <bean class="com.cloud.mina.component.unit_a.sport.
                    Sport LogoutParser" />
            </list>
        </property>
</bean>
<bean id="unitABPComponent" class="com.cloud.mina.component.filter.
    Unit ABPComponent">
      <property name="list">
          <list>
              <bean class="com.cloud.mina.component.unit_a.bp.
                  BPNo1Login Parer" />
              <bean class="com.cloud.mina.component.unit_a.bp.
                  BPNo4Syns TimeParer" />
              <bean class="com.cloud.mina.component.unit_a.bp.
                  BPNo8Data Parer" />
          </list>
      </property>
</bean>
<!-- 数据包解码器-->
<bean id="codec" class="com.cloud.mina.component.filter.
    ComponentIOFilter">
     <constructor-arg index="0">
         <bean class="com.cloud.mina.component.filter.MHRootComponent">
             <property name="list">
                 <list>
                     <ref bean="unitASportsComponent"></ref>
                     <ref bean="unitABPComponent"></ref>
                 </list>
             </property>
         </bean>
     </constructor-arg>
</bean>
<!-- 多线程处理过滤器,为后面的操作开启多线程,一般放在编解码过滤器之后,开始业务逻辑处理 -->
```

```xml
<bean id="executors" class="org.apache.mina.filter.executor.
    ExecutorFilter" />
<!-- Mina自带日志过滤器,默认级别为debug -->
<bean id="loggerFilter" class="org.apache.mina.filter.logging.
    LoggingFilter">
    <property name="messageReceivedLogLevel" ref="info"></property>
    <property name="exceptionCaughtLogLevel" ref="info"></property>
</bean>
<!-- 枚举类型,依赖注入,需要先通过此类进行类型转换 -->
<bean id="info"
    class="org.springframework.beans.factory.config.
        FieldRetrievingFactoryBean">
    <property name="staticField" value="org.apache.mina.filter.logging.
        LogLevel.INFO" />
</bean>
<bean id="filterChainBuilder"
    class="org.apache.mina.core.filterchain.DefaultIoFilterChainBuilder">
    <property name="filters">
        <map>
            <entry key="codec" value-ref="codec" />
            <entry key="logger" value-ref="loggerFilter" />
            <entry key="executors" value-ref="executors" />
        </map>
    </property>
</bean>
<bean id="defaultLocalAddress" class="java.net.InetSocketAddress">
    <constructor-arg index="0" value="${tcpPort}"></constructor-arg>
</bean>
<!-- session config -->
<bean id="sessionConfig" factory-bean="ioAcceptor" factory-method=
    "getSessionConfig">
    <property name="readerIdleTime" value="40" />
    <property name="minReadBufferSize" value="512" />
    <property name="maxReadBufferSize" value="10240" />
    <!--<property name="readBufferSize" value="20480"/> --><!--<property
        name="receiveBufferSize" value="5000"/> -->
</bean>
<bean id="ioAcceptor" class="org.apache.mina.transport.socket.nio.
    NioSocketAcceptor"
    init-method="bind" destroy-method="unbind">
    <!-- 默认启用的线程个数是CPU的核数+1,-->
    <constructor-arg index="0" value="10"></constructor-arg>
    <property name="defaultLocalAddress" ref="defaultLocalAddress" />
    <property name="handler" ref="handler" />
    <property name="filterChainBuilder" ref="filterChainBuilder" />
```

```
        </bean>
</beans>
```

（4）Maven 项目核心配置文件 pom.xml，主要描述项目相关信息以及指定该项目中用到的所有相关 jar 包。配置实现如下：

```xml
<project xmlns="http://Maven.apache.org/POM/4.0.0" xmlns:xsi="http://
    www.w3.org/2001/XMLSchema-instance"
    xsi:schemaLocation="http://Maven.apache.org/POM/4.0.0 http://Maven.
    apache.org/xsd/Maven-4.0.0.xsd">
<modelVersion>4.0.0</modelVersion>
<groupId>com.cloud.mina</groupId>
<artifactId>BD_AggregateServer_Maven</artifactId>
<version>0.0.1-SNAPSHOT</version>
<packaging>war</packaging>
<properties>
    <spring.version>4.3.7.RELEASE</spring.version>
</properties>
<build>
    <plugins>
        <plugin>
            <groupId>org.apache.Maven.plugins</groupId>
            <artifactId>Maven-compiler-plugin</artifactId>
            <version>3.3</version>
            <configuration>
                <source>1.7</source>
                <target>1.7</target>
            </configuration>
        </plugin>
    </plugins>
</build>
<dependencies>
    <!-- spring 常用配置 -->
    <dependency>
        <groupId>org.springframework</groupId>
        <artifactId>spring-core</artifactId>
        <version>${spring.version}</version>
    </dependency>

    <dependency>
        <groupId>org.springframework</groupId>
        <artifactId>spring-beans</artifactId>
        <version>${spring.version}</version>
    </dependency>

    <dependency>
```

```xml
        <groupId>org.springframework</groupId>
        <artifactId>spring-context</artifactId>
        <version>${spring.version}</version>
</dependency>

<dependency>
        <groupId>org.springframework</groupId>
        <artifactId>spring-web</artifactId>
        <version>${spring.version}</version>
</dependency>
<!-- mina 配置 -->
<dependency>
        <groupId>org.apache.mina</groupId>
        <artifactId>mina-core</artifactId>
        <version>2.0.4</version>
</dependency>
<!-- json 配置 -->
<dependency>
        <groupId>net.sf.json-lib</groupId>
        <artifactId>json-lib</artifactId>
        <version>2.3</version>
        <classifier>jdk15</classifier>
</dependency>
<dependency>
        <groupId>org.apache.commons</groupId>
        <artifactId>commons-lang3</artifactId>
        <version>3.0.1</version>
</dependency>
<!-- 添加数据库驱动 -->
<dependency>
        <groupId>mysql</groupId>
        <artifactId>mysql-connector-java</artifactId>
        <version>5.1.30</version>
</dependency>
<!-- c3p0 -->
<dependency>
        <groupId>c3p0</groupId>
        <artifactId>c3p0</artifactId>
        <version>0.9.1.2</version>
</dependency>
<!-- dbcp -->
<dependency>
        <groupId>commons-dbcp</groupId>
        <artifactId>commons-dbcp</artifactId>
        <version>1.3</version>
```

```xml
        </dependency>
        <dependency>
            <groupId>org.slf4j</groupId>
            <artifactId>slf4j-log4j12</artifactId>
            <version>1.6.6</version>
        </dependency>
        <dependency>
            <groupId>commons-httpclient</groupId>
            <artifactId>commons-httpclient</artifactId>
            <version>3.1</version>
        </dependency>

        <!--https://mvnrepository.com/artifact/commons-logging/commons-
           logging -->
        <!-- https://mvnrepository.com -->
        <dependency>
            <groupId>commons-logging</groupId>
            <artifactId>commons-logging</artifactId>
            <version>1.1.3</version>
        </dependency>
        <dependency>
            <groupId>log4j</groupId>
            <artifactId>log4j</artifactId>
            <version>1.2.17</version>
        </dependency>
    </dependencies>
</project>
```

2.3.2 Spring Boot 微服务构建

在大数据和互联网高速发展时期，平台系统如何满足需求变化和用户增长快的通用需求？从系统架构设计的角度来说，构建灵活、易扩展的系统来应对日新月异的需求变化；从系统质量特性的角度来说，构建可伸缩性、高可用性系统才能满足用户快速增长的需求。微架构通过组件化和服务化的设计思想，可以解决独立部署和快速迭代开发的变化需求。Spring Boot 是 Java 领域最优秀的微服务架构代表，就是基于 Spring 开发，助力开发者快速、敏捷地开发新一代基于 Spring 框架的应用程序。也就是说，它不是用来替代 Spring 的解决方案，而是和 Spring 框架紧密结合的，同时集成了大量的第三方库配置（如 Redis、MongoDB、Jpa、RabbitMQ、Quartz、Mina 等），Spring Boot 集成第三方库可以"插拔式"方式使用，让开发者减少配置和版本兼容性考虑，专注于业务逻辑设计和开发。下面开始 Spring Boot 的微服务构建，具体步骤如下。

（1）配置 Spring Boot 的 pom.xml 文件，用于描述该项目 Maven 信息以及项目中用到的 jar 包依赖，对比 Spring 版本会发现，Spring boot 相关依赖较少，其实不然。上面有提到，Spring Boot 是基于 Spring 的，故想要使用 Spring Boot，则需要先引入 spring。但是由于是 Maven 项目，我们只需要指定 Spring Boot 的版本，则 Maven 会自动下载其对应的 Spring 依赖包，这也就是 Maven 的一大特色。配置实现如下：

```xml
<?xml version="1.0" encoding="UTF-8"?>
<project xmlns="http://Maven.apache.org/POM/4.0.0" xmlns:xsi="http://
    www.w3.org/2001/XMLSchema-instance"
xsi:schemaLocation="http://Maven.apache.org/POM/4.0.0 http://Maven.apache.
    org/xsd/Maven-4.0.0.xsd">
<modelVersion>4.0.0</modelVersion>
<groupId>com.cloud.bigdata</groupId>
<artifactId>BD_AggregateSever_B</artifactId>
<version>0.0.1-SNAPSHOT</version>
<packaging>jar</packaging>

<name>service-hi</name>
<description>Demo project for Spring Boot</description>
<parent>
    <groupId>org.springframework.boot</groupId>
    <artifactId>spring-boot-starter-parent</artifactId>
    <version>1.5.2.RELEASE</version>
    <relativePath/><!-- lookup parent from repository -->
</parent>
<properties>
    <project.build.sourceEncoding>UTF-8</project.build.sourceEncoding>
    <project.reporting.outputEncoding>UTF-8</project.reporting.
       output Encoding>
    <java.version>1.8</java.version>
</properties>
<dependencies>
    <dependency>
        <groupId>org.springframework.cloud</groupId>
        <artifactId>spring-cloud-starter-eureka</artifactId>
    </dependency>
    <dependency>
        <groupId>org.springframework.cloud</groupId>
        <artifactId>spring-cloud-starter-ribbon</artifactId>
    </dependency>
    <dependency>
        <groupId>org.springframework.boot</groupId>
        <artifactId>spring-boot-starter-web</artifactId>
    </dependency>
    <dependency>
        <groupId>org.springframework.boot</groupId>
        <artifactId>spring-boot-starter-test</artifactId>
```

```xml
        <scope>test</scope>
    </dependency>
    <!-- mina 配置 -->
    <dependency>
        <groupId>org.apache.mina</groupId>
        <artifactId>mina-core</artifactId>
        <version>2.0.4</version>
    </dependency>
    <!-- json 配置 -->
    <dependency>
        <groupId>net.sf.json-lib</groupId>
        <artifactId>json-lib</artifactId>
        <version>2.3</version>
        <classifier>jdk15</classifier>
    </dependency>
    <dependency>
        <groupId>org.apache.commons</groupId>
        <artifactId>commons-lang3</artifactId>
        <version>3.0.1</version>
    </dependency>
    <dependency>
        <groupId>commons-httpclient</groupId>
        <artifactId>commons-httpclient</artifactId>
        <version>3.1</version>
    </dependency>
    <!-- c3p0 -->
    <dependency>
        <groupId>c3p0</groupId>
        <artifactId>c3p0</artifactId>
        <version>0.9.1.2</version>
    </dependency>
    <dependency>
        <groupId>org.apache.httpcomponents</groupId>
        <artifactId>httpclient</artifactId>
        <version>4.5.5</version>
    </dependency>
    <!-- poi -->
    <dependency>
        <groupId>org.apache.poi</groupId>
        <artifactId>poi</artifactId>
        <version>3.7</version>
    </dependency>
    <!-- 添加数据库驱动 -->
    <dependency>
        <groupId>mysql</groupId>
```

```xml
            <artifactId>mysql-connector-java</artifactId>
        </dependency>
        <dependency>
            <groupId>org.springframework.boot</groupId>
            <artifactId>spring-boot-configuration-processor</artifactId>
            <optional>true</optional>
        </dependency>
</dependencies>
<dependencyManagement>
    <dependencies>
        <dependency>
            <groupId>org.springframework.cloud</groupId>
            <artifactId>spring-cloud-dependencies</artifactId>
            <version>Dalston.RC1</version>
            <type>pom</type>
            <scope>import</scope>
        </dependency>
    </dependencies>
</dependencyManagement>
<build>
    <plugins>
        <plugin>
            <groupId>org.springframework.boot</groupId>
            <artifactId>spring-boot-Maven-plugin</artifactId>
        </plugin>
    </plugins>
</build>
<repositories>
    <repository>
        <id>spring-milestones</id>
        <name>Spring Milestones</name>
        <url>https://repo.spring.io/milestone</url>
        <snapshots>
            <enabled>false</enabled>
        </snapshots>
    </repository>
</repositories>
</project>
```

（2）在项目编译路径 resources 文件夹下配置 Spring Boot 的核心文件 application.properties，该文件主要用于指定服务的端口、访问路径、mina 的参数配置以及 eureka 注册中心相关配置。配置实现如下：

```
server.port:8086
#server.context-path:/boot_aggregate
```

```
spring.application.name=boot-aggregate
#springCloud注册中心服务地址
eureka.client.serviceUrl.defaultZone=http://localhost:8761/eureka/
#restTemplate饿汉式加载服务
ribbon.eager-load.enabled=true
ribbon.eager-load.clients=boot-dispatch
#mina的相关配置
mina.ip=127.0.0.1
mina.port=8888
mina.readerIdleTime=600
mina.minReadBufferSize=512
mina.maxReadBufferSize=102400
```

（3）Spring Boot 的启动类 AggregateBootSatrter（@Spring BootApplication 注解声明该类为 Spring Boot 的入口），主要实现 Spring 容器的初始化以及服务器的开启。代码实现如下：

```java
package com.cloud.mina.Spring Boot;

import org.springframework.boot.SpringApplication;
import org.springframework.boot.autoconfigure.Spring BootApplication;
import org.springframework.cloud.client.discovery.EnableDiscoveryClient;
import org.springframework.context.annotation.ComponentScan;

/**
 * Spring Boot 的启动器
 *
 * @author changyaobin
 *
 */
@Spring BootApplication//spring boot 启动类
@ComponentScan(basePackages = { "com.cloud.mina" })//扫描该包下的所有注解,
                                        //初始化到spring容器中
@EnableDiscoveryClient
public class AggregateBootSatrter {
  public static void main(String[] args){
      SpringApplication.run(AggregateBootSatrter.class,args);
  }
}
```

（4）mina 的核心配置类 MinaConfig，主要实现自定义解码器的组装、mina 以及 restTemplate（微服务之间相互调用）配置，用于替代项目 Spring 版本的 minaContext.xml 文件。代码实现如下：

```java
package com.cloud.mina.Spring Boot;

import java.io.IOException;
import java.net.InetSocketAddress;
import javax.annotation.Resource;
import org.apache.mina.core.filterchain.IoFilter;
import org.apache.mina.core.service.IoAcceptor;
import org.apache.mina.core.session.IdleStatus;
import org.apache.mina.core.session.IoSessionConfig;
import org.apache.mina.filter.executor.ExecutorFilter;
import org.apache.mina.filter.logging.LoggingFilter;
import org.apache.mina.transport.socket.nio.NioSocketAcceptor;
import org.springframework.boot.context.properties.ConfigurationProperties;
import org.springframework.cloud.client.loadbalancer.LoadBalanced;
import org.springframework.context.annotation.Bean;
import org.springframework.context.annotation.Configuration;
import org.springframework.web.client.RestTemplate;
import com.cloud.mina.component.filter.ComponentIOFilter;
import com.cloud.mina.component.filter.MHRootComponent;
import com.cloud.mina.component.filter.UnitASportComponent;
import com.cloud.mina.component.unit_a.sport.No8OneWayParser;
import com.cloud.mina.component.unit_a.sport.No8ThreeWayParser;
import com.cloud.mina.component.unit_a.sport.No8TwoWayParser;
import com.cloud.mina.component.unit_a.sport.SportLoginParser;
import com.cloud.mina.component.unit_a.sport.SportLogoutParser;
import com.cloud.mina.unit_a.strategy.StrategyFactroyHandler;
import com.cloud.mina.util.Logger;

/**
 * mina 的核心配置文件
 *
 * @author changyaobin
 *
 */
@SuppressWarnings(value = "all")
@Configuration
@ConfigurationProperties(prefix = "mina")
public class MinaConfig {
    private String ip;
    private int port;
    private int readerIdleTime;
    private int minReadBufferSize;
    private int maxReadBufferSize;
    // 智能终端运动的底层解析包
```

```java
@Resource(name = "sportLoginParser")
private SportLoginParser sportLoginParser;
@Resource(name = "no8OneWayParser")
private No8OneWayParser no8OneWayParser;
@Resource(name = "no8TwoWayParser")
private No8TwoWayParser no8TwoWayParser;
@Resource(name = "no8ThreeWayParser")
private No8ThreeWayParser no8ThreeWayParser;
@Resource(name = "sportLogoutParser")
private SportLogoutParser sportLogoutParser;

/**
 * 设置mina的ioHandler自定义处理类
 *
 * @return
 */
@Bean
public StrategyFactroyHandler getIoHandler(){
    StrategyFactroyHandler strategyFactroyHandler = new
        StrategyFactroy Handler();
    strategyFactroyHandler.setRestTemplate(restTemplate());
    return strategyFactroyHandler;
}
/**
 * Mina的IoAccptor设置
 *
 * @return
 * @throws IOException
 */
@Bean
public IoAcceptor getIoAccptor()throws IOException {
    IoAcceptor acceptor = new NioSocketAcceptor();
    IoSessionConfig sessionConfig = acceptor.getSessionConfig();
    sessionConfig.setIdleTime(IdleStatus.READER_IDLE,readerIdleTime);
    sessionConfig.setMinReadBufferSize(minReadBufferSize);
    sessionConfig.setMaxReadBufferSize(maxReadBufferSize);
    acceptor.setDefaultLocalAddress(getInetAddress());
    acceptor.getFilterChain().addLast("codec",getIOFilter());
    acceptor.getFilterChain().addLast("logger",getLogFilter());
    acceptor.getFilterChain().addLast("executors",getExecutorFilter());
    acceptor.setHandler(getIoHandler());
    acceptor.bind();
    Logger.writeLog("监听端口" + port + "....");
    return acceptor;
}
```

```java
/**
 * 业务树解码器的根类(注册到spring容器)
 *
 * @author changyaobin
 *
 */
@Bean
public MHRootComponent getMHRootComponent(){
    MHRootComponent mHRootComponent = new MHRootComponent();
    mHRootComponent.add(getUnitASportComponent());
    return mHRootComponent;
}
/**
 * unitA 智能终端运动解码器组装
 *
 * @return
 */
@Bean
public UnitASportComponent getUnitASportComponent(){
    UnitASportComponent unitASportComponent = new UnitASportComponent();
    unitASportComponent.add(sportLoginParser);
    unitASportComponent.add(no8OneWayParser);
    unitASportComponent.add(no8TwoWayParser);
    unitASportComponent.add(no8ThreeWayParser);
    unitASportComponent.add(sportLogoutParser);
    return unitASportComponent;
}
/**
 * springCloud 消息交互模板
 *
 * @return
 */
@Bean
@LoadBalanced
RestTemplate restTemplate(){
    return new RestTemplate();
}
@Bean
public InetSocketAddress getInetAddress(){
    return new InetSocketAddress(port);
}

/**
 * Mina 的日志过滤器
```

```java
 *
 * @return
 */
@Bean
public LoggingFilter getLogFilter(){
    return new LoggingFilter();
}
/**
 * Mina的业务多线程处理过滤器
 *
 * @return
 */
@Bean
public ExecutorFilter getExecutorFilter(){
    return new ExecutorFilter();
}
/**
 * Mina的IoFileter设置
 *
 * @return
 */
@Bean
public IoFilter getIOFilter(){
    ComponentIOFilter componentIOFilter = new ComponentIOFilter
      (getMHRootComponent());
    return componentIOFilter;
}
public String getIp(){
    return ip;
}
public void setIp(String ip){
    this.ip = ip;
}
public int getPort(){
    return port;
}
public void setPort(int port){
    this.port = port;
}
public int getReaderIdleTime(){
    return readerIdleTime;
}
public void setReaderIdleTime(int readerIdleTime){
    this.readerIdleTime = readerIdleTime;
}
```

```
public int getMinReadBufferSize(){
    return minReadBufferSize;
}
public void setMinReadBufferSize(int minReadBufferSize){
    this.minReadBufferSize = minReadBufferSize;
}
public int getMaxReadBufferSize(){
    return maxReadBufferSize;
}
public void setMaxReadBufferSize(int maxReadBufferSize){
    this.maxReadBufferSize = maxReadBufferSize;
}
}
```

2.3.3 数据包定义和实现

数据包定义主要包括：定义抽象类包 PackageData，抽象出数据包的公共字段，也就是每一个厂家的智能终端运动数据包必带的字段头，如 UnitA 厂家的智能终端运动数据包的名称和类型等。定义 UnitA 厂家的智能终端运动包有 5 个子类，分别是登录包 LoginPacket、1 号包 No8OneWayPacket、2 号包 No8TwoWayPacket、3 号包 No8ThreeWayPacket 和退出包 LogoutPacket。每一个包的字段定义完全参照设备接口规范进行设计，代码实现如下。

（1）数据包抽象类 PackageData，声明数据包名称、类型、设备 id、app 类型等通用属性。代码实现如下：

```
package com.cloud.mina.unit_a.sportpackage;

/**
 * UnitA 智能终端运动数据包抽象父类
 *
 * @author changyaobin
 *
 */
public abstract class PackageData {
protected String name = "";
protected String type = "";
protected String deviceID = "";
protected String patientID = "";
protected String company = "";
protected String password = "";
protected String appType = "";

public String getName(){
    return name;
```

```java
    }

    public void setName(String name){
        this.name = name;
    }

    public String getType(){
        return type;
    }

    public void setType(String type){
        this.type = type;
    }

    public String getDeviceID(){
        return deviceID;
    }

    public void setDeviceID(String deviceID){
        this.deviceID = deviceID;
    }

    public String getPatientID(){
        return patientID;
    }

    public void setPatientID(String patientID){
        this.patientID = patientID;
    }

    public String getCompany(){
        return company;
    }

    public void setCompany(String company){
        this.company = company;
    }

    public String getPassword(){
        return password;
    }

    public void setPassword(String password){
        this.password = password;
    }
```

```java
public String getAppType(){
    return appType;
}

public void setAppType(String appType){
    this.appType = appType;
}

@Override
public String toString(){
    return "PackageData [name=" + name + ",type=" + type + ",deviceID="
        + deviceID + ",patientID=" + patientID + ",company=" + company
        + ",password=" + password + ",appType= " + appType + "]";
}

}
```

（2）数据登录包 LoginPacket，定义登录数据包信息。代码实现如下：

```java
package com.cloud.mina.unit_a.sportpackage;

/**
 * UnitA 智能终端运动登录数据包
 *
 * @author changyaobin
 *
 */
public class LoginPacket extends PackageData {
    private String loginTime = "";

    public String getLoginTime(){
        return loginTime;
    }

    public void setLoginTime(String loginTime){
        this.loginTime = loginTime;
    }
}
```

（3）数据 1 号包 No8OneWayPacket，定义 1 号包的类属性，具体参考传输协议。代码实现如下：

```java
package com.cloud.mina.unit_a.sportpackage;
```

```java
/**
 * UnitA 智能终端运动数据包(1 号包)
 *
 * @author changyaobin
 *
 */
public class No8OneWayPacket extends PackageData
/** */
{
    // 电量
    private int battery = 0;
    // 体重
    private int weight;
    // 步幅
    private int stride;
    // 卡路里
    private long kcal;
    // 总步数
    private long step;
    // 距离
    private long distance = 0;
    // 智能终端运动等级
    private int level1;
    // 智能终端运动等级
    private int level2;
    // 智能终端运动等级
    private int level3;
    // 智能终端运动等级
    private int level4;
    // 自动发送 0,手动发送 1
    private int tran_type;
    // 有效步数
    private long effective_step;
    private String stepdate = "";              // 数据真实时间
    private String firmware_version = "";      // 固件版本
    private String prefix = "";                // 设备前缀

    public No8OneWayPacket(){
        this.name = "sports";
        this.type = "No8-1";
    }

    public String getStepdate(){
        return stepdate;
    }
```

```java
public void setStepdate(String stepdate){
    this.stepdate = stepdate;
}

public int getBattery(){
    return battery;
}

public void setBattery(int battery){
    this.battery = battery;
}

public int getWeight(){
    return weight;
}

public void setWeight(int weight){
    this.weight = weight;
}

public int getStride(){
    return stride;
}

public void setStride(int stride){
    this.stride = stride;
}

public long getKcal(){
    return kcal;
}

public void setKcal(long kcal){
    this.kcal = kcal;
}

public long getStep(){
    return step;
}

public void setStep(long step){
    this.step = step;
}
```

```java
    public long getDistance(){
        return distance;
    }

    public void setDistance(long distance){
        this.distance = distance;
    }

    public int getLevel1(){
        return level1;
    }

    public void setLevel1(int level1){
        this.level1 = level1;
    }

    public int getLevel2(){
        return level2;
    }

    public void setLevel2(int level2){
        this.level2 = level2;
    }

    public int getLevel3(){
        return level3;
    }

    public void setLevel3(int level3){
        this.level3 = level3;
    }

    public int getLevel4(){
        return level4;
    }

    public void setLevel4(int level4){
        this.level4 = level4;
    }

    public int getTran_type(){
        return tran_type;
    }

    public void setTran_type(int tran_type){
```

```
        this.tran_type = tran_type;
}

public long getEffective_step(){
        return effective_step;
}

public void setEffective_step(long effective_step){
        this.effective_step = effective_step;
}

public String getFirmware_version(){
        return firmware_version;
}

public void setFirmware_version(String firmware_version){
        this.firmware_version = firmware_version;
}

public String getPrefix(){
        return prefix;
}

public void setPrefix(String prefix){
        this.prefix = prefix;
}
}
```

（4）数据 2 号包 No8TwoWayPacket，定义 2 号包的类属性，可以参考传输协议。代码实现如下：

```
package com.cloud.mina.unit_a.sportpackage;

import java.util.ArrayList;
import java.util.List;

/**
 * UnitA 智能终端运动数据包(2 号包)
 *
 * @author changyaobin
 *
 */
public class No8TwoWayPacket extends PackageData {
private List<String> stepcount2data = new ArrayList<String>();
private List<String> stepdate = new ArrayList<String>();
        /**
```

```
 * 根据1分钟数据计算出的有效步数包
 */
private List<String> stepEffective = new ArrayList<String>();

public No8TwoWayPacket(){
    this.name = "sports";
    this.type = "No8-2";
}

public List<String> getStepcount2data(){
    return stepcount2data;
}

public void setStepcount2data(List<String> stepcount2data){
    this.stepcount2data = stepcount2data;
}

public List<String> getStepdate(){
    return stepdate;
}

public void setStepdate(List<String> stepdate){
    this.stepdate = stepdate;
}

public List<String> getStepEffective(){
    return stepEffective;
}

public void setStepEffective(List<String> stepEffective){
    this.stepEffective = stepEffective;
}
}
```

（5）数据3号包No8ThreeWayPacket，定义3号包的类属性。代码实现如下：

```
package com.cloud.mina.unit_a.sportpackage;
/**
 * UnitA 智能终端运动数据包(3号包)
 *
 * @author changyaobin
 *
 */
public class No8ThreeWayPacket extends PackageData {
private int battery;
private int weight;
```

```java
private int stride;
private long kcal;
private long step;
private long distance;
private int level1;
private int level2;
private int level3;
private int level4;
private int tran_type;
private long effective_step;
private String stepdate = "";              // 数据真实时间
private String firmware_version = "";      // 固件版本
private String prefix = "";                // 设备前缀
private String locationInfoStr;

public No8ThreeWayPacket(){
    this.name = "sports";
    this.type = "No8-3";
}

public String getStepdate(){
    return stepdate;
}

public void setStepdate(String stepdate){
    this.stepdate = stepdate;
}

public int getBattery(){
    return battery;
}

public void setBattery(int battery){
    this.battery = battery;
}

public int getWeight(){
    return weight;
}

public void setWeight(int weight){
    this.weight = weight;
}

public int getStride(){
```

```java
        return stride;
    }

    public void setStride(int stride){
        this.stride = stride;
    }

    public long getKcal(){
        return kcal;
    }

    public void setKcal(long kcal){
        this.kcal = kcal;
    }

    public long getStep(){
        return step;
    }

    public void setStep(long step){
        this.step = step;
    }

    public long getDistance(){
        return distance;
    }

    public void setDistance(long distance){
        this.distance = distance;
    }

    public int getLevel1(){
        return level1;
    }

    public void setLevel1(int level1){
        this.level1 = level1;
    }

    public int getLevel2(){
        return level2;
    }

    public void setLevel2(int level2){
        this.level2 = level2;
```

```java
}

public int getLevel3(){
    return level3;
}

public void setLevel3(int level3){
    this.level3 = level3;
}

public int getLevel4(){
    return level4;
}

public void setLevel4(int level4){
    this.level4 = level4;
}

public int getTran_type(){
    return tran_type;
}

public void setTran_type(int tran_type){
    this.tran_type = tran_type;
}

public long getEffective_step(){
    return effective_step;
}

public void setEffective_step(long effective_step){
    this.effective_step = effective_step;
}

public String getLocationInfoStr(){
    return locationInfoStr;
}

public void setLocationInfoStr(String locationInfoStr){
    this.locationInfoStr = locationInfoStr;
}

public String getFirmware_version(){
    return firmware_version;
}
```

```java
public void setFirmware_version(String firmware_version){
    this.firmware_version = firmware_version;
}

public String getPrefix(){
    return prefix;
}

public void setPrefix(String prefix){
    this.prefix = prefix;
}
}
```

（6）数据退出登录包LogoutPacket，继承父类PackageData的通用属性。代码实现如下：

```java
package com.cloud.mina.unit_a.sportpackage;

/**
 * UnitA 智能终端运动退出登录数据包
 *
 * @author changyaobin
 *
 */
public class LogoutPacket extends PackageData {
//可以不实现
}
```

2.3.4 业务树构建和实现

业务树构建是按照设备类型和数据包类型进行的，目的是从字节流解码转换为对象的过程。同时为了保证业务的高度扩展性，本次设计通过工厂、组合、迭代的设计模式实现。实现过程如下。

（1）解码器接口 Component，主要完成解码器 Component 的定义，包括添加子解码器、移除子解码器和解析数据包等。代码实现如下：

```java
package com.cloud.mina.component.filter;

import org.apache.mina.core.buffer.IoBuffer;
import com.cloud.mina.unit_a.sportpackage.PackageData;
/**
 * 解码器组件
 *
 * @author changyaobin
 *
 */
```

```java
public interface Component {
    // 添加子解码器
    public void add(Component t);
    // 移除子解码器
    public void remove(Component t);
    // 解析数据包
    public PackageData getDataFromBuffer(IoBuffer buffer);
    // 从 IOBuffer 中解析数据包
    public PackageData generateRealPackageData(IoBuffer buffer);
}
```

（2）自定义过滤器 ComponentIOFilter，主要功能是重写 mina 的 ioFilter 核心方法 messageReceived，将我们组装的解码器注入，调用自定义解码器完成数据的解析以及将数据包传递到下一个 ioFilter。代码实现如下：

```java
package com.cloud.mina.component.filter;

import org.apache.mina.core.buffer.IoBuffer;
import org.apache.mina.core.filterchain.IoFilterAdapter;
import org.apache.mina.core.session.IoSession;
import com.cloud.mina.unit_a.sportpackage.PackageData;
/**
 * mina 的 IOFilter 自定义扩展类
 *
 * @author changyaobin
 *
 */
public class ComponentIOFilter extends IoFilterAdapter {
    public Component component;

    public ComponentIOFilter(Component component){
        super();
        this.component = component;
    }

    public ComponentIOFilter(){
        super();
    }

    // 数据接收转换核心方法
    @Override
    public void messageReceived(NextFilter nextFilter,IoSession session,
        Object message)throws Exception {
        // 1.调用接口 component 实现字节流转为 Java 对象
        // data = component.getDataFromBuffer(ioBuffer);
```

```
        // 2.递归调用 messageReceived,处理下一个设备
        // nextFilter.message Received(session, data);
        packageHandle(nextFilter,session,message);
    }
    private void packageHandle(NextFilter nextFilter,IoSession session,
      Object message){
        PackageData data = null;
        // 1.判断 message 是字节流还是 Java 对象 PackageData
        // 如登录包被解析后,message 换为 LoginPacket,这个时候进入 if(data == null),
        // 此时的 nextFilter 是 unitABPComponent,但是没有内容结束程序
        if(message instanceof IoBuffer){
            IoBuffer ioBuffer =(IoBuffer)message;
            ioBuffer.setAutoExpand(true);
            data = component.getDataFromBuffer(ioBuffer);
        }
        String appType =(String)session.getAttribute("appType");
        if(data == null){
            // 2.Filter 就是 unitASportsComponent 和 unitABPComponent,因为
            // unit ABPComponent 的 nextfilter=null,结束程序
            // 登录包过来后,IoFilterAdapter 的 messageReceived 方法,执行
            // next Filter.messageReceived(session,data)之后,递归进入packageHandle,
            //   在下面方法结束程序
            nextFilter.messageReceived(session,message);
        } else {
            nextFilter.messageReceived(session,data);
        }
    }
}
```

（3）解码器抽象父类 PacketFilterComponent，主要功能是实现接口定义方法，供子类直接调用；利用组合迭代设计模式完成 iobuffer 字节数据到数据包 javaBean 的转换。代码实现如下：

```
package com.cloud.mina.component.filter;

import java.util.ArrayList;
import java.util.Iterator;
import java.util.List;
import org.apache.log4j.Logger;
import org.apache.mina.core.buffer.IoBuffer;
import com.cloud.mina.unit_a.sportpackage.PackageData;
/**
 * 解码器的抽象父类
 *
 * @author changyaobin
```

```
 *
 */
public abstract class PacketFilterComponent implements Component {

public static Logger log = Logger.getLogger(PacketFilterComponent.class);
public List<Component> list = new ArrayList<Component>();

public void add(Component t){
    this.list.add(t);
}

public void remove(Component t){
    this.list.add(t);
}

/**
 * 解析 iobuffer 中的数据,看是否符合要求
 *
 * @param buffer
 * @return
 */
public abstract boolean check(IoBuffer buffer);

// 迭代模式:叠加递归算法进行编码
public PackageData getDataFromBuffer(IoBuffer buffer){
    return createTreeData(buffer);
}

private PackageData createTreeData(IoBuffer buffer){
    // 没有子节点,该节点为叶子节点,直接生成 data
    if(list.size()== 0){
        return generateRealPackageData(buffer);
    }
    // 非叶子节点,调用叶子节点的方法生成 data
    Iterator<Component> iterator = list.iterator();
    while(iterator.hasNext()){
        PacketFilterComponent filter =(PacketFilterComponent)iterator.
            next();
        if(filter.check(buffer)){
            return filter.getDataFromBuffer(buffer);
        }
    }
    return null;
}
```

```java
public List<Component> getList(){
    return list;
}

public void setList(List<Component> list){
    this.list = list;
}

}
```

（4）设备A的根解码器MHRootComponent，无实际功能，主要作为解码器的根，具体功能由其子类去实现。代码实现如下：

```java
package com.cloud.mina.component.filter;
import org.apache.mina.core.buffer.IoBuffer;
import com.cloud.mina.unit_a.sportpackage.PackageData;
/**
 * 业务树解码器的根类
 *
 * @author changyaobin
 *
 */
public class MHRootComponent extends PacketFilterComponent {

public PackageData generateRealPackageData(IoBuffer buffer){
    return null;
}
@Override
public boolean check(IoBuffer buffer){
    return false;
}
}
```

（5）智能终端运动数据包解码器UnitASportComponent，主要功能是实现字节流的数据头检查，具体解析数据功能由其各个子类去实现。代码实现如下：

```java
package com.cloud.mina.component.filter;
import org.apache.mina.core.buffer.IoBuffer;
import com.cloud.mina.unit_a.sportpackage.PackageData;

/**
 * unitA公司的智能终端运动解码器
 *
 * @author changyaobin
```

```
 *
 */
public class UnitASportComponent extends MHRootComponent {

    @Override
    public boolean check(IoBuffer buffer){
        log.info("byte[0]=" + buffer.get(0)+ " byte[1]=" + buffer.get(1)+
          " byte[2]=" + buffer.get(2)+ " byte[3]=" + buffer.get(3));
        log.info("byte[4]=" + buffer.get(4)+ " byte[5]=" + buffer.get(5)+
          " byte[6]=" + buffer.get(6)+ " byte[7]=" + buffer.get(7));
        log.info("byte[8]=" + buffer.get(8)+ " byte[9]=" + buffer.get(9));
        if((buffer.get(0)== -89)&&(buffer.get(1)== -72)&&(buffer.get(2)==
          0)&&(buffer.get(3)== 1)){
            log.info("buffer.length=" + buffer.array().length);
            log.info(this.getClass().getSimpleName()+".check()return true");
            return true;
        }
        log.info(this.getClass().getSimpleName()+ ".check()return false");
        return false;
    }

    @Override
    public PackageData generateRealPackageData(IoBuffer buffer){
        return null;
    }
}
```

（6）数据登录数据包，继承 UnitASportComponent 类，主要功能是检查登录数据包头信息，将登录字节解析为登录数据包对象，解析规则可以参考设备接口规范。代码实现如下：

```
package com.cloud.mina.component.unit_a.sport;

import org.apache.mina.core.buffer.IoBuffer;
import org.springframework.stereotype.Component;
import com.cloud.mina.component.filter.UnitASportComponent;
import com.cloud.mina.unit_a.sportpackage.LoginPacket;
import com.cloud.mina.unit_a.sportpackage.PackageData;
import com.cloud.mina.util.DateUtil;
import com.cloud.mina.util.DeviceIDResolver;
import com.cloud.mina.util.Logger;

/**
 * unitA 公司智能终端运动登录数据包解码器
```

```java
 *
 * @author changyaobin
 *
 */
@Component
public class SportLoginParser extends UnitASportComponent {
@Override
public boolean check(IoBuffer buffer){
    if(buffer.get(8)== 1 && buffer.get(9)== -128){
        return true;
    }
    return false;
}

@Override
public PackageData generateRealPackageData(IoBuffer buffer){
    log.info(this.getClass().getSimpleName()+
      ".generateRealPackageData()begin...");
    LoginPacket data = new LoginPacket();
    data.setDeviceID(DeviceIDResolver.getDeviceIDFromBytes(buffer.
      array(),10));
    data.setPatientID(DeviceIDResolver.searchPatientidByDeviceid
      (data.getDeviceID()));
    data.setAppType(DeviceIDResolver.searchAppTypeByDeviced(data.
      getDeviceID()));
    data.setLoginTime(DateUtil.getCurrentTime());
    data.setName("sports");
    data.setType("login");
    if(data.getDeviceID()!= null && data.getDeviceID().length()> 4){
data.setCompany(DeviceIDResolver.searchCompanyByDeviceid(data.
  getDeviceID()));
    }
    Logger.writeLog("NO.1 package handled device_id:" + data.
      getDevice ID()+ " patientID:" + data.getPatientID()+ " company:"
      + data.getCompany());
    log.info(this.getClass().getSimpleName()+
      ".generateRealPackage Data()end.");
    return data;
    }
}
```

（7）数据 1 号包解码器 No8OneWayParser，继承 UnitASportComponent 类，主要功能是检查简要包数据头，将简要包字节解析为数据包对象，解析规则可以参考设备接口规范。代码实现如下：

```java
package com.cloud.mina.component.unit_a.sport;

import org.apache.mina.core.buffer.IoBuffer;
import org.springframework.stereotype.Component;
import com.cloud.mina.component.filter.UnitASportComponent;
import com.cloud.mina.unit_a.sportpackage.No8OneWayPacket;
import com.cloud.mina.unit_a.sportpackage.PackageData;
import com.cloud.mina.util.DataTypeChangeHelper;
import com.cloud.mina.util.DateUtil;
import com.cloud.mina.util.DeviceIDResolver;
/**
 * unitA 公司智能终端运动数据包(1 号数据包)解码器
 *
 * @author changyaobin
 *
 */
@Component
public class No8OneWayParser extends UnitASportComponent {
@Override
public boolean check(IoBuffer buffer){
    if(buffer.get(8)== 8 && buffer.get(9)== 1){
        return true;
    }
    return false;
}

public PackageData generateRealPackageData(IoBuffer buffer){
    log.info(this.getClass().getSimpleName()+
      ".generateRealPackage Data()begin...");
    No8OneWayPacket packet = new No8OneWayPacket();
    byte kcal_b[] = new byte[4];
    byte step_b[] = new byte[4];
    byte effective_step_b[] = new byte[4];
    byte distance_b[] = new byte[4];
    byte level1_b[] = new byte[2];
    byte level2_b[] = new byte[2];
    byte level3_b[] = new byte[2];
    byte level4_b[] = new byte[2];
    int tran_type = buffer.get(10);
    int year = buffer.get(11);
    int month = buffer.get(12);
    int day = buffer.get(13);
    effective_step_b[0] = buffer.get(14);
    effective_step_b[1] = buffer.get(15);
    effective_step_b[2] = buffer.get(16);
```

```java
effective_step_b[3] = buffer.get(17);
int battery = buffer.get(22);
int weight = buffer.get(23);
int stride = buffer.get(24);
kcal_b[0] = buffer.get(25);
kcal_b[1] = buffer.get(26);
kcal_b[2] = buffer.get(27);
kcal_b[3] = buffer.get(28);
step_b[0] = buffer.get(29);
step_b[1] = buffer.get(30);
step_b[2] = buffer.get(31);
step_b[3] = buffer.get(32);
distance_b[0] = buffer.get(33);
distance_b[1] = buffer.get(34);
distance_b[2] = buffer.get(35);
distance_b[3] = buffer.get(36);
level1_b[1] = buffer.get(37);
level1_b[0] = buffer.get(38);
level2_b[1] = buffer.get(39);
level2_b[0] = buffer.get(40);
level3_b[1] = buffer.get(41);
level3_b[0] = buffer.get(42);
level4_b[1] = buffer.get(43);
level4_b[0] = buffer.get(44);
long kcal = DataTypeChangeHelper.unsigned4BytesToInt(kcal_b,0);
long step = DataTypeChangeHelper.unsigned4BytesToInt(step_b,0);
long effective_step = DataTypeChangeHelper.unsigned4BytesToInt
   (effective_step_b,0);
long distance = DataTypeChangeHelper.unsigned4BytesToInt
   (distance_b,0);
int level1 = DataTypeChangeHelper.byte2int(level1_b)* 2;
int level2 = DataTypeChangeHelper.byte2int(level2_b)* 2;
int level3 = DataTypeChangeHelper.byte2int(level3_b)* 2;
int level4 = DataTypeChangeHelper.byte2int(level4_b)* 2;
String firmware_version = DeviceIDResolver.getFirmwareVersion(buffer.
   array(),18);
String prefix = DeviceIDResolver.getDeviceIDPrefix(buffer.
   array(),45);
String deviceID = DeviceIDResolver.getDeviceIDFromBytes(buffer.
   array(),50);
String patientID = DeviceIDResolver.searchPatientidByDeviceid
   (deviceID);
String company = DeviceIDResolver.searchCompanyByDeviceid(deviceID);
String stepdate = DateUtil.getStepdate(year,month,day);
```

```
        packet.setStepdate(stepdate);
        packet.setBattery(battery);
        packet.setWeight(weight);
        packet.setStride(stride);
        packet.setKcal(kcal);
        packet.setStep(step);
        packet.setDistance(distance);
        packet.setEffective_step(effective_step);
        packet.setLevel1(level1);
        packet.setLevel2(level2);
        packet.setLevel3(level3);
        packet.setLevel4(level4);
        packet.setTran_type(tran_type);
        packet.setDeviceID(deviceID);
        packet.setPatientID(patientID);
        packet.setCompany(company);
        packet.setFirmware_version(firmware_version);
        packet.setPrefix(prefix);
        log.info(this.getClass().getSimpleName()+
          ".generateRealPackageData()end.");
        return packet;
    }
}
```

（8）数据 2 号包解码器 No8TwoWayParser，继承 UnitASportComponent 类，主要功能是检查详细包数据头，将详细包字节数据解析成数据包对象，解析规则可以参考设备接口规范。代码实现如下：

```
package com.cloud.mina.component.unit_a.sport;
import org.apache.mina.core.buffer.IoBuffer;
import org.springframework.stereotype.Component;
import com.cloud.mina.component.filter.UnitASportComponent;
import com.cloud.mina.unit_a.sportpackage.No8TwoWayPacket;
import com.cloud.mina.unit_a.sportpackage.PackageData;
import com.cloud.mina.util.DataTypeChangeHelper;
import com.cloud.mina.util.DateUtil;
import com.cloud.mina.util.DeviceIDResolver;

/**
 * unitA 公司智能终端运动数据包(2 号数据包)解码器
 *
 * @author changyaobin
 *
 */
```

```java
@Component
public class No8TwoWayParser extends UnitASportComponent {
    @Override
    public boolean check(IoBuffer buffer){
        if(buffer.get(8)== 8 && buffer.get(9)== 2){
            return true;
        }
        return false;
    }
    @Override
    public PackageData generateRealPackageData(IoBuffer buffer){
        log.info(this.getClass().getSimpleName()+
            ".generateRealPackageData()begin...");
        String prefix = DeviceIDResolver.getNo8PackageDevicePrefix
            (buffer.array());
        No8TwoWayPacket packet = null;
        packet = handle5MinutesDetailPacket(buffer);
        log.info(this.getClass().getSimpleName()+
            ".generateRealPackageData()end.");
        return packet;
    }
    private No8TwoWayPacket handle5MinutesDetailPacket(IoBuffer buffer){
        No8TwoWayPacket packet = new No8TwoWayPacket();
        int number = 0;
        byte length[] = new byte[4];
        byte year[] = new byte[2];
        byte stepcount[] = new byte[2];
        byte stepkcal[] = new byte[2];
        byte data[] = new byte[2];
        int year_u[] = new int[24];
        int month_u[] = new int[24];
        int day_u[] = new int[24];
        int Hour[] = new int[24];
        int hourdata[][] = new int[24][72];
        int hourdata_real[][] = new int[24][72];
        length[0] = buffer.get(4);
        length[1] = buffer.get(5);
        length[2] = buffer.get(6);
        length[3] = buffer.get(7);
        long lengthvalue = DataTypeChangeHelper.unsigned4BytesToInt
            (length,0);
        String deviceID = DeviceIDResolver.getDeviceIDFromBytes(buffer.
            array(),(int)(lengthvalue - 18));
        String patientID = DeviceIDResolver.searchPatientidByDeviceid
            (deviceID);
```

```
String company = DeviceIDResolver.searchCompanyByDeviceid(deviceID);
packet.setDeviceID(deviceID);
packet.setPatientID(patientID);
packet.setCompany(company);
int unUsedataNum = 12;
if((lengthvalue - 12)% 114 == 0){
    unUsedataNum = 12;
} else {
    unUsedataNum = 33;
}
long max_times_tran =(lengthvalue - unUsedataNum)/ 114;
for(int i = 0;i <= 24;i++){
    if(number ==(lengthvalue - unUsedataNum)){
        break;
    }
    year[1] = buffer.get(10 + number);
    year[0] = buffer.get(11 + number);
    year_u[i] = DataTypeChangeHelper.byte2int(year);
    month_u[i] = buffer.get(13 + number);
    day_u[i] = buffer.get(14 + number);
    Hour[i] = buffer.get(15 + number);
    for(int j = 0;j < 12;j++){
        stepcount[0] = buffer.get(16 + j * 2 + number);
        stepcount[1] = buffer.get(17 + j * 2 + number);
        hourdata[i][j] = DataTypeChangeHelper.byte2int(stepcount);
        hourdata_real[i][j] = DataTypeChangeHelper.byte2int(stepcount);
    }
    for(int j = 0;j < 12;j++){
        stepkcal[0] = buffer.get(40 + j * 2 + number);
        stepkcal[1] = buffer.get(41 + j * 2 + number);
        hourdata[i][j + 12] = DataTypeChangeHelper.byte2int(stepkcal);
    }
    for(int j = 0;j < 12;j++){
        if(buffer.get(64 + j + number)< 0){
            hourdata[i][j + 24] = buffer.get(64 + j + number)+ 256;
            hourdata[i][j + 24] = hourdata[i][j + 24] * 2;
        } else {
            hourdata[i][j + 24] = buffer.get(64 + j + number);
            hourdata[i][j + 24] = hourdata[i][j + 24] * 2;
        }
        if(buffer.get(76 + j + number)< 0){
            hourdata[i][j + 36] = buffer.get(76 + j + number)+ 256;
            hourdata[i][j + 36] = hourdata[i][j + 36] * 2;
        } else {
            hourdata[i][j + 36] = buffer.get(76 + j + number);
```

```java
                hourdata[i][j + 36] = hourdata[i][j + 36] * 2;
            }
            if(buffer.get(88 + j + number)< 0){
                hourdata[i][j + 48] = buffer.get(88 + j + number)+ 256;
                hourdata[i][j + 48] = hourdata[i][j + 48] * 2;
            } else {
                hourdata[i][j + 48] = buffer.get(88 + j + number);
                hourdata[i][j + 48] = hourdata[i][j + 48] * 2;
            }
        }
        for(int j = 0;j < 12;j++){
            data[0] = buffer.get(100 + j * 2 + number);
            data[1] = buffer.get(101 + j * 2 + number);
            hourdata[i][j + 60] = DataTypeChangeHelper.byte2int(data);
        }
        number = number + 114;
    }
    for(int times_tran = 0;times_tran < max_times_tran;times_tran++){
        StringBuffer stepcount2data = new StringBuffer();
        stepcount2data = stepcount2data.append("{\"data\":{\"datatype\":
            \"STEPCOUNT2\",");
        stepcount2data = stepcount2data.append("\"hour\"" + ":\"" +
            String.valueOf(Hour[times_tran])+ "\"," + "\"datavalue\":
            [{\"snp5\":");
        stepcount2data.append("\"");
        for(int i = 0;i < 12;i++){
            if(i == 11){
                stepcount2data.append(String.valueOf(hourdata[times_
                    tran][i]));
            } else {
                stepcount2data.append(String.valueOf(hourdata[times_
                    tran][i])).append(",");
            }
        }
        stepcount2data.append("\"");
        stepcount2data.append("},{\"knp5\":");
        stepcount2data.append("\"");
        for(int i = 0;i < 12;i++){
            if(i == 11){
                stepcount2data.append(String.valueOf(hourdata[times_
                    tran][i + 12]));
            } else {
                stepcount2data.append(String.valueOf(hourdata[times_
                    tran][i + 12])).append(",");
            }
        }
```

```
}
stepcount2data.append("\"");
stepcount2data.append("},{\"level2p5\":");
stepcount2data.append("\"");
for(int i = 0;i < 12;i++){
    if(i == 11){
        stepcount2data.append(String.valueOf(hourdata[times_
            tran][i + 24]));
    } else {
        stepcount2data.append(String.valueOf(hourdata[times_
            tran][i + 24])).append(",");
    }
}
stepcount2data.append("\"");
stepcount2data.append("},{\"level3p5\":");
stepcount2data.append("\"");
for(int i = 0;i < 12;i++){
    if(i == 11){
        stepcount2data.append(String.valueOf(hourdata[times_
            tran][i + 36]));
    } else {
        stepcount2data.append(String.valueOf(hourdata[times_
            tran][i + 36])).append(",");
    }
}
stepcount2data.append("\"");
stepcount2data.append("},{\"level4p5\":");
stepcount2data.append("\"");
for(int i = 0;i < 12;i++){
    if(i == 11){
        stepcount2data.append(String.valueOf(hourdata[times_
            tran][i + 48]));
    } else {
        stepcount2data.append(String.valueOf(hourdata[times_
            tran][i + 48])).append(",");
    }
}
stepcount2data.append("\"");
stepcount2data.append("},{\"yuanp5\":");
stepcount2data.append("\"");
for(int i = 0;i < 12;i++){
    if(i == 11){
        stepcount2data.append(String.valueOf(hourdata[times_
            tran][i + 60]));
    } else {
```

```
                    stepcount2data.append(String.valueOf(hourdata[times_
                        tran][i + 60])).append(",");
                }
            }
            stepcount2data.append("\"");
            stepcount2data.append("}]}}");
            String stepdate = DateUtil.format(year_u[times_tran] + "-" +
                month_u[times_tran] + "-" + day_u[times_tran]);
            stepdate = stepdate.replaceAll("-","");
            packet.getStepcount2data().add(stepcount2data.toString());
            packet.getStepdate().add(stepdate);
        }
        return packet;
    }
}
```

（9）数据 3 号包解码器 No8ThreeWayParser，继承 UnitASportComponent 类，主要功能是检测历史简要包字节包头信息，将字节数据解析成历史简要包数据，解析规则可以参考设备接口规范。代码实现如下：

```
package com.cloud.mina.component.unit_a.sport;
import org.apache.mina.core.buffer.IoBuffer;
import org.springframework.stereotype.Component;
import com.cloud.mina.component.filter.UnitASportComponent;
import com.cloud.mina.unit_a.sportpackage.No8ThreeWayPacket;
import com.cloud.mina.unit_a.sportpackage.PackageData;
import com.cloud.mina.util.DataTypeChangeHelper;
import com.cloud.mina.util.DateUtil;
import com.cloud.mina.util.DeviceIDResolver;
/**
 * unitA 公司智能终端运动数据包(3 号包)解码器
 *
 * @author changyaobin
 *
 */
@Component
public class No8ThreeWayParser extends UnitASportComponent {
 @Override
 public boolean check(IoBuffer buffer){
     if(buffer.get(8)== 8 && buffer.get(9)== 3){
         return true;
     }
     return false;
 }
```

```java
@Override
public PackageData generateRealPackageData(IoBuffer buffer){
    log.info(this.getClass().getSimpleName()+ ".generateRealPackageData()
      begin...");

    No8ThreeWayPacket packet = packetPacking(buffer);
    log.info(this.getClass().getSimpleName()+ ".generateRealPackageData()
      end.");
    return packet;
}
private No8ThreeWayPacket packetPacking(IoBuffer buffer){
    No8ThreeWayPacket packet = new No8ThreeWayPacket();
    byte kcal_b[] = new byte[4];
    byte step_b[] = new byte[4];
    byte effective_step_b[] = new byte[4];
    byte distance_b[] = new byte[4];
    byte level1_b[] = new byte[2];
    byte level2_b[] = new byte[2];
    byte level3_b[] = new byte[2];
    byte level4_b[] = new byte[2];
    int tran_type = buffer.get(10);
    int year = buffer.get(11);
    int month = buffer.get(12);
    int day = buffer.get(13);
    effective_step_b[0] = buffer.get(14);
    effective_step_b[1] = buffer.get(15);
    effective_step_b[2] = buffer.get(16);
    effective_step_b[3] = buffer.get(17);
    int battery = buffer.get(22);
    int weight = buffer.get(23);
    int stride = buffer.get(24);
    kcal_b[0] = buffer.get(25);
    kcal_b[1] = buffer.get(26);
    kcal_b[2] = buffer.get(27);
    kcal_b[3] = buffer.get(28);
    step_b[0] = buffer.get(29);
    step_b[1] = buffer.get(30);
    step_b[2] = buffer.get(31);
    step_b[3] = buffer.get(32);
    distance_b[0] = buffer.get(33);
    distance_b[1] = buffer.get(34);
    distance_b[2] = buffer.get(35);
    distance_b[3] = buffer.get(36);
    level1_b[1] = buffer.get(37);
    level1_b[0] = buffer.get(38);
```

```java
            level2_b[1] = buffer.get(39);
            level2_b[0] = buffer.get(40);
            level3_b[1] = buffer.get(41);
            level3_b[0] = buffer.get(42);
            level4_b[1] = buffer.get(43);
            level4_b[0] = buffer.get(44);
            long kcal = DataTypeChangeHelper.unsigned4BytesToInt(kcal_b,0);
            long step = DataTypeChangeHelper.unsigned4BytesToInt(step_b,0);
            long effective_step = DataTypeChangeHelper.unsigned4BytesToInt
                (effective_step_b,0);
            long distance = DataTypeChangeHelper.unsigned4BytesToInt(distance_
                b,0);
            int level1 = DataTypeChangeHelper.byte2int(level1_b)* 2;
            int level2 = DataTypeChangeHelper.byte2int(level2_b)* 2;
            int level3 = DataTypeChangeHelper.byte2int(level3_b)* 2;
            int level4 = DataTypeChangeHelper.byte2int(level4_b)* 2;
            String firmware_version = DeviceIDResolver.getFirmwareVersion
                (buffer.array(),18);
            String prefix = DeviceIDResolver.getDeviceIDPrefix(buffer.array(),
                65);
            String deviceID = DeviceIDResolver.getDeviceIDFromBytes(buffer.
                array(),70);
            String patientID = DeviceIDResolver.searchPatientidByDeviceid(deviceID);
            String company = DeviceIDResolver.searchCompanyByDeviceid(deviceID);
            packet.setDeviceID(deviceID);
            packet.setPatientID(patientID);
            packet.setCompany(company);
            packet.setFirmware_version(firmware_version);
            packet.setPrefix(prefix);
            String stepdate = DateUtil.getStepdate(year,month,day);
            packet.setBattery(battery);
            packet.setStepdate(stepdate);
            packet.setWeight(weight);
            packet.setStride(stride);
            packet.setKcal(kcal);
            packet.setStep(step);
            packet.setDistance(distance);
            packet.setEffective_step(effective_step);
            packet.setLevel1(level1);
            packet.setLevel2(level2);
            packet.setLevel3(level3);
            packet.setLevel4(level4);
            packet.setTran_type(tran_type);
            return packet;
    }
}
```

(10) 数据退出包解码器 SportLogoutParser, 主要功能是检查退出包数据头信息, 将字节数据解析成退出数据包, 解析规则可以参考设备接口规范。代码实现如下:

```java
package com.cloud.mina.component.unit_a.sport;

import org.apache.mina.core.buffer.IoBuffer;
import org.springframework.stereotype.Component;
import com.cloud.mina.component.filter.UnitASportComponent;
import com.cloud.mina.unit_a.sportpackage.LogoutPacket;
import com.cloud.mina.unit_a.sportpackage.PackageData;
/**
 * unitA 公司智能终端运动退出解码器
 *
 * @author changyaobin
 *
 */
@Component
public class SportLogoutParser extends UnitASportComponent {
    @Override
    public boolean check(IoBuffer buffer){
        if(buffer.get(8)== 1 && buffer.get(9)== 3){
            return true;
        }
        return false;
    }
    @Override
    public PackageData generateRealPackageData(IoBuffer buffer){
        log.info(this.getClass().getSimpleName()+
          ".generateRealPackage Data()begin...");
        LogoutPacket packet = new LogoutPacket();
        packet.setName("sports");
        packet.setType("logout");
        log.info(this.getClass().getSimpleName()+
          ".generateRealPackage Data()end.");
        return packet;
    }
}
```

2.3.5 数据包状态进行解析实现

业务数据包处理是按照厂家和数据类型进行设计的, 不同厂家的设备协议不同, 采用策略模式实现, 不同数据类型的数据包, 采用状态模式实现, 主要可以实现不同厂家和不同数据类型的可扩展实现。首先定义总的策略 MHDataPacketHandleStrategy, 不同厂家就是不同的策略, unitA 厂家包括 UnitASportsPacketHandleStrategy 和

UnitABloodPressurePacketHandleStrategy 策略，当设备的数据包到达平台时，每个数据包头通过名字来标识，如 sports 和 bloodpressure。unitA 厂家的不同数据包是按照状态模式实现的，总的状态定义为 SportsPacketHandleState，包括子类有登录包 SportNo1LoginState、1 号包 SportNo8OneWayState、2 号包 SportNo8TwoWayState、3 号包 SportNo8ThreeWayState 和退出包 SportLogoutState。登录包和退出包不会入库，登录包实现数据包的合法验证，退出包实现数据传输结束。数据包完成解析后按照数据对象进行入库，入库调用统一的数据接口 SaveSportsNo8PacketUtil。

（1）项目包 com.cloud.mina.unit_a.strategy 的 StrategyFactroyHandler，继承 mina 提供的 IoHandlerAdapter 适配器类，主要功能是重写数据处理核心方法 messageReceived，完成业务数据入库；根据数据包不同的类型，利用 Java 反射出不同的策略处理类；在 SpringBoot 版本会调用 restTemplate 将数据转发到 dispatch 服务端。代码实现如下。

① Spring MVC 版本：

```java
package com.cloud.mina.unit_a.strategy;

import java.util.ArrayList;
import java.util.HashMap;
import java.util.List;
import java.util.Map;
import org.apache.commons.httpclient.NameValuePair;
import org.apache.commons.lang.StringUtils;
import org.apache.mina.core.service.IoHandlerAdapter;
import org.apache.mina.core.session.IoSession;
import com.cloud.mina.unit_a.sportpackage.PackageData;
import com.cloud.mina.util.HttpClientUtil;
import com.cloud.mina.util.Log;
import com.cloud.mina.util.PropertiesReader;
/**
 * mina 的 Iohandler 自定义扩展类
 *
 * @author Changyaobin
 *
 */
public class StrategyFactroyHandler extends IoHandlerAdapter {
    // 定义变量区域
    public MHDataPacketHandleStrategy chain = null;
    PackageData packet = null;
    static Map<String,Class> classMap = new HashMap<String,Class>();
    static {
        /**
         * 不同厂家就是不同的策略,unitA 的 sport/BP 通过数据包的名字来匹配类名,如
         *   sports 和 bloodpressure
```

```
        */
        classMap.put("sports",UnitASportsPacketHandleStrategy.class);
        classMap.put("bloodpressure",UnitABloodPressurePacketHandle Strategy.
           class);
    }

    public void messageReceived(IoSession session,Object message)throws 
       Exception {
        // 1.参数验证
        if(message != null && message instanceof PackageData){
            packet =(PackageData)message;
            // 2.调用具体设备处理
            chain=(MHDataPacketHandleStrategy)classMap.get
               (packet.get Name()).newInstance();
            if(chain != null){
                chain.handle(session,message);
                String dispatchPath = PropertiesReader.getProp("DISPATCH_
                   SERVER_PATH");
                if(StringUtils.isNotBlank(dispatchPath)){
                    // 参数拼接
                    List<NameValuePair> urlParameters = new ArrayList<>();
                    urlParameters.add(new NameValuePair("appType",packet.
                       getAppType()));
                    urlParameters.add(new NameValuePair("dataType",packet.
                       getType()));
                    // 数据发送给转发服务
                    boolean sendSuccess = HttpClientUtil.sendHttpData(this.
                       getClass().getName(),dispatchPath,urlParameters.
                       toArray(new NameValuePair[url
                       Parameters.size()]));
                    if(!sendSuccess){
                        Log.error("数据发送到转发服务失败,转发服务路径为" +
                           dispatchPath + ",数据为" + packet.toString());
                    }
                }else {
                    Log.error("转发服务路径没配置");
                }
            }
        }
    }
}
```

② Spring Boot 版本：

```
package com.cloud.mina.unit_a.strategy;
```

```java
import java.util.HashMap;
import java.util.Map;
import org.apache.mina.core.service.IoHandlerAdapter;
import org.apache.mina.core.session.IoSession;
import org.springframework.web.client.RestTemplate;
import com.cloud.mina.bean.Message;
import com.cloud.mina.unit_a.sportpackage.PackageData;
import com.cloud.mina.util.Logger;
/**
 * mina 的 Iohandler 自定义扩展类
 *
 * @author changyaobin
 *
 */
public class StrategyFactroyHandler extends IoHandlerAdapter {
    // 定义变量区域
    public MHDataPacketHandleStrategy chain = null;
    PackageData packet = null;
    // springCloud 消息转发模板
    private RestTemplate restTemplate;
    static Map<String,Class> classMap = new HashMap<String,Class>();

    static {
        /**
         * 不同厂家就是不同的策略,unitA 的 sport/BP 通过数据包的名字来匹配类名,如
         *   sports 和 bloodpressure
         */
        classMap.put("sports",UnitASportsPacketHandleStrategy.class);
        classMap.put("bloodpressure",UnitABloodPressurePacketHandle Strategy.
            class);
    }
    public void messageReceived(IoSession session,Object message)throws
      Exception {
        // 1.参数验证
        if(message != null && message instanceof PackageData){
            packet =(PackageData)message;
            // 2.调用具体设备处理
            chain=(MHDataPacketHandleStrategy)classMap.get(packet.getName()).
              newInstance();
            if(chain != null){
                chain.handle(session,message);
            }
            // 发送给转发服务
            try {
```

```
            Message responseMes = restTemplate.getForObject("http://
                BOOT-DISPATCH/sendData?appType=" + packet.getAppType()
                + "&dataType=" + packet.get Type(),Message.class);
            if(responseMes != null && responseMes.getCode()== 1001){
                // 发送成功
                Logger.writeLog("发送数据成功");
            } else {
                Logger.writeLog("发送数据失败");
            }
        } catch(Exception e){
            Logger.errorLog("连接转发服务异常,转发服务地址为http://
                BOOT-DISPATCH");
        }
    }
}
public RestTemplate getRestTemplate(){
    return restTemplate;
}
public void setRestTemplate(RestTemplate restTemplate){
    this.restTemplate = restTemplate;
}
}
```

（2）数据包处理接口 MHDataPacketHandleStrategy，主要功能是定义数据厂家策略接口，以及数据处理方法。如果有其他的厂家接入时，要实现这个接口，重写数据处理方法即可，这样可以提高项目的扩展性。代码实现如下：

```
package com.cloud.mina.unit_a.strategy;
import org.apache.mina.core.session.IoSession;

/**
 * UnitA 公司业务处理策略接口
 *
 * @author changyaobin
 *
 */
public interface MHDataPacketHandleStrategy {
    // UnitA 数据处理方法
    public void handle(IoSession session,Object message);
}
```

（3）智能终端运动数据包具体策略类 UnitASportsPacketHandleStrategy 代表 unitA 厂家的运动设备，继承 MHDataPacketHandleStrategy 总策略，主要功能是实现 unitA 厂家的智能终端运动数据包处理，按照数据包头的类型进行匹配属于哪个状态后，调用具体状态类进行数据处理。代码实现如下：

```java
package com.cloud.mina.unit_a.strategy;

import java.util.HashMap;
import java.util.Map;
import org.apache.mina.core.session.IoSession;
import com.cloud.mina.unit_a.sportpackage.PackageData;
import com.cloud.mina.unit_a.sportstate.SportLogoutState;
import com.cloud.mina.unit_a.sportstate.SportNo1LoginState;
import com.cloud.mina.unit_a.sportstate.SportNo8OneWayState;
import com.cloud.mina.unit_a.sportstate.SportNo8ThreeWayState;
import com.cloud.mina.unit_a.sportstate.SportNo8TwoWayState;
import com.cloud.mina.unit_a.sportstate.SportsPacketHandleState;
/**
 * unitA 公司智能终端运动数据处理策略类(策略模式)
    UnitASportsPacketHandle Strategy 就是 Context(状态模式)
 */
public class UnitASportsPacketHandleStrategy implements
   MHDataPacket HandleStrategy {
static Map<String,Class> classMap = new HashMap<String,Class>();
// 定义变量区域
static {
    classMap.put("login",SportNo1LoginState.class);
    classMap.put("logout",SportLogoutState.class);
    classMap.put("No8-1",SportNo8OneWayState.class);
    classMap.put("No8-2",SportNo8TwoWayState.class);
    classMap.put("No8-3",SportNo8ThreeWayState.class);
}
SportsPacketHandleState state = null;
public void setState(SportsPacketHandleState state){
    this.state = state;
}
public void handle(IoSession session,Object message){
    // 根据数据包的头,调用具体的状态类
    if(message != null && message instanceof PackageData){
        PackageData packageData =(PackageData)message;
        try {
            setState((SportsPacketHandleState)classMap.get(packageData.
              getType()).newInstance());
            state.handlePacket(session,message);
        } catch(InstantiationException e){
            e.printStackTrace();
        } catch(IllegalAccessException e){
```

```
                e.printStackTrace();
            }
        }
    }
}
```

（4）数据登录包状态类 SportNo1LoginState，实现 SportsPacketHandleState 总状态，主要功能是智能终端运动登录包数据处理，存入 mina 的 ioSession 会话中，供其他业务处理调用，但不入库。代码实现如下：

```
package com.cloud.mina.unit_a.sportstate;
import java.util.Calendar;
import org.apache.mina.core.buffer.IoBuffer;
import org.apache.mina.core.session.IoSession;
import com.cloud.mina.unit_a.sportpackage.LoginPacket;
import com.cloud.mina.util.DataTypeChangeHelper;
import com.cloud.mina.util.Logger;
import com.cloud.mina.util.MLinkCRC;
/**
 * unitA 公司智能终端运动数据包(1 号包)登录状态处理类
 *
 * @author changyaobin
 *
 */
public class SportNo1LoginState implements SportsPacketHandleState {
    public boolean handlePacket(IoSession session,Object message){
        LoginPacket packet = null;
        if(message != null && message instanceof LoginPacket){
            packet =(LoginPacket)message;
            if(packet.getPatientID()== null || "".equals(packet.getPatientID().
               trim())){
                return false;
            }
            session.setAttribute("patientId",packet.getPatientID());
            session.setAttribute("deviceId",packet.getDeviceID());
            session.setAttribute("company",packet.getCompany());
            session.setAttribute("loginTime",packet.getLoginTime());
            session.setAttribute("appType",packet.getAppType());
            // 回复ACK
            responseToClient(session);
            return true;
        } else {
            // 回复NAK
            // responseToClient(session);
            return false;
```

```java
        }
    }
    private void responseToClient(IoSession session){
        byte[] ack = responsePacking(session);
        session.write(IoBuffer.wrap(ack));
    }
    private byte[] responsePacking(IoSession session){
        byte[] ack = new byte[19];
        byte[] crc_c = new byte[4];
        ack[0] = -89;
        ack[1] = -72;
        ack[2] = 0;
        ack[3] = 1;
        ack[4] = 0;
        ack[5] = 0;
        ack[6] = 0;
        ack[7] = 19;
        ack[8] = 1;
        ack[9] = 1;
        Logger.writeLog("patientID" + session.getAttribute("patientId")+
          " has no param data to send to stepcounter");
        Calendar cal = Calendar.getInstance();
        int year = cal.get(Calendar.YEAR);
        int month = cal.get(Calendar.MONTH)+ 1;
        int day = cal.get(Calendar.DAY_OF_MONTH);
        int hour = cal.get(Calendar.HOUR_OF_DAY);
        int minute = cal.get(Calendar.MINUTE);
        int second = cal.get(Calendar.SECOND);
        byte[] year_b = DataTypeChangeHelper.int2byte(year);
        byte[] month_b = DataTypeChangeHelper.int2byte(month);
        byte[] day_b = DataTypeChangeHelper.int2byte(day);
        byte[] hour_b = DataTypeChangeHelper.int2byte(hour);
        byte[] minute_b = DataTypeChangeHelper.int2byte(minute);
        byte[] second_b = DataTypeChangeHelper.int2byte(second);
        ack[10] = year_b[1];
        ack[11] = year_b[0];
        ack[12] = month_b[0];
        ack[13] = day_b[0];
        ack[14] = hour_b[0];
        ack[15] = minute_b[0];
        ack[16] = second_b[0];
        crc_c = MLinkCRC.crc16(ack);
        ack[17] = crc_c[0];
        ack[18] = crc_c[1];
```

```
        return ack;
    }
}
```

（5）数据 1 号包状态类 SportNo8OneWayState，实现 SportsPacketHandleState 总状态，主要功能是 SportNo1LoginState 处理数据后，就会切换到该状态类，该类调用 SaveSports No8PacketUtil 将简要包数据入库。代码实现如下：

```
package com.cloud.mina.unit_a.sportstate;

import org.apache.mina.core.session.IoSession;
import com.cloud.mina.unit_a.sportpackage.No8OneWayPacket;
import com.cloud.mina.util.Log;
import com.cloud.mina.util.SaveSportsNo8PacketUtil;
/**
 * unitA 公司智能终端运动数据包(1 号包)状态处理类
 *
 * @author changyaobin
 *
 */
public class SportNo8OneWayState implements SportsPacketHandleState {

    public boolean handlePacket(IoSession session,Object message){
        // 定义变量
        No8OneWayPacket packet = null;
        // 1.验证参数
        if(message != null && message instanceof No8OneWayPacket){
            packet =(No8OneWayPacket)message;
            session.setAttribute("patientId",packet.getPatientID());
            session.setAttribute("deviceId",packet.getDeviceID());
            session.setAttribute("company",packet.getCompany());
            // 2.数据包入库
            boolean result = SaveSportsNo8PacketUtil.saveNewSportHistory
              (session,packet);

            // 3.给设备应答
            if(result){
                SaveSportsNo8PacketUtil.sendNo8Ack(session,result,1);
            } else {
                Log.error("数据保存失败!");
            }
        }
        return false;
    }
}
```

（6）数据 2 号包状态类 SportNo8TwoWayState，实现 SportsPacketHandleState 总状态，主要功能是 SportNo8OneWayState 处理数据后，切换到该状态类，调用 SaveSportsNo8PacketUtil 将详细包数据入库。代码实现如下：

```java
package com.cloud.mina.unit_a.sportstate;
import org.apache.mina.core.session.IoSession;
import com.cloud.mina.unit_a.sportpackage.No8TwoWayPacket;
import com.cloud.mina.util.SaveSportsNo8PacketUtil;
/**
 * unitA 公司智能终端运动数据包(2 号包)状态处理类
 *
 * @author changyaobin
 *
 */
public class SportNo8TwoWayState implements SportsPacketHandleState {
 public boolean handlePacket(IoSession session,Object message){
    No8TwoWayPacket packet = null;
    if(message != null && message instanceof No8TwoWayPacket){
        packet =(No8TwoWayPacket)message;
        if(packet.getPatientID()!= null && !"".equals(packet.
          getPatient ID())){
           session.setAttribute("patientId",packet.getPatientID());
           session.setAttribute("deviceId",packet.getDeviceID());
           session.setAttribute("company",packet.getCompany());
        }
        // 数据存储入库
        for(int i = 0;i < packet.getStepcount2data().size();i++){
            SaveSportsNo8PacketUtil.saveNewSportDetail(session,
               packet.getStepcount2data().get(i), packet.
               getStepdate().get(i));
        }
        SaveSportsNo8PacketUtil.sendNo8Ack(session,true,2);
        return true;
    }
    return false;
 }
}
```

（7）数据 3 号包状态类 SportNo8ThreeWayState，实现 SportsPacketHandleState 总状态，主要功能是 SportNo8TwoWayState 处理数据后，切换到该状态类，调用 SaveSportsNo8PacketUtil 将有效包数据入库。代码实现如下：

```java
package com.cloud.mina.unit_a.sportstate;

import org.apache.mina.core.session.IoSession;
```

```java
import com.cloud.mina.unit_a.sportpackage.No8ThreeWayPacket;
import com.cloud.mina.util.SaveSportsNo8PacketUtil;
/**
 * unitA 公司智能终端运动数据包(3号包)状态处理类
 *
 * @author changyaobin
 *
 */
public class SportNo8ThreeWayState implements SportsPacketHandleState {
    public boolean handlePacket(IoSession session,Object message){
        No8ThreeWayPacket packet = null;
        if(message != null && message instanceof No8ThreeWayPacket){
            packet =(No8ThreeWayPacket)message;
            if(packet.getPatientID()!= null && !"".equals(packet.
              getPatient ID())){
                session.setAttribute("patientId",packet.getPatientID());
                session.setAttribute("deviceId",packet.getDeviceID());
                session.setAttribute("company",packet.getCompany());
            }
            boolean result = false;
            result = SaveSportsNo8PacketUtil.saveNewSportSimple(session,
              packet);
            if(result){
                SaveSportsNo8PacketUtil.sendNo8Ack(session,result,3);
            }
            return true;
        }
        return false;
    }
}
```

（8）项目包 com.cloud.mina.unit_a.sportstate 的 SportLogoutState 类是运动设备的数据包退出类，实现 SportsPacketHandleState 总状态，主要功能是 SportNo8ThreeWayState 处理数据后，调用该类进行退出，关闭 session。代码实现如下：

```java
package com.cloud.mina.unit_a.sportstate;

import org.apache.mina.core.buffer.IoBuffer;
import org.apache.mina.core.session.IoSession;
import com.cloud.mina.unit_a.sportpackage.LogoutPacket;
import com.cloud.mina.util.Logger;
import com.cloud.mina.util.MLinkCRC;
/**
 * unitA 公司智能终端运动退出登录处理状态类
 *
```

```java
 * @author changyaobin
 *
 */
public class SportLogoutState implements SportsPacketHandleState {
  public boolean handlePacket(IoSession session,Object message){
      if(message != null && message instanceof LogoutPacket){
          Logger.writeLog("logout package be handled patientID:" + session.
              getAttribute("patientId")+ " company:" + session.getAttribute
              ("company")+" device_id:" + session.getAttribute("deviceId"));
          handleLogoutData(session);
          session.close(true);
          return true;
      }
      return false;
  }
  private void handleLogoutData(IoSession session){
      byte[] ack = new byte[12];
      byte[] crc_c = new byte[2];
      ack[0] = -89;
      ack[1] = -72;
      ack[2] = 0;
      ack[3] = 1;
      ack[4] = 0;
      ack[5] = 0;
      ack[6] = 0;
      ack[7] = 12;
      ack[8] = 1;
      ack[9] = 3;
      crc_c = MLinkCRC.crc16(ack);
      ack[10] = crc_c[0];
      ack[11] = crc_c[1];
      session.write(IoBuffer.wrap(ack));
      Logger.writeLog("in method handleLogoutData end the ack:" +
         "-89 -72 0 1 0 0 0 12 1 3 " + crc_c[0] + " " + crc_c[1]);
  }
}
```

2.3.6 按照通用方式进行高并发入库实现

（1）SaveSportsNo8PacketUtil 通用类是根据业务封装，引入核心工具类 C3P0Util 操作数据库，将各种数据包存入 MySQL 数据库中。C3P0 的配置文件主要包括 MySQL 的驱动、地址，用户名、密码、各种性能参数等。配置 xml 文件和工具类的代码实现如下：

```xml
<?xml version="1.0" encoding="UTF-8"?>
<c3p0-config>
<!-- 默认配置,如果没有指定则使用这个配置 -->
<default-config>
    <property name="driverClass">com.mysql.jdbc.Driver</property>
    <property name="jdbcUrl">
        <![CDATA[jdbc:mysql://localhost:3306/aggregate?useUnicode=
            true&characterEncoding=UTF-8]]>
    </property>
    <property name="user">root</property>
    <property name="password">root</property>
    <property name="acquireIncrement">3</property><!-- 如果池中数据连接
        不够时一次增长多少个 -->
    <property name="initialPoolSize">3</property>
    <property name="minPoolSize">3</property>
    <property name="maxPoolSize">20</property>
    <property name="maxStatements">0</property><!-- 一次向数据库最多可以
        发多少个 Sql 指令 -->
    <property name="idleConnectionTestPeriod">3600</property><!--每
        3600 秒检查所有连接池中的空闲连接。Default:0 -->
    <property name="maxIdleTime ">60</property><!-- seconds -->
        <!-- default:0 -->
    <property name="testConnectionOnCheckin">true</property>
    <property name="acquireRetryAttempts">10</property>
    <property name="acquireRetryDelay">1000</property><!--两次连接中间隔
        时间,单位毫秒。Default:1000 -->
    <property name="breakAfterAcquireFailure">false</property>
    <property name="checkoutTimeout">3000</property>
</default-config>
</c3p0-config>
```

（2）C3P0 工具类，用于操作 MySQL 库，封装了对数据库常用的操作。代码实现如下：

```java
package com.cloud.mina.util;

import java.sql.Connection;
import java.sql.PreparedStatement;
import java.sql.ResultSet;
import java.sql.ResultSetMetaData;
import java.sql.SQLException;
import java.sql.Statement;
import java.util.ArrayList;
import java.util.Arrays;
import java.util.HashMap;
import java.util.Iterator;
```

```java
import java.util.List;
import java.util.Map;
import javax.sql.DataSource;
import org.apache.log4j.Logger;
import com.mchange.v2.c3p0.ComboPooledDataSource;
/**
 * C3P0 工具类,用于操作 MySQL 库
 *
 * @author changyaobin
 *
 */
public class C3P0Util {
    private static Logger log = Logger.getLogger(C3P0Util.class);
    private static DataSource ds = null;

    static {
        // 默认读取 classpath 下的 c3p0-config.xml
        ds = new ComboPooledDataSource();
    }

    /**
     * 获取一个数据库连接
     *
     * @return Connection
     */
    public static Connection getConnection(){
        try {
            return ds.getConnection();
        } catch(SQLException e){
            e.printStackTrace();
            log.error("从C3P0连接池获取数据库连接失败!");
        }
        return null;
    }
    public static int getCount(String sql){
        Connection conn = null;
        Statement st = null;
        ResultSet rs = null;
        int i = 0;
        try {
            conn = getConnection();
            st = conn.createStatement();
            rs = st.executeQuery(sql);
            if(rs.next()){
                i = Integer.parseInt(rs.getString(1));
```

```java
            }
        } catch(Exception e){
            log.error(e.getMessage());
            e.printStackTrace();
        } finally {
            releaseResource(conn,st,rs);
        }
        return i;
    }
    public static boolean executeUpdate(String sql){
        Connection conn = null;
        Statement st = null;
        boolean ret = true;
        try {
            conn = getConnection();
            st = conn.createStatement();
            st.executeUpdate(sql);
            System.out.println(sql);
        } catch(Exception e){
            log.error(e.getMessage());
            e.printStackTrace();
            ret = false;
        } finally {
            releaseResource(conn,st,null);
        }
        return ret;
    }
    /**
    * 可插入任意个数 value 参数,但 sql 中的问号个数务必与 value 个数相同
    * @param sql
    * @param value
    * @return
    * @throws SQLException
    *         boolean
    */
    public static boolean insertOrUpdateData(String sql,String... value){
        log.info(sql + Arrays.toString(value));
        Connection conn = null;
        PreparedStatement prst = null;
        boolean ret = false;
        int num = 0;
        try {
            conn = getConnection();
            prst = conn.prepareStatement(sql);
            for(int i = 0;i < value.length;i++){
```

```java
                prst.setString(i + 1,value[i]);
            }
            num = prst.executeUpdate();
            if(num > 0){
                ret = true;
            }
        } catch(Exception e){
            log.error(e.getMessage());
            e.printStackTrace();
            ret = false;
        } finally {
            releaseResource(conn,prst,null);
        }
        return ret;
    }
    /**
     * 执行删除操作,并返回影响的行数
     *
     * @param sql
     * @return
     * @throws SQLException
     */
    public static int executeDelete(String sql)throws SQLException {
        Connection conn = null;
        Statement st = null;
        int result = 0;
        try {
            conn = getConnection();
            st = conn.createStatement();
            result = st.executeUpdate(sql);
            System.out.println(sql);
        } catch(Exception e){
            log.error(e.getMessage());
            e.printStackTrace();
        } finally {
            releaseResource(conn,st,null);
        }
        return result;
    }
    /**
     * JDBC 获取数据,结果组织成 HashMap 形式,key 为列名,value 为数据,一条 HashMap 对
       应于查询到的一条数据
     *
     * @param sql
     * @return
```

```java
 * @throws Exception
 */
public static List<HashMap<String,String>> getData(String sql){
    Connection conn = null;
    Statement st = null;
    ResultSet rs = null;
    List<HashMap<String,String>> result = new ArrayList<HashMap<String,
      String>>();
    try {
        conn = getConnection();
        st = conn.createStatement();
        rs = st.executeQuery(sql);
        ResultSetMetaData rsmd = rs.getMetaData();
        while(rs.next()){
            HashMap<String,String> map = new HashMap<String,String>();
            for(int i = 0;i < rsmd.getColumnCount();i++){
                map.put(rsmd.getColumnLabel(i + 1),rs.getString(i + 1));
            }
            result.add(map);
        }
    } catch(Exception e){
        log.error(e.getMessage());
        e.printStackTrace();
    } finally {
        releaseResource(conn,st,rs);
    }
    return result;
}
public static List<HashMap<String,String>> getScollData(String sql,int
  pageno,int pagesize){
    Connection conn = null;
    PreparedStatement pstat = null;
    ResultSet rs = null;
    List<HashMap<String,String>> result = new ArrayList<HashMap<String,
      String>>();
    try {
        // conn.prepareStatement(sql,游标类型,能否更新记录);
        // 游标类型:
        // ResultSet.TYPE_FORWORD_ONLY:只进游标
        // ResultSet.TYPE_SCROLL_INSENSITIVE:可滚动。但是不受其他用户对数
        // 据库更改的影响
        // ResultSet.TYPE_SCROLL_SENSITIVE:可滚动。当其他用户更改数据库时这
        // 个记录也会改变
        // 能否更新记录:
        // ResultSet.CONCUR_READ_ONLY,只读
```

```java
            // ResultSet.CONCUR_UPDATABLE,可更新
            conn = getConnection();
            pstat = conn.prepareStatement(sql,ResultSet.TYPE_SCROLL_
                INSENSITIVE,ResultSet.CONCUR_READ_ONLY);
            // 最大查询的记录条数
            pstat.setMaxRows(pageno * pagesize);
            rs = pstat.executeQuery();
            // 将游标移动到第一条记录
            rs.first();
            // 游标移动到要输出的第一条记录
            rs.relative((pageno - 1) * pagesize - 1);
            ResultSetMetaData rsmd = rs.getMetaData();
            while(rs.next()){
                HashMap<String,String> map = new HashMap<String,String>();
                for(int i = 0;i < rsmd.getColumnCount();i++){
                    map.put(rsmd.getColumnLabel(i + 1),rs.getString(i + 1));
                }
                result.add(map);
            }
        } catch(Exception e){
            e.printStackTrace();
        } finally {
            releaseResource(conn,pstat,rs);
        }
        return result;
    }
    /**
     * 释放数据库资源
     *
     * @param Connection
     *          conn
     * @param PreparedStatement
     *          pstmt
     * @param ResultSet
     *          rs
     */
    public static void releaseResource(Connection conn,PreparedStatement
      pstmt,ResultSet rs){
        try {// 关闭顺序:rs,pstmt,conn
            if(rs != null)
                rs.close();
            if(pstmt != null)
                pstmt.close();
            if(conn != null)
                conn.close();
```

```
        } catch(Exception e){
            log.error(e.getMessage());
        }
    }
    /**
     * 重载释放数据库资源
     *
     * @param Connection
     *            conn
     * @param Statement
     *            stmt
     * @param ResultSet
     *            rs
     */
    public static void releaseResource(Connection conn,Statement stmt,
      ResultSet rs){
        try {// 关闭顺序:rs,stmt,conn
            if(rs != null)
                rs.close();
            if(stmt != null)
                stmt.close();
            if(conn != null)
                conn.close();
        } catch(Exception e){
            log.error(e.getMessage());
        }
    }
    /**
     * 判断 map 是否包含 key
     *
     * @param map
     * @param key
     * @return 包含返回 value,不包含返回 null
     */
    public static Object getMapElement(Map map,String key){
        if(map != null){
            for(Iterator ite = map.entrySet().iterator();ite.hasNext();){
                Map.Entry entry =(Map.Entry)ite.next();
                if((key.trim()).equals(entry.getKey()))
                    return entry.getValue();
            }
        }
        return null;
    }
}
```

（3）数据入库操作的通用类 SaveSportsNo8PacketUtil，调用 C3P0 操作数据库。如果业务上有更多的数据包需要处理，直接在 SaveSportsNo8PacketUtil 中增加相应的接口即可。代码实现如下：

```java
package com.cloud.mina.util;

import java.util.Arrays;
import org.apache.mina.core.buffer.IoBuffer;
import org.apache.mina.core.session.IoSession;
import com.cloud.mina.unit_a.sportpackage.No8OneWayPacket;
import com.cloud.mina.unit_a.sportpackage.No8ThreeWayPacket;
import net.sf.json.JSONArray;
import net.sf.json.JSONObject;
/**
 * Tcp 协议智能终端运动设备 8 号包数据入库通用方法 类名称:SaveSportsNo8PacketUtil
 *
 *
 * @version
 */
public class SaveSportsNo8PacketUtil {
/**
 * 存储智能终端运动数据-历史包
 *
 * @param session
 * @param packet
 * @return
 */
public static boolean saveNewSportHistory(IoSession session,
    No8OneWay Packet packet){
      Logger.writeLog("存储-历史包数据!");
      JSONArray jsArr = new JSONArray();
      JsonUtil.addEntryToJsonArray(jsArr,"stepSum",packet.getStep()+ "");
      JsonUtil.addEntryToJsonArray(jsArr,"calSum",packet.getKcal()+ "");
      JsonUtil.addEntryToJsonArray(jsArr,"distanceSum",packet.
        getDistance()+ "");
      JsonUtil.addEntryToJsonArray(jsArr,"yxbsSum",packet.getEffective_
        step()+ "");
      JsonUtil.addEntryToJsonArray(jsArr,"weight",packet.getWeight()+ "");
      JsonUtil.addEntryToJsonArray(jsArr,"stride",packet.getStride()+ "");
      JsonUtil.addEntryToJsonArray(jsArr,"degreeOne",packet.getLevel1()+ "");
      JsonUtil.addEntryToJsonArray(jsArr,"degreeTwo",packet.getLevel2()+ "");
      JsonUtil.addEntryToJsonArray(jsArr,"degreeThree",packet.
        getLevel3()+ "");
      JsonUtil.addEntryToJsonArray(jsArr,"degreeFour",packet.
        getLevel4()+ "");
```

```java
        JsonUtil.addEntryToJsonArray(jsArr,"uploadType",packet.getTran_
            type()+ "");
        JsonUtil.addEntryToJsonArray(jsArr,"measureTime",DateUtil.
            formatRestfulDate(packet.getStepdate()));

        String sql = "insert into sports(phone,deviceId,apptype,dataType,
            realTime,datavalue,pname,receiveTime)values(?,?,?,?,?,?,?,now())";
        boolean ret = C3P0Util.insertOrUpdateData(sql,(String)session.
            getAttribute("patientId"),
                (String)session.getAttribute("deviceId"),(String)session.
                    getAttribute("appType"),
                PropertiesReader.getProp("DATATYPE_STEPCOUNT"),DateUtil.
                    format(packet.getStepdate()),jsArr.toString(),
                "No8-1");
        return ret;
    }
    /**
     * 存储详细包
     *
     * @param session
     * @param data
     * @param stepdate
     */
    public static boolean saveNewSportDetail(IoSession session,String data,
        String stepdate){
        Logger.writeLog("以新协议格式存储-详细包数据!");
        JSONObject jo = JSONObject.fromObject(data);
        JSONObject dataJson = jo.getJSONObject("data");
        String hour = dataJson.getString("hour");
        JSONArray dataValue = dataJson.getJSONArray("datavalue");
        JsonUtil.addEntryToJsonArray(dataValue,"hour",hour);
        JsonUtil.addEntryToJsonArray(dataValue,"measureTime",DateUtil.
            format(stepdate));
        String sql = "insert into sports(phone,deviceId,apptype,dataType,
            realTime,datavalue,pname,receiveTime)values(?,?,?,?,?,?,?,now())";
        boolean ret = C3P0Util.insertOrUpdateData(sql,(String)session.
            getAttribute("patientId"),
                (String)session.getAttribute("deviceId"),(String)session.
                    getAttribute("appType"),
                PropertiesReader.getProp("changyaobin"),DateUtil.format
                    (stepdate),dataValue.toString(),
                "No8-2");
        return ret;
    }
```

```java
/**
 * 存储智能终端运动数据-简要包
 *
 * @param session
 * @param packet
 * @return
 */
public static boolean saveNewSportSimple(IoSession session,
    No8ThreeWay Packet packet){
    Logger.writeLog("存储-简要包数据!");
    JSONArray jsArr = new JSONArray();
    JsonUtil.addEntryToJsonArray(jsArr,"stepSum",packet.getStep()+ "");
    JsonUtil.addEntryToJsonArray(jsArr,"calSum",packet.getKcal()+ "");
    JsonUtil.addEntryToJsonArray(jsArr,"distanceSum",packet.getDistance()+
        "");
    JsonUtil.addEntryToJsonArray(jsArr,"yxbsSum",packet.getEffective_
        step()+ "");
    JsonUtil.addEntryToJsonArray(jsArr,"weight",packet.getWeight()+
        "");
    JsonUtil.addEntryToJsonArray(jsArr,"stride",packet.getStride()+
        "");
    JsonUtil.addEntryToJsonArray(jsArr,"degreeOne",packet.getLevel1()+
        "");
    JsonUtil.addEntryToJsonArray(jsArr,"degreeTwo",packet.getLevel2()+
        "");
    JsonUtil.addEntryToJsonArray(jsArr,"degreeThree",packet.getLevel3()+
        "");
    JsonUtil.addEntryToJsonArray(jsArr,"degreeFour",packet.getLevel4()+
        "");
    JsonUtil.addEntryToJsonArray(jsArr,"uploadType",packet.getTran_
        type()+ "");
    JsonUtil.addEntryToJsonArray(jsArr,"measureTime",DateUtil.
        format RestfulDate(packet.getStepdate()));
    String sql = "insert into sports(phone,deviceId,apptype,dataType,
        realTime,datavalue,pname,receiveTime)values(?,?,?,?,?,?,?,now())";
    boolean ret = C3P0Util.insertOrUpdateData(sql,(String)session.
        getAttribute("patientId"),
            (String)session.getAttribute("deviceId"),(String)session.
                getAttribute("appType"),
            PropertiesReader.getProp("DATATYPE_STEPCOUNT"),DateUtil.
                format(packet.getStepdate()),jsArr.toString(),
            "No8-3");
    return ret;
}
```

```java
/**
 * unitA 智能终端运动数据返回 ack 给客户端
 *
 * @param out
 * @param result
 * @param type
 */
public static void sendNo8Ack(IoSession out,boolean result,int type){
    byte[] ack = new byte[13];
    byte[] crc_c = new byte[2];
    ack[0] = -89;
    ack[1] = -72;
    ack[2] = 0;
    ack[3] = 1;
    ack[4] = 0;
    ack[5] = 0;
    ack[6] = 0;
    ack[7] = 13;
    ack[8] = 8;
    ack[9] =(byte)type;
    if(result)
        ack[10] = 14;          // 成功
    else
        ack[10] = 15;          // 失效
    crc_c = MLinkCRC.crc16(ack);
    ack[11] = crc_c[0];
    ack[12] = crc_c[1];
    out.write(IoBuffer.wrap(ack));
    Logger.writeLog("return No8-" + type + " ACK end" + Arrays.toString
        (ack));
}
}
```

（4）配置读取操作的工具类 PropertiesReader，通过 io 可以读取业务配置文件 SysConf.properties。SysConf.properties 中定义了 mina 的端口、业务信息等。代码实现如下：

```java
package com.cloud.mina.util;

import java.io.InputStream;
import java.util.Properties;
import org.springframework.core.io.ClassPathResource;
/**
 * 读取配置文件流的工具类
 *
 * @author changyaobin
 *
```

```java
*/
public class PropertiesReader {
 private static Properties prop = new Properties();
 static {
    try {
        InputStream SystemIn = new ClassPathResource("com/Config/
           SysConf.properties").getInputStream();
        prop.load(SystemIn);
    } catch(Exception e){
        e.printStackTrace();
    }
 }
 public static String getProp(String name){
    if(prop != null && prop.containsKey(name)){
        return prop.getProperty(name,"");
    } else {
        return "";
    }
 }
}
```

（5）项目业务配置文件，主要用于定义本项目需要用到的业务配置，例如智能终端运动的数据类型，mina 的端口（spring boot 配置在 application.properties）等。具体配置如下：

```
##constants DATATYPE
changyaobin=stepDetail
DATATYPE_STEPEFFECTIVE=stepEffective
DATATYPE_STEPCOUNT=stepCount
DATATYPE_BLOODPRESSURE=bloodPressure
#mina 监听端口
tcpPort=8888
DISPATCH_SERVER_PATH=http://localhost:8080/BD_DispatchServer_Maven/
  sendData
```

（6）编写 DataTypeChangeHelper 工具类，可以实现数据类型的转换，包括将一个单字节的 byte 转换成十六进制的数，将 16 位的 short 转换成 byte 数组，将 32 位整数转换成长度为 4 的 byte 数组。代码实现如下：

```java
/**
 * 数据协议使用工具类
 *
 * @version
 */
public class DataTypeChangeHelper {
```

```java
/**
 * 将一个单字节的 byte 转换成 32 位的 int
 *
 * @param b
 *          byte
 * @return convert result
 */
public static int unsignedByteToInt(byte b){
    return(int)b & 0xFF;
}
/**
 * 将一个单字节的 byte 转换成十六进制的数
 *
 * @param b
 *          byte
 * @return convert result
 */
public static String byteToHex(byte b){
    int i = b & 0xFF;
    return Integer.toHexString(i);
}
/**
 * 将一个 4byte 的数组转换成 32 位的 int
 *
 * @param buf
 *          bytes buffer
 * @param byte[]中开始转换的位置
 * @return convert result
 */
public static long unsigned4BytesToInt(byte[] buf,int pos){
    int firstByte = 0;
    int secondByte = 0;
    int thirdByte = 0;
    int fourthByte = 0;
    int index = pos;
    firstByte =(0x000000FF &((int)buf[index]));
    secondByte =(0x000000FF &((int)buf[index + 1]));
    thirdByte =(0x000000FF &((int)buf[index + 2]));
    fourthByte =(0x000000FF &((int)buf[index + 3]));
    index = index + 4;
    return((long)(firstByte << 24 | secondByte << 16 | thirdByte << 8 |
        fourthByte))& 0xFFFFFFFFL;
}
/**
 * 将 16 位的 short 转换成 byte 数组
```

```java
 *
 * @param s
 *            short
 * @return byte[] 长度为2
 */
public static byte[] shortToByteArray(short s){
    byte[] targets = new byte[2];
    for(int i = 0;i < 2;i++){
        int offset =(targets.length - 1 - i)* 8;
        targets[i] =(byte)((s >>> offset)& 0xff);
    }
    return targets;
}
/**
 * 将32位整数转换成长度为4的byte数组
 *
 * @param s
 *            int
 * @return byte[]
 */
public static byte[] intToByteArray(int s){
    byte[] targets = new byte[2];
    for(int i = 0;i < 4;i++){
        int offset =(targets.length - 1 - i)* 8;
        targets[i] =(byte)((s >>> offset)& 0xff);
    }
    return targets;
}
/**
 * long to byte[]
 *
 * @param s
 *            long
 * @return byte[]
 */
public static byte[] longToByteArray(long s){
    byte[] targets = new byte[2];
    for(int i = 0;i < 8;i++){
        int offset =(targets.length - 1 - i)* 8;
        targets[i] =(byte)((s >>> offset)& 0xff);
    }
    return targets;
}
/** 32位int转byte[] */
public static byte[] int2byte(int res){
```

```
        byte[] targets = new byte[4];
        targets[0] =(byte)(res & 0xff);              // 最低位
        targets[1] =(byte)((res >> 8)& 0xff);        // 次低位
        targets[2] =(byte)((res >> 16)& 0xff);       // 次高位
        targets[3] =(byte)(res >>> 24);              // 最高位,无符号右移
        return targets;
}
/**
 * 将长度为2的byte数组转换为16位int
 *
 * @param res
 *            byte[]
 * @return int
 */
public static int byte2int(byte[] res){
    // res = InversionByte(res);
    // 一个byte数据左移24位变成0x??000000,再右移8位变成0x00??0000
    int targets =(res[0] & 0xff)|((res[1] << 8)& 0xff00);// | 表示安位或
    return targets;
}
}
```

（7）日期处理工具类，封装了一些常用的日期操作。代码实现如下：

```
package com.cloud.mina.util;

import java.text.ParseException;
import java.text.SimpleDateFormat;
import java.util.Calendar;
import java.util.Date;
/**
 * 日期处理工具类
 *
 * @author changyaobin
 *
 */
public class DateUtil {
 /**
  * yyyy-MM-dd
  */
 public static SimpleDateFormat ft = new SimpleDateFormat("yyyy-MM-dd");
 /**
  * yyyy-MM-dd HH:mm:ss
  */
 public static SimpleDateFormat formatter = new SimpleDateFormat("yyyy-
    MM-dd HH:mm:ss");
```

```java
/**
 * yyyyMMddHHmmss
 */
public static SimpleDateFormat compact_formatter = new SimpleDateFormat
    ("yyyyMMddHHmmss");
/**
 * 获得当前的日期 格式:yyyy-MM-dd
 */
public static String getToday(){
    Calendar cal = Calendar.getInstance();
    synchronized(ft){
        return ft.format(cal.getTime());
    }
}
/**
 * 获得当前时间 格式:yyyy-MM-dd HH:mm:ss
 */
public static String getCurrentTime(){
    Calendar cal = Calendar.getInstance();
    synchronized(formatter){
        return formatter.format(cal.getTime());
    }
}
/**
 * 将给定格式的时间字符串格式化为:yyyy-MM-dd,例如,2013-1-1 经格式化后为:
   2013-01-01
 *
 * @param str
 *          格式:yyyy-M-d
 * @return 格式:yyyy-MM-dd
 */
public static String formatDate(String str){
    try {
        synchronized(ft){
            return ft.format(ft.parse(str));
        }
    } catch(ParseException e){
        e.printStackTrace();
    }
    return null;
}
/**
 * 将给定格式的时间字符串格式化 <br>
 * 例如,2013-1-1 1:1:0 经格式化后为:2013-01-01 01:01:00 <br>
 * 2013-1-1 经格式化后为:2013-01-01 <br>
```

```java
 * 20130101 经格式化后为:2013-01-01
 *
 * @param str
 *            可接受格式:yyyy-M-d H:m:s || yyyy-M-d || yyyyMMdd
 * @return  标准格式:yyyy-MM-dd HH:mm:ss || yyyy-MM-dd || yyyy-MM-dd
 */
public static synchronized String format(String str){
    if(str == null){
        return null;
    }
    try {
        if(str.contains(":")){
            return formatter.format(formatter.parse(str));
        } else if(str.contains("-")){
            return ft.format(ft.parse(str));
        } else if(str.length()== 8){
            String date = str.substring(0,4)+ "-" + str.substring(4,6)+
                "-" + str.substring(6,8);
            return date;
        } else if(str.length()== 14){
            return formatRestfulDate(str);
        }
    } catch(ParseException e){
        e.printStackTrace();
    }
    return null;
}
/**
 * 将给定格式的时间字符串格式化 <br>
 * 例如,2013-12-1 13:14:0 经格式化后为:20131201131400
 *
 * @param str
 *            可接受格式:yyyy-M-d H:m:s
 * @return  标准格式:yyyyMMddHHmmss
 */
public static synchronized String getCompactDatetime(String str){
    try {
        return compact_formatter.format(formatter.parse(str));
    } catch(ParseException e){
        e.printStackTrace();
    }
    return null;
}
/**
 * 指定时间指定分钟偏移量后的时间
```

```java
 *
 * @param str
 *          格式:yyyy-MM-dd HH:mm:ss
 * @param add
 * @return 格式:yyyy-MM-dd HH:mm:ss
 */
public static synchronized String addMinutes(String str,int add){
    try {
        Calendar cal = Calendar.getInstance();
        cal.setTime(formatter.parse(str));
        cal.add(Calendar.MINUTE,add);
        return formatter.format(cal.getTime());
    } catch(ParseException e){
        e.printStackTrace();
    }
    return null;
}
/**
 * 指定时间指定小时偏移量后的时间
 *
 * @param str
 *          格式:yyyy-MM-dd HH:mm:ss
 * @param add
 * @return 格式:yyyy-MM-dd HH:mm:ss
 */
public static synchronized String addHour(String str,int add){
    try {
        Calendar cal = Calendar.getInstance();
        cal.setTime(formatter.parse(str));
        cal.add(Calendar.HOUR,add);
        return formatter.format(cal.getTime());
    } catch(ParseException e){
        e.printStackTrace();
    }
    return null;
}
/**
 * @param year
 * @param month
 * @param day
 * @param hour
 * @param minute
 * @param second
 * @return stepdate
 */
```

```java
public static String getStepdate(int year,int month,int day){
    try {
        if(year < 100){
            year += 2000;
        }
        String date = year + "-" + month + "-" + day;
        date = DateUtil.format(date);
        if(date.equals(DateUtil.getToday())){
            Calendar cal = Calendar.getInstance();
            synchronized(compact_formatter){
                return compact_formatter.format(cal.getTime());
            }
        } else {
            date = date.replace("-","");
            return date + "235959";
        }
    } catch(Exception e){
        e.printStackTrace();
        return null;
    }
}
/**
 * 获取简要包数据时间,若不是今天则默认为 23 点 59 分 59 秒
 *
 * @param date
 *            yyyy-MM-dd
 * @return stepdate
 */
public static String getStepdate(String date){
    try {
        date = formatDate(date);
        if(date.equals(getToday())){
            Calendar cal = Calendar.getInstance();
            synchronized(compact_formatter){
                return compact_formatter.format(cal.getTime());
            }
        } else {
            date = date.replace("-","");
            return date + "235959";
        }
    } catch(Exception e){
        e.printStackTrace();
        return null;
    }
}
```

```java
/**
 * 将给定的数字年、月、日格式化为 YYYY-MM-DD
 *
 * @param year
 * @param month
 * @param day
 * @return
 */
public static String getDate(int year,int month,int day){
    Calendar cal = Calendar.getInstance();
    if(year < 100)
        year = year + 2000;
    cal.set(year,month - 1,day);
    synchronized(ft){
        return ft.format(cal.getTime());
    }
}
public static Date getDate(String str){
    try {
        synchronized(ft){
            return ft.parse(str);
        }
    } catch(ParseException e){
        e.printStackTrace();
        return null;
    }
}
public static String getTime(int hour,int minute){
    StringBuffer sb = new StringBuffer();
    if(hour < 10){
        sb.append("0");
    }
    sb.append(hour).append(":");
    if(minute < 10){
        sb.append("0");
    }
    sb.append(minute);
    return sb.toString();
}
/**
 * 用于格式化 restful 协议中的时间传输格式的解析
 *
 * @param date
 *           yyyyMMddHHmmss
 * @return 符合规范则返回 yyyy-MM-dd HH:mm:ss 类型数据,不符合规范的不做处理
```

```java
 */
public static String formatRestfulDate(String date){
    // 判断时间是否为空,长度是否满足解析要求
    if(null != date && date.length()== 14){
        StringBuffer sb = new StringBuffer();
        sb.append(date.substring(0,4)).append('-').append(date.substring
            (4,6)).append('-')
                .append(date.substring(6,8)).append(' ').append(date.
                    substring(8,10)).append(':')
                .append(date.substring(10,12)).append(':').append
                    (date.substring(12,14));
        return sb.toString();
    }
    return date;
}
/**
 * 获得指定格式的日期的指定偏移量后的日期
 *
 * @param dateString
 *         格式:yyyy-MM-dd
 * @param num
 *
 * @return  格式:yyyy-MM-dd
 */
public static synchronized String getDayAdd(String dateString,int num){
    Calendar cal = Calendar.getInstance();
    try {
        cal.setTime(ft.parse(dateString));
    } catch(ParseExceptione){
        e.printStackTrace();
    }
    cal.add(Calendar.DAY_OF_MONTH,num);
    return ft.format(cal.getTime());
}
/**
 * 将给定格式的时间字符串转换为Calendar对象
 *
 * @param currentTime
 *         yyyy-MM-dd HH:mm:ss
 * @return
 */
public static synchronized Calendar getCalendar(String currentTime){
    Calendar cal = Calendar.getInstance();
    try {
        cal.setTime(formatter.parse(currentTime));
```

```
        } catch(ParseException e){
            e.printStackTrace();
        }
        return cal;
    }
}
```

（8）http 请求数据发送类，主要通过 httpClient 来给指定路径服务发送参数。代码实现如下：

```
package com.cloud.mina.util;

import java.util.Arrays;
import org.apache.commons.httpclient.HttpClient;
import org.apache.commons.httpclient.NameValuePair;
import org.apache.commons.httpclient.methods.PostMethod;
import org.apache.commons.httpclient.params.HttpMethodParams;
/**
 * http 客户端
 * @author Changyaobin
 *
 */
public class HttpClientUtil {
    public static boolean sendHttpData(String className,String url,
        NameValuePair[] parameter){
        HttpClient client = new HttpClient();
        PostMethod post = new PostMethod(url);
        boolean isSuccess=true;
        try {
            post.getParams().setParameter(HttpMethodParams.HTTP_CONTENT_
                CHARSET,"UTF-8");
            post.setRequestBody(parameter);
            Log.info(className+" send data:"+Arrays.deepToString(parameter));
            //设置连接超时时间
            client.getHttpConnectionManager().getParams().
                setConnection Timeout(3000);
            //设置响应超时时间
            client.getHttpConnectionManager().getParams().setSoTimeout
                (15000);
            int returnFlag = client.executeMethod(post);
            if(returnFlag!=200){
                isSuccess=false;
            }
            Log.info(className+" success receive form post:" +
                post.get StatusLine().toString()+ ",returnFlag="+returnFlag);
        } catch(Exception e){
```

```
            e.printStackTrace();
            isSuccess=false;
            Log.info(className+" fail receive form post:" +e.getMessage());
        }finally{
            if(post!=null){
                post.releaseConnection();
                Log.info(className+" post.releaseConnection()" + "is coming");
            }
        }
        return isSuccess;
    }
}
```

（9）设备 id 处理类，主要用于操作用户的手机号码与设备信息。代码实现如下：

```
package com.cloud.mina.util;

import java.sql.Connection;
import java.sql.PreparedStatement;
import java.sql.ResultSet;
import java.sql.ResultSetMetaData;
import java.sql.Statement;
import java.util.HashMap;
import org.apache.log4j.Logger;
import com.cloud.mina.util.DataTypeChangeHelper;
/**
 * 设备id处理工具类
 *
 * @author changyaobin
 *
 */
public class DeviceIDResolver {
 private static Logger log = Logger.getLogger(DeviceIDResolver.class);

 /**
  * 通过用户的设备号获取用户的手机号
  */

 public static String searchPatientidByDeviceid(String device_id){
     String patientID = "";
     // 查询
 Connection conn = null;
 PreparedStatement pst = null;
 String sql = "SELECT patientID FROM usertbl WHERE deviceID=? and
    deviceUseFlag='1' ";
     ResultSet rs = null;
```

```java
    try {
        conn = C3P0Util.getConnection();
        // 查询数据
        pst = conn.prepareStatement(sql);
        pst.setString(1,device_id);
        rs = pst.executeQuery();
        if(rs != null && rs.next()){
            patientID = rs.getString("patientID");
        }
    } catch(Exception e){
        log.error("数据库查询异常,获取patientID失败!!!" + e.getMessage());
    } finally {
        C3P0Util.releaseResource(conn,pst,rs);
    }
    return patientID;
}
/**
 * 通过用户的设备号和设备类型获取用户的手机号(目前不区分设备类型)
 *
 * @param deviceID
 * @param deviceType
 * @return String
 */
public static String searchPatientIDByDeviceidAndDeviceType(String
   deviceID,String deviceType){
    String patientID = "";
    // 查询
Connection conn = null;
PreparedStatement pst = null;
String sql = "SELECT patientID FROM usertbl WHERE deviceID=? and
   deviceUseFlag='1' ";
    ResultSet rs = null;
    try {
        conn = C3P0Util.getConnection();
        pst = conn.prepareStatement(sql);
        pst.setString(1,deviceID);
        rs = pst.executeQuery();
        if(rs != null && rs.next()){
            patientID = rs.getString("patientID");
        }
    } catch(Exception e){
        log.error("数据库查询异常,获取patientID失败!!!");
        log.error(e.getMessage());
    } finally {
        C3P0Util.releaseResource(conn,pst,rs);
```

```java
        return patientID;
}
/**
 * 通过用户的设备号获取对应用户信息(目前不区分设备类型)
 *
 * @param deviceID
 * @param deviceType
 * @return String
 */
public static HashMap<String,String> searchPatientInfoByDeviceid
   (String deviceID){
    HashMap<String,String> map = new HashMap<String,String>();
    // 查询
    Connection conn = null;
    PreparedStatement pst = null;
    String sql = "SELECT * FROM usertbl WHERE deviceID=? and
       deviceUseFlag='1' ";
    ResultSet rs = null;
    try {
        conn = C3P0Util.getConnection();
        pst = conn.prepareStatement(sql);
        pst.setString(1,deviceID);
        rs = pst.executeQuery();
        if(rs != null && rs.next()){
            ResultSetMetaData rsmd = rs.getMetaData();
            for(int i = 0;i < rsmd.getColumnCount();i++){
                map.put(rsmd.getColumnLabel(i + 1),rs.getString(i + 1));
            }
        }
    } catch(Exception e){
        log.error("数据库查询异常,获取patientInfo失败!!!");
        log.error(e.getMessage());
    } finally {
        C3P0Util.releaseResource(conn,pst,rs);
    }
    return map;
}
/**
 * 通过用户的设备号和设备类型获取用户的单位代号
 *
 * @param deviceID
 * @param deviceType
 * @return String
 */
```

```java
public static String searchCompanyByDeviceidAndDeviceType(String
    deviceID,String deviceType){
    String company = "";
    // 查询
    Connection conn = null;
    PreparedStatement pst = null;
    String sql = "SELECT company FROM usertbl WHERE deviceID=? and
        deviceType=? and deviceUseFlag='1' ";
    ResultSet rs = null;
    try {
        conn = C3P0Util.getConnection();
        pst = conn.prepareStatement(sql);
        pst.setString(1,deviceID);
        pst.setString(2,deviceType);
        rs = pst.executeQuery();
        if(rs != null && rs.next()){
            company = rs.getString("company");
        }
    } catch(Exception e){
        log.error("数据库查询异常,获取company失败!!!");
        log.error(e.getMessage());
    } finally {
        C3P0Util.releaseResource(conn,pst,rs);
    }
    return company;
}
/**
 * 通过用户的设备号判断用户所在单位
 */
public static String searchCompanyByDeviceid(String device_id){
    String company = "";
    if(device_id != null){
        Connection conn = null;
        Statement st = null;
        ResultSet rs = null;
        try {
            conn = C3P0Util.getConnection();
            // 查询数据
            st = conn.createStatement();
            String query = "SELECT company FROM usertbl WHERE
                deviceUseFlag='1' and deviceID='" + device_id + "'";
            rs = st.executeQuery(query);
            if(rs != null && rs.next()){
                company = rs.getString("company");
            }
```

```
            } catch(Exception e){
                System.out.println(e.getMessage());
            } finally {
                C3P0Util.releaseResource(conn,st,rs);
            }
        }
        return company;
    }
    /**
     * 从字节数组中解析出设备号,忽略 null、空格、回车、换行 null,ASCII 码 0 换
       行,ASCII 码 10 回车,ASCII 码 13
     * 空格,ASCII 码 32
     *
     * @param b
     *           字节数组
     * @param deviceIDBeginIndex
     *           设备号开始下标
     * @return String
     */
    public static String getDeviceIDFromBytes(byte[] b,int
      deviceIDBegin Index){
        StringBuffer sbid = new StringBuffer();
        for(int j = deviceIDBeginIndex;j < deviceIDBeginIndex + 16;j++){
            // 空格和回车忽略
            if(b[j] != 0 && b[j] != 13 && b[j] != 10 && b[j] != 32){
                sbid.append((char)b[j]);
            }
        }
        log.info("The deviceID is " + sbid.toString());
        return sbid.toString();
    }
    /**
     * 从字节数组中解析出设备号,忽略 null、空格、回车、换行 null,ASCII 码 0 换行,
       ASCII 码 10 回车,ASCII 码 13
     * 空格,ASCII 码 32
     *
     * @param b
     *           字节数组
     * @param passwordBeginIndex
     *           开始下标
     * @return String
     */
    public static String getPasswordFromBytes(byte[] b,int
      passwordBegin Index){
        StringBuffer sbid = new StringBuffer();
```

```java
        for(int j = passwordBeginIndex;j < passwordBeginIndex + 16;j++){
            // 空格和回车忽略
            if(b[j] != 0 && b[j] != 13 && b[j] != 10 && b[j] != 32){
                sbid.append((char)b[j]);
            }
        }
        log.info("The password is " + sbid.toString());
        return sbid.toString();
    }
    /**
     * 从字节数组中解析出设备前缀
     *
     * @param b
     *            字节数组
     * @param prefixBeginIndex
     *            设备前缀开始下标
     * @return String
     */
    public static String getDeviceIDPrefix(byte[] b,int prefixBeginIndex){
        StringBuffer prefix = new StringBuffer();
        for(int i = prefixBeginIndex;i <(prefixBeginIndex + 5);i++){
            // 空格忽略
            if(b[i] != 0){
                prefix.append((char)b[i]);
            }
        }
        log.info("The deviceID prefix is " + prefix.toString());
        return prefix.toString();
    }
    /**
     * 从字节数组中解析出固件版本
     *
     * @param b
     *            字节数组
     * @param versionBeginIndex
     *            固件版本开始下标
     * @return String
     */
    public static String getFirmwareVersion(byte[] b,int versionBeginIndex){
        StringBuffer version = new StringBuffer();
        for(int i = versionBeginIndex;i <(versionBeginIndex + 4);i++){
            version.append(b[i]);
        }
        log.info("The firmware version is " + version.toString());
        return version.toString();
```

```java
}
/**
 * add by RCM on 2014/02/13 获取 8 号包方式二的设备号前缀
 * @param b
 * @return
 */
public static String getNo8PackageDevicePrefix(byte[] b){
    byte length[] = new byte[4];
    length[0] = b[4];
    length[1] = b[5];
    length[2] = b[6];
    length[3] = b[7];
    long lengthvalue = DataTypeChangeHelper.unsigned4BytesToInt(length,0);
    return getDeviceIDPrefix(b,(int)(lengthvalue - 23));
}
/**
 * 获取用户 appType
 * @param deviceId
 * @return
 */
public static String searchAppTypeByDeviced(String deviceId){
    String appType = "";
    // 查询
Connection conn = null;
PreparedStatement pst = null;
String sql = "SELECT appType FROM usertbl WHERE deviceID=? and
  device UseFlag='1' ";
    ResultSet rs = null;
    try {
        conn = C3P0Util.getConnection();
        pst = conn.prepareStatement(sql);
        pst.setString(1,deviceId);
        // pst.setString(2,deviceType);
        rs = pst.executeQuery();
        if(rs != null && rs.next()){
            appType = rs.getString("appType");
        }
    } catch(Exception e){
        log.error("数据库查询异常,获取 appType 失败!!!");
        log.error(e.getMessage());
    } finally {
        C3P0Util.releaseResource(conn,pst,rs);
    }
```

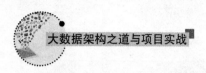

```
        return appType;
    }
}
```

（10）Json工具类，封装了一些对json格式数据的常用操作。代码实现如下：

```
package com.cloud.mina.util;

import java.beans.IntrospectionException;
import java.beans.Introspector;
import java.beans.PropertyDescriptor;
import java.util.Date;
import java.util.List;
import net.sf.json.JSONArray;
import net.sf.json.JSONObject;
import org.apache.commons.lang.StringUtils;
/**
 * json操作类
 */
public class JsonUtil {
    /**
     *
     * @param json
     * @return
     */
    public static boolean isBadJson(String json){
        return !isGoodJson(json);
    }
    /**
     * 校验字符串是否符合json格式
     *
     * @param json
     * @return
     */
    public static boolean isGoodJson(String json){
        if(StringUtils.isBlank(json)){
            return false;
        }
        try {
            JSONObject.fromObject(json);
        } catch(Exception e){
            return false;
        }
        return true;
    }
    /**
```

```java
 * 功能描述:传入一个javabean对象生成一个json格式的字符串
 *
 * @param bean
 * @return String
 */
public static String beanToJson(Object bean){
    StringBuilder json = new StringBuilder();
    json.append("{");
    PropertyDescriptor[] props = null;
    try {
        props = Introspector.getBeanInfo(bean.getClass(),Object.class).
            getPropertyDescriptors();
    } catch(IntrospectionException e){
    }
    if(props != null){
        for(int i = 0;i < props.length;i++){
            try {
                String name = objectToJson(props[i].getName());
                String value = objectToJson(props[i].getReadMethod().
                    invoke(bean));
                json.append(name);
                json.append(":");
                json.append(value);
                json.append(",");
            } catch(Exception e){
                e.printStackTrace();
            }
        }
        json.setCharAt(json.length()- 1,'}');
    } else {
        json.append("}");
    }
    return json.toString();
}
/**
 * 功能描述:传入任意一个Object对象生成一个json格式的字符串
 *
 * @param object
 *          任意对象
 * @return String
 */
@SuppressWarnings("rawtypes")
public static String objectToJson(Object object){
    StringBuilder json = new StringBuilder();
    if(object == null){
```

```java
                json.append("\"\"");// 输出双引号
            } else if(object instanceof String || object instanceof Integer ||
                    object instanceof Long || object instanceof Boolean ||
                    object instanceof Date || object instanceof java.sql.Date){
                json.append("\"").append(object.toString()).append("\"");
            } else if(object instanceof List){
                json.append(listToJson((List)object));
            } else {
                json.append(beanToJson(object));
            }
        return json.toString();
    }
    /**
     * 功能描述:通过传入一个列表对象,调用指定方法将列表中的数据生成一个JSON规格指定字
     符串
     *
     * @param list
     *          列表对象
     * @return String
     */
    public static String listToJson(List<?> list){
        StringBuilder json = new StringBuilder();
        json.append("[");
        if(list != null && list.size()> 0){
            for(Object obj:list){
                json.append(objectToJson(obj));
                json.append(",");
            }
            json.setCharAt(json.length()- 1,']');
        } else {
            json.append("]");
        }
        return json.toString();
    }
    /**
     * get jsonobject's field value
     *
     * @param jo
     * @param field
     */
    public static JSONArray getJsonParamterArray(JSONObject jo,String field){
        JSONArray re_str = new JSONArray();
        if(jo != null && jo.containsKey(field)){
            re_str = jo.getJSONArray(field);
        }
```

```java
        return re_str;
}
/**
 * 从json对象中根据key获取某一个字段值
 *
 * @param jo
 * @param field
 */
public static JSONObject getJsonParamterObject(JSONObject jo,String field){
    JSONObject re_str = new JSONObject();
    if(jo != null && jo.containsKey(field)){
        re_str = jo.getJSONObject(field);
    }
    return re_str;
}
/**
 * 从json对象中根据key获取某一个字段值(返回值为字符串)
 *
 * @param jo
 * @param field
 */
public static String getJsonParamterString(JSONObject jo,String field){
    String re_str = "";
    if(jo != null && jo.containsKey(field)){
        re_str = jo.getString(field);
    }
    return re_str;
}
/**
 * 从json对象中根据key获取某一个字段值(返回值为int)
 *
 * @param jo
 * @param field
 */
public static int getJsonParamterInteger(JSONObject jo,String field){
    int re_str = 0;
    if(jo != null && jo.containsKey(field)){
        re_str = jo.getInt(field);
    }
    return re_str;
}
/**
 * 从json对象中根据key获取某一个字段值(返回值为long)
 * @param jo
 * @param field
```

```java
     */
    public static long getJsonParamterLong(JSONObject jo,String field){
        long re_str = 0l;
        if(jo != null && jo.containsKey(field)){
            re_str = jo.getLong(field);
        }
        return re_str;
    }
    /**
     * 从json对象中根据key获取某一个字段值(返回值为double)
     *
     * @param jo
     * @param field
     */
    public static double getJsonParamterDouble(JSONObject jo,String field){
        double re_str = 0d;
        if(jo != null && jo.containsKey(field)){
            re_str = jo.getDouble(field);
        }
        return re_str;
    }
    /**
     * 添加一个json对象(Key/Value)到JSON数组末尾
     *
     * @param arr
     * @param key
     * @param value
     * @return JSONArray
     */
    public static JSONArray addEntryToJsonArray(JSONArray arr,String key,
      String value){
        JSONObject jo = new JSONObject();
        jo.put(key,value);
        arr.add(jo);
        return arr;
    }
}
```

(11) 日志处理通用类，主要用于记录日志信息到文件中。代码实现如下：

```java
package com.cloud.mina.util;

public class Logger {
    static org.apache.log4j.Logger logger = null;
    static {
        logger = org.apache.log4j.Logger.getLogger(Logger.class);
    }
```

```
/**
 * @param log
 */
public static void writeLog(String log){
    logger.info(log);
}
/**
 * @param log
 */
public static void errorLog(String log){
    logger.error(log);
}
```

（12）协议专用类，主要用于根据协议，返回规定的格式给客户端。代码实现如下：

```
package com.cloud.mina.util;
/**
 * MLink 协议专用 类名称 修改人:修改时间:修改备注:
 *
 * @version
 */
public final class MLinkCRC {
private static final int[] crcTable = { 0x0000,0x1189,0x2312,0x329b,
   0x4624,0x57ad,0x6536,0x74bf,0x8c48,
        0x9dc1,0xaf5a,0xbed3,0xca6c,0xdbe5,0xe97e,0xf8f7,0x1081,0x0108,
           0x3393,0x221a,0x56a5,0x472c,
        0x75b7,0x643e,0x9cc9,0x8d40,0xbfdb,0xae52,0xdaed,0xcb64,0xf9ff,
           0xe876,0x2102,0x308b,0x0210,
        0x1399,0x6726,0x76af,0x4434,0x55bd,0xad4a,0xbcc3,0x8e58,0x9fd1,
           0xeb6e,0xfae7,0xc87c,0xd9f5,
        0x3183,0x200a,0x1291,0x0318,0x77a7,0x662e,0x4b5,0x453c,0xbdcb,
           0xac42,0x9ed9,0x8f50,0xfbef,
        0xea66,0xd8fd,0xc974,0x4204,0x538d,0x6116,0x709f,0x0420,0x15a9,
           0x2732,0x36bb,0xce4c,0xdfc5,
        0xed5e,0xfcd7,0x8868,0x99e1,0xab7a,0xbaf3,0x5285,0x430c,0x7197,
           0x601e,0x14a1,0x0528,0x37b3,
        0x263a,0xdecd,0xcf44,0xfddf,0xec56,0x98e9,0x8960,0xbbfb,0xaa72,
           0x6306,0x728f,0x4014,0x519d,
        0x2522,0x34ab,0x0630,0x17b9,0xef4e,0xfec7,0xcc5c,0xddd5,0xa96a,
           0xb8e3,0x8a78,0x9bf1,0x7387,
        0x620e,0x5095,0x411c,0x35a3,0x242a,0x16b1,0x0738,0xffcf,0xee46,
           0xdcdd,0xcd54,0xb9eb,0xa862,
        0x9af9,0x8b70,0x8408,0x9581,0xa71a,0xb693,0xc22c,0xd3a5,0xe13e,
           0xf0b7,0x0840,0x19c9,0x2b52,
        0x3adb,0x4e64,0x5fed,0x6d76,0x7cff,0x9489,0x8500,0xb79b,0xa612,
```

```
            0xd2ad,0xc324,0xf1bf,0xe036,
        0x18c1,0x0948,0x3bd5,0x2a5a,0x5ee5,0x4f6c,0x7df7,0x6c7e,0xa50a,
            0xb483,0x8618,0x9791,0xe32e,
        0xf2a7,0xc03c,0xd1b5,0x2942,0x38cb,0x0a50,0x1bd9,0x6f66,0x7eef,
            0x4c74,0x5dfd,0xb58b,0xa402,
        0x9699,0x8710,0xf3af,0xe226,0xd0bd,0xc134,0x39c3,0x284a,0x1ad1,
            0x0b58,0x7fe7,0x6e6e,0x5cf5,
        0x4d7c,0xc60c,0xd785,0xe51e,0xf497,0x8028,0x91a1,0xa33a,0xb2b3,
            0x4a44,0x5bcd,0x6956,0x78df,
        0x0c60,0x1de9,0x2f72,0x3efb,0xd68d,0xc704,0xf59f,0xe416,0x90a9,
            0x8120,0xb3bb,0xa232,0x5ac5,
        0x4b4c,0x79d7,0x685e,0x1ce1,0x0d68,0x3ff3,0x2e7a,0xe70e,0xf687,
            0xc41c,0xd595,0xa12a,0xb0a3,
        0x8238,0x93b1,0x6b46,0x7acf,0x4854,0x59dd,0x2d62,0x3ceb,0x0e70,
            0x1ff9,0xf78f,0xe606,0xd49d,
        0xc514,0xb1ab,0xa022,0x92b9,0x8330,0x7bc7,0x6a4e,0x58d5,0x495c,
            0x3de3,0x2c6a,0x1ef1,0x0f78 };

public static byte[] crc16(byte[] ba){
    byte[] crc = new byte[2];
    int crc16 = 0x0000;
    for(byte b:ba){
        crc16 =(crc16 >>> 8)^ crcTable[(crc16 ^ b)& 0xff];
    }
    crc[0] =(byte)((crc16 >> 0)& 0xFF);
    crc[1] =(byte)((crc16 >> 8)& 0xFF);
    return crc;
}
}
```

2.3.7 客户端模拟器工具类进行高并发测试

模拟器采用多线程，模拟上千个设备同时并发测试，因为是对字节流的测试，最好采用自主研发模拟器。代码实现如下：

```
package com.simulator.client;
/**
 * 模拟器多线程客户端
 * @author changyaobin
 *
 */
public class ThreadClient{
public static void main(String args[])throws Exception {
    String deviceId="0526";
    for(int i = 0;i <1000;i++){
        new Thread(new StepcountPackageThread(Integer.parseInt
```

```java
            (deviceId+i))).start();
        }
    }
}
package com.simulator.client;

import java.io.IOException;
import java.io.InputStream;
import java.io.OutputStream;
import java.net.InetAddress;
import java.net.Socket;
import java.util.Arrays;
import java.util.Calendar;
import org.apache.log4j.Logger;
import com.simulator.utils.ByteUtil;
import com.simulator.utils.MLinkCRC;
/**
 * 智能终端运动数据包模拟器多线程处理类
 * @author changyaobin
 *
 */
public class StepcountPackageThread implements Runnable {
    private static Logger log = Logger.getLogger(StepcountPackageThread.
        class);
    private int deviceId;
    private String device;
    public StepcountPackageThread(int deviceId){
        this.deviceId = deviceId;
        device = "0" + Integer.toString(deviceId);
    }
    @Override
    public void run(){
        byte[] serverIp = new byte[4];
        serverIp[0] =(byte)127;
        serverIp[1] =(byte)0;
        serverIp[2] =(byte)0;
        serverIp[3] =(byte)1;
        InputStream in = null;
        OutputStream out = null;
        byte[] b = new byte[1024];
        try {
            InetAddress address = InetAddress.getByAddress(serverIp);
            Socket client = new Socket(address,8888);
            in = client.getInputStream();
```

```
            out = client.getOutputStream();
            client.setKeepAlive(true);
            sentLoginPackage(out);
            Thread.sleep(3000);
            in.read(b);
            log.info("login:" + Arrays.toString(b));
            sentPackage8One(out);
            Thread.sleep(1000);
            in.read(b);
            log.info("1:" + Arrays.toString(b));
            System.out.println("1:" + Arrays.toString(b));
            sendPackage8Two(out);
            Thread.sleep(1000);
            in.read(b);
            log.info("2:" + Arrays.toString(b));
            System.out.println("2:" + Arrays.toString(b));
            sendPackage8Three(out);
            Thread.sleep(3000);
            in.read(b);
            log.info("3:" + Arrays.toString(b));
            System.out.println("3:" + Arrays.toString(b));
            sentLogoutPackage(out);
            Thread.sleep(1000);
            in.read(b);
            log.info("logout:" + Arrays.toString(b));
            // 数据发送完毕后,关闭流和socket连接
            out.close();
            client.close();
        } catch(Exception e){
            e.printStackTrace();
        }
    }
    /**
     * 发送登录数据包
     *
     * @param out
     * @throws IOException
     */
    private void sentLoginPackage(OutputStream out)throws IOException {
        log.info("sentLoginPackage... " + device);
        byte[] sendData = null;
        sendData = new byte[44];
        sendData[0] = -89;
        sendData[1] = -72;
```

```java
            sendData[2] = 0;
            sendData[3] = 1;
            ByteUtil.putIntByLarge(sendData,44,4);
            sendData[8] = 1;
            sendData[9] = -128;
            String deviceIdStr = this.device;
            char[] array = deviceIdStr.toCharArray();
            for(int i = 0;i < array.length;i++){
                sendData[10 + i] =(byte)array[i];
            }
            // crc
            sendData[42] = 0;
            sendData[43] = 0;
            out.write(sendData);
            out.flush();
}
/**
 * 发送1号数据包
 *
 * @param out
 * @throws IOException
 * @throws InterruptedException
 */
private void sentPackage8One(OutputStream out)throws IOException,
    InterruptedException {
        log.info("sentPackage8One... " + device);
        byte[] sendData = null;
        sendData = new byte[68];
        sendData[0] = -89;
        sendData[1] = -72;
        sendData[2] = 0;
        sendData[3] = 1;
        ByteUtil.putIntByLarge(sendData,68,4);
        sendData[8] = 8;
        sendData[9] = 1;
        sendData[10] = 1;
        Calendar c = Calendar.getInstance();
        sendData[11] =(byte)(c.get(Calendar.YEAR)- 2000);
        sendData[12] =(byte)(c.get(Calendar.MONTH)+ 1);
        sendData[13] =(byte)c.get(Calendar.DATE);
        sendData[18] = 3;
        sendData[19] = 5;
        sendData[20] = 0;
        sendData[21] = 3;
```

```java
        sendData[22] = 60;
        sendData[23] = 70;
        sendData[24] = 70;
        ByteUtil.putIntByLarge(sendData,(this.deviceId + 1)* 1000,25);
        ByteUtil.putIntByLarge(sendData,(this.deviceId + 1)* 1000,29);
        ByteUtil.putIntByLarge(sendData,(this.deviceId + 1) * (int)(10 *
            Math.random()),33);
        ByteUtil.putShortByLarge(sendData,(short)((this.deviceId + 1)* 10),
            37);
        ByteUtil.putShortByLarge(sendData,(short)((this.deviceId + 1)* 10),
            39);
        ByteUtil.putShortByLarge(sendData,(short)((this.deviceId + 1)* 10),
            41);
        ByteUtil.putShortByLarge(sendData,(short)((this.deviceId + 1)* 10),
            43);
        sendData[45] = 'D';
        sendData[46] = 'E';
        sendData[47] = 'V';
        sendData[48] = 'I';
        sendData[49] = 'D';
        String deviceIdStr = this.device;
        char[] array = deviceIdStr.toCharArray();
        for(int i = 0;i < array.length;i++){
            sendData[50 + i] =(byte)array[i];
        }
        // crc
        sendData[66] = 0;
        sendData[67] = 0;
        out.write(sendData);
        out.flush();
    }
    /**
     * 发送2号数据包
     *
     * @param out
     * @throws IOException
     */
    private void sendPackage8Two(OutputStream out)throws IOException {
        log.info("sendPackage8Two..." + device);
        System.out.println("sendPackage8Two..." + device);
        byte[] sendData = null;
        int hours = 8;
        sendData = new byte[33 + 114 * hours];
        // Header(4)
```

```
        sendData[0] = -89;
        sendData[1] = -72;
        sendData[2] = 0;
        sendData[3] = 1;
        // Length(4)
        ByteUtil.putIntByLarge(sendData,33 + 114 * hours,4);
        sendData[8] = 8;
        sendData[9] = 2;
        // USEDATA(114/h)
        Calendar c = Calendar.getInstance();
        for(int j = 0;j < hours;j++){
            // Year(2)
            ByteUtil.putShortByLarge(sendData,(short)(c.get(Calendar.
               YEAR)),j * 114 + 10);
            sendData[j * 114 + 12] = 0;
            sendData[j * 114 + 13] =(byte)(c.get(Calendar.MONTH)+ 1);
            sendData[j * 114 + 14] =(byte)c.get(Calendar.DATE);
            sendData[j * 114 + 15] =(byte)j;
            for(int i = 0;i < 12;i++){
                ByteUtil.putShortByLarge(sendData,(short)(i * 10 + 3),
                   j * 114 + 16 + i * 2);
            }
            for(int i = 0;i < 12;i++){
                ByteUtil.putShortByLarge(sendData,(short)(i * 100 + 3),
                   j * 114 + 39 + i * 2);
            }
            for(int i = 0;i < 12;i++){
                sendData[j * 114 + 63 + i] = 5;
            }
        }
        // deviceID
        String deviceIdStr = this.device;
        char[] array = deviceIdStr.toCharArray();
        for(int i = 0;i < array.length;i++){
            sendData[15 + 114 * hours + i] =(byte)array[i];
        }
        sendData[145] = 0;
        sendData[146] = 0;
        out.write(sendData);
        out.flush();
    }
    /**
     * 发送 3 号数据包
     *
```

```java
 * @param out
 * @throws IOException
 */
private void sendPackage8Three(OutputStream out)throws IOException {
    log.info("sendPackage8Three..." + device);
    byte[] sendData = null;
    sendData = new byte[88];
    // Header(4)
    sendData[0] = -89;
    sendData[1] = -72;
    sendData[2] = 0;
    sendData[3] = 1;
    // Length(4)
    ByteUtil.putIntByLarge(sendData,88,4);
    // Type(2)
    sendData[8] = 8;
    sendData[9] = 3;
    // USEDATA
    sendData[10] = 1;
    Calendar c = Calendar.getInstance();
    sendData[11] =(byte)(c.get(Calendar.YEAR)- 2000);
    sendData[12] =(byte)(c.get(Calendar.MONTH)+ 1);
    sendData[13] =(byte)c.get(Calendar.DATE);
    sendData[18] = 3;
    sendData[19] = 5;
    sendData[20] = 0;
    sendData[21] = 3;
    sendData[22] = 60;
    sendData[23] = 70;
    sendData[24] = 70;
    ByteUtil.putIntByLarge(sendData,(this.deviceId + 1) * 1000,25);
    ByteUtil.putIntByLarge(sendData,(this.deviceId + 1) * 1000,29);
    ByteUtil.putIntByLarge(sendData,(this.deviceId + 1) * (int)(10 *
       Math.random()),33);
    ByteUtil.putShortByLarge(sendData,(short)((this.deviceId + 1)*
       10),37);
    ByteUtil.putShortByLarge(sendData,(short)((this.deviceId + 1)*
       10),39);
    ByteUtil.putShortByLarge(sendData,(short)((this.deviceId + 1)*
       10),41);
    ByteUtil.putShortByLarge(sendData,(short)((this.deviceId + 1)*
       10),43);
    String deviceIdStr = this.device;
    sendData[65] = 'D';
```

```java
        sendData[66] = 'E';
        sendData[67] = 'V';
        sendData[68] = 'I';
        sendData[69] = 'D';
        char[] array = deviceIdStr.toCharArray();
        for(int i = 0;i < array.length;i++){
            sendData[70 + i] =(byte)array[i];
        }
        // crc
        sendData[86] = 0;
        sendData[87] = 0;
        out.write(sendData);
        out.flush();
}
/**
 * 发送退出数据包
 *
 * @param out
 * @throws IOException
 */
private void sentLogoutPackage(OutputStream out)throws IOException {
    log.info("sentLogoutPackage... ");
    byte[] sendData = null;
    sendData = new byte[12];
    byte[] crc_c = new byte[2];
    sendData[0] = -89;
    sendData[1] = -72;
    sendData[2] = 0;
    sendData[3] = 1;
    ByteUtil.putIntByLarge(sendData,12,4);
    sendData[8] = 1;
    sendData[9] = 3;
    crc_c = MLinkCRC.crc16(sendData);
    // crc
    sendData[10] = crc_c[0];
    sendData[11] = crc_c[1];
    out.write(sendData);
    out.flush();
}
```

（1）测试验证结果如图 2-15 所示，可以实现单机 3000TPS 以上的高并发测试要求，如果是服务器可以实现上万个并发测试。

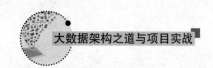

图 2-15 数据入库结果

（2）数据表的设计。

① 数据表 usertbl，包括用户 id（名称）、deviceID（设备号）、patientID（手机号）、dataType（数据类型）、appType（应用类型）等主要字段。

② 数据产品表 product，包括 appID（应用 ID）、appName（应用名称）、appSendFlag（应用发送标志）、appUrl（应用的目标地址）。

③ 数据产品表 sports，包括 id、phone（手机号）、deviceID（设备编号）、dataType（数据类型）、appType（应用类型）、pname（用户姓名）、sendFlag（发送标识）、dataValue（数据内容）等。

④ 本章用到的表的创建参考如下 sql 语句：

```
/*!40101 SET NAMES utf8 */;
/*!40101 SET SQL_MODE=''*/;
/*!40014 SET @OLD_UNIQUE_CHECKS=@@UNIQUE_CHECKS,UNIQUE_CHECKS=0 */;
/*!40014 SET @OLD_FOREIGN_KEY_CHECKS=@@FOREIGN_KEY_CHECKS,FOREIGN_KEY_
  CHECKS=0 */;
/*!40101 SET @OLD_SQL_MODE=@@SQL_MODE,SQL_MODE='NO_AUTO_VALUE_ON_ZERO'
  */;
/*!40111 SET @OLD_SQL_NOTES=@@SQL_NOTES,SQL_NOTES=0 */;
CREATE DATABASE /*!32312 IF NOT EXISTS*/'aggregate' /*!40100 DEFAULT
  CHARACTER SET utf8 COLLATE utf8_unicode_ci */;
USE 'aggregate';
/*Table structure for table 'datatype' */
DROP TABLE IF EXISTS 'datatype';
CREATE TABLE 'datatype'(
  'datatypeID' int(11)NOT NULL AUTO_INCREMENT,
  'dataDesc' varchar(50)COLLATE utf8_bin DEFAULT NULL,
  'dataTypeName' varchar(50)COLLATE utf8_bin DEFAULT NULL,
```

```sql
  'tableName' varchar(50)COLLATE utf8_bin DEFAULT NULL,
  'updateTime' varchar(20)COLLATE utf8_bin DEFAULT NULL,
  PRIMARY KEY('datatypeID'),
  UNIQUE KEY 'datatypeID'('datatypeID')
)ENGINE=InnoDB AUTO_INCREMENT=12 DEFAULT CHARSET=utf8 COLLATE=utf8_bin;

/*Table structure for table 'product' */
DROP TABLE IF EXISTS 'product';
CREATE TABLE 'product' (
  'appID' int(11)NOT NULL AUTO_INCREMENT,
  'appDesc' varchar(50)COLLATE utf8_bin DEFAULT NULL,
  'appName' varchar(50)COLLATE utf8_bin DEFAULT NULL,
  'appSendFlag' varchar(50)COLLATE utf8_bin DEFAULT NULL,
  'appToggle' varchar(50)COLLATE utf8_bin DEFAULT NULL,
  'appUrl' varchar(100)COLLATE utf8_bin DEFAULT NULL,
  'updateTime' varchar(20)COLLATE utf8_bin DEFAULT NULL,
  'appQueueName' varchar(50)COLLATE utf8_bin DEFAULT NULL,
  PRIMARY KEY('appID'),
  UNIQUE KEY 'appID' ('appID')
)ENGINE=InnoDB AUTO_INCREMENT=6 DEFAULT CHARSET=utf8 COLLATE=utf8_bin;
/*Table structure for table 'product_datatype' */
DROP TABLE IF EXISTS 'product_datatype';
CREATE TABLE 'product_datatype' (
  'id' int(11)NOT NULL AUTO_INCREMENT,
  'dataTypeID' int(11)DEFAULT NULL,
  'productID' int(11)DEFAULT NULL,
  'toggle' varchar(20)COLLATE utf8_bin DEFAULT NULL COMMENT 'on/off',
  'updatetTime' varchar(20)COLLATE utf8_bin DEFAULT NULL,
  PRIMARY KEY('id'),
  UNIQUE KEY 'id'('id'),
  KEY 'FK7DF1B3449BED0BB'('dataTypeID'),
  KEY 'FK7DF1B3425633361'('productID'),
  CONSTRAINT 'FK7DF1B3425633361' FOREIGN KEY('productID')REFERENCES
    'product'('appID'),
  CONSTRAINT 'FK7DF1B3449BED0BB' FOREIGN KEY('dataTypeID')REFERENCES
    'datatype' ('datatypeID')
)ENGINE=InnoDB AUTO_INCREMENT=17 DEFAULT CHARSET=utf8 COLLATE=utf8_bin;
/*Table structure for table 'sports' */
DROP TABLE IF EXISTS 'sports';
CREATE TABLE 'sports' (
  'id' int(10)NOT NULL AUTO_INCREMENT COMMENT 'id',
  'phone' varchar(20)COLLATE utf8_bin DEFAULT NULL COMMENT '手机号',
  'deviceID' varchar(50)COLLATE utf8_bin DEFAULT NULL COMMENT '设备编号',
  'dataType' varchar(50)COLLATE utf8_bin DEFAULT NULL COMMENT '数据类型',
```

```sql
  'appType' varchar(100)COLLATE utf8_bin DEFAULT NULL COMMENT '应用类型',
  'pname' varchar(50)COLLATE utf8_bin DEFAULT NULL COMMENT '用户姓名',
  'sendFlag' char(1)COLLATE utf8_bin DEFAULT '0' COMMENT '发送标识',
  'receiveTime' varchar(20)COLLATE utf8_bin DEFAULT NULL COMMENT '接收
    时间',
  'realTime' varchar(20)COLLATE utf8_bin DEFAULT NULL COMMENT '数据真实
    时间',
  'sendTime' varchar(20)COLLATE utf8_bin DEFAULT NULL COMMENT '转发时间',
  'deviceType' varchar(50)COLLATE utf8_bin DEFAULT NULL COMMENT '设备类
    型:手机计步为 PHONE',
  'dataValue' varchar(1500)COLLATE utf8_bin DEFAULT NULL COMMENT '数据
    内容',
  'AppA_flag' char(1)COLLATE utf8_bin DEFAULT '0' COMMENT 'AppA 发送标识',
  'AppB_flag' char(1)COLLATE utf8_bin DEFAULT '0' COMMENT 'AppB 发送标识',
  'AppC_flag' char(1)COLLATE utf8_bin DEFAULT '0' COMMENT 'AppC 发送标识',
  'AppD_flag' char(1)COLLATE utf8_bin DEFAULT '0' COMMENT 'AppD 发送标识',
  PRIMARY KEY('id')
)ENGINE=InnoDB AUTO_INCREMENT=191 DEFAULT CHARSET=utf8 COLLATE=utf8_
  bin;
/*Table structure for table 'user' */
DROP TABLE IF EXISTS 'user';
CREATE TABLE 'user' (
  'id' int(11)NOT NULL AUTO_INCREMENT,
  'apptype' varchar(255)COLLATE utf8_bin DEFAULT NULL,
  'email' varchar(255)COLLATE utf8_bin DEFAULT NULL,
  'idcard' varchar(255)COLLATE utf8_bin DEFAULT NULL,
  'mark' varchar(255)COLLATE utf8_bin DEFAULT NULL,
  'name' varchar(255)COLLATE utf8_bin DEFAULT NULL,
  'password' varchar(255)COLLATE utf8_bin DEFAULT NULL,
  'phone' varchar(255)COLLATE utf8_bin DEFAULT NULL,
  'realname' varchar(255)COLLATE utf8_bin DEFAULT NULL,
  PRIMARY KEY('id')
)ENGINE=InnoDB AUTO_INCREMENT=4 DEFAULT CHARSET=utf8 COLLATE=utf8_bin;
/*Table structure for table 'userbaseinfo' */
DROP TABLE IF EXISTS 'userbaseinfo';
CREATE TABLE 'userbaseinfo' (
  'user_email' varchar(100)COLLATE utf8_unicode_ci NOT NULL,
  'user_passwd' varchar(100)COLLATE utf8_unicode_ci NOT NULL,
  'user_realName' varchar(100)COLLATE utf8_unicode_ci DEFAULT NULL,
  'user_birth' datetime DEFAULT NULL,
  'user_sex' bigint(1)DEFAULT NULL,
  'user_phone' varchar(50)COLLATE utf8_unicode_ci NOT NULL,
  'user_registerTime' datetime DEFAULT NULL,
  'user_pic' varchar(100)COLLATE utf8_unicode_ci DEFAULT NULL,
```

```sql
  'activateflag' char(1)COLLATE utf8_unicode_ci DEFAULT '0' COMMENT
    '0 表示未激活,1 表示已经激活',
  'shortmsgflag' char(1)COLLATE utf8_unicode_ci DEFAULT '1' COMMENT
    '0 表示不发送短信提醒,1 表示发送短信提醒',
  PRIMARY KEY('user_email'),
  KEY 'ind_userbaseinfo_userpasswd' ('user_passwd'),
  KEY 'ind_userbaseinfo_userphone' ('user_phone')
)ENGINE=InnoDB DEFAULT CHARSET=utf8 COLLATE=utf8_unicode_ci;

/*Table structure for table 'userparaminfo_gateway' */
DROP TABLE IF EXISTS 'userparaminfo_gateway';
CREATE TABLE 'userparaminfo_gateway' (
  'phone' varchar(50)NOT NULL,
  'weight' varchar(50)DEFAULT NULL,
  'age' varchar(50)DEFAULT NULL COMMENT '年龄段',
  'height' varchar(50)DEFAULT NULL COMMENT '身高',
  'sex' char(1)DEFAULT NULL COMMENT '性别,1 表示男,0 表示女',
  'changetime' datetime DEFAULT NULL COMMENT '更改时间',
  'needsend' char(1)DEFAULT NULL COMMENT '0 表示没有下推到智能终端运动设备,
    1 表示已经下推到智能终端运动设备',
  'datafromip' varchar(50)DEFAULT NULL,
  'datafromdomain' varchar(50)DEFAULT NULL,
  'port' varchar(50)DEFAULT NULL,
  'autouploadtime' varchar(50)DEFAULT NULL,
  PRIMARY KEY('phone')
)ENGINE=InnoDB DEFAULT CHARSET=gbk;

/*Table structure for table 'usertbl' */
DROP TABLE IF EXISTS 'usertbl';
CREATE TABLE 'usertbl' (
  'id' int(10)NOT NULL AUTO_INCREMENT,
  'deviceID' varchar(100)COLLATE utf8_unicode_ci NOT NULL DEFAULT ''
    COMMENT '设备编号',
  'patientID' varchar(100)COLLATE utf8_unicode_ci DEFAULT NULL COMMENT
    '手机号',
  'deviceType' varchar(100)COLLATE utf8_unicode_ci DEFAULT NULL COMMENT
    '设备类型',
  'appType' varchar(100)COLLATE utf8_unicode_ci DEFAULT NULL COMMENT
    '应用类型,多个之间以分号分隔',
  'deviceUseFlag' varchar(100)COLLATE utf8_unicode_ci DEFAULT '1'
    COMMENT '1 正常使用,0 停用',
  'company' varchar(100)COLLATE utf8_unicode_ci DEFAULT NULL COMMENT
    '单位代号',
  'pname' varchar(100)COLLATE utf8_unicode_ci DEFAULT NULL COMMENT '姓名',
  'email' varchar(100)COLLATE utf8_unicode_ci DEFAULT NULL COMMENT '邮箱',
```

```
    'teamName' varchar(100)COLLATE utf8_unicode_ci DEFAULT NULL COMMENT
        '班组名',
    'companyName' varchar(100)COLLATE utf8_unicode_ci DEFAULT NULL,
    'isActivate' varchar(10)COLLATE utf8_unicode_ci DEFAULT '0' COMMENT
        '0 未激活,1 已激活',
    'lastTime' varchar(100)COLLATE utf8_unicode_ci DEFAULT NULL COMMENT
        '最后一次上传数据时间',
    'modifyTime' varchar(20)COLLATE utf8_unicode_ci DEFAULT NULL COMMENT
        '用户信息最新修改时间',
    'ywId' varchar(20)COLLATE utf8_unicode_ci DEFAULT NULL COMMENT '用户
        在业务系统的 id',
    PRIMARY KEY('id'),
    UNIQUE KEY 'index_usertbl_diviceid' ('deviceID'),
    UNIQUE KEY 'userId' ('ywId'),
    KEY 'index_usertbl_patientid' ('patientID'),
    KEY 'index_usertbl_company' ('company'),
    KEY 'index_usertbl_useflag' ('deviceUseFlag')
)ENGINE=InnoDB AUTO_INCREMENT=81 DEFAULT CHARSET=utf8 COLLATE=utf8_
    unicode_ci;

/*!40101 SET SQL_MODE=@OLD_SQL_MODE */;
/*!40014 SET FOREIGN_KEY_CHECKS=@OLD_FOREIGN_KEY_CHECKS */;
/*!40014 SET UNIQUE_CHECKS=@OLD_UNIQUE_CHECKS */;
/*!40111 SET SQL_NOTES=@OLD_SQL_NOTES */;
```

⑤ 数据库表设计结果如图 2-16 所示。

图 2-16　数据库设计

2.4 项目小结

1. Mina 框架核心接口 IoFilter、IoHandler 总结

IoFilter 和 IoHandler 是 Mina 最为核心的两个接口。当 Mina 接收到数据以后，首先会传到 IoFilter 链中。在 IoFilter 链中，可以实现编解码、黑白名单过滤、日志记录、开启业务处理线程池等功能。在本章中，使用组合迭代模式，将解码器进行组装，然后在自定义 IoFilter 类 ComponentIOFilter 的核心方法 messageReceived 中调用组装解码器进行解码操作。当解码成功后，将解码后的 javaBean 传给下一个 IoFilter。当解码失败后，将接受到的 IoBuffer 数据同样传递到下一个解码器进行解码。当数据在 IoFilter 链中传递完毕后，会传递给 IoHandler 中。

IoHandler 主要功能就是业务数据处理。在本章中，编写了 IoHandler 自定义扩展类 StrategyFactroyHandler，重写业务处理核心方法 messageReceived，对接收到的 message 数据进行分策略、分状态入库处理。

2. 性能调优

当业务代码完成时，就需要考虑服务的性能问题了，性能调优是一个优秀程序员必备的素质。在本章中，数据接收采用的是 Mina 框架。Mina 会提供一些参数来让使用者根据业务数据量、服务器配置的不同来调整服务性能。性能调优方法如下。

（1）Session 的关闭与空闲：对于长连接来说，一旦客户端与服务端连接上，就会一直保持连接，服务端会一直保存会话 session。但是对于短连接来说，数据发送完毕后，就应该及时关闭连接。这样，可以减少服务器压力。Mina 的接口 IoSession 就是用来管理 session 的。IoSeesion 有个重要的方法 session.close（false）就是用来关闭 session 的。那么何时需要关闭 session 呢。首先，当模拟器传上退出登录包时，就需要关闭 session。其次，当读写通道在一段时间内无任何操作（也就是服务端与客户端没有数据交互）时，则需要关闭 session。这里，可以通过 acceptor.getSessionConfig().setIdleTime（IdleStatus. BOTH_IDLE，40）来设置读写通道空闲时间来关闭 session。

（2）Mina 线程数：Mina 是一个多线程高并发数据处理框架。既然是多线程，必然可以设置线程的启动个数。在创建 Mina 服务端时，需要用到 new NioSocketAcceptor（int processorcount）方法，构造函数里的参数就是要开启线程的个数。Mina 默认开启的线程数是计算机 cpu 的核数+1。这里的参数没有固定的数值，必须是程序员根据服务器的性能来经过大量的测试调整完成。

3. 业务树构建方法

考虑到业务的扩展性和程序的耦合性，本章中解码器的组装采用了组合和迭代模式。具体组装过程可以参考代码如下：

```xml
<bean id="unitASportsComponent" class="com.cloud.mina.component.filter.
    UnitASportComponent">
    <property name="list">
        <list>
            <bean class="com.cloud.mina.component.unit_a.sport.
                SportLoginParser" />
            <bean class="com.cloud.mina.component.unit_a.sport.
                No8OneWayParser" />
            <bean class="com.cloud.mina.component.unit_a.sport.
                No8TwoWayParser" />
            <bean class="com.cloud.mina.component.unit_a.sport.
                No8ThreeWayParser" />
            <bean class="com.cloud.mina.component.unit_a.sport.
                SportLogoutParser" />
        </list>
    </property>
</bean>
<bean id="mHRootComponent" class="com.cloud.mina.component.filter.
    MHRootComponent">
    <property name="list">
        <list>
            <ref bean="unitASportsComponent"></ref>
            <ref bean="unitABPComponent"></ref>
        </list>
    </property>
</bean>
```

从上述代码不难看出，采用 List 集合的方式来存储各个解码器的层级关系。对于本章业务来说，智能终端运动数据包分登录包、1 号包、2 号包、3 号包和退出包，同时智能终端运动数据是 unitA 下的一个运动子业务，这样就存在着层级关系。所以，unitA 业务对应一个解码器，这个解码器包含智能终端运动父解码器，同时智能终端运动父解码器包含各个数据包对应的底层解码器。当进行解码时，永远都是由最外层找到最内层的解码器。例如，多个文件归类到同一个文件夹下。当你想要去找某一个文件时，就首先会进入目录，递归找每个子文件和子文件中的文件，直至找到目标文件。

第 3 章

大数据灵活转发微服务引擎

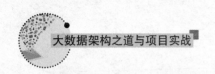

架构之道分享之三：孙子兵法的《谋攻篇》论述了"不战而屈人之兵"的战略指导思想，以最小代价赢得战争，并揭示了"知己知彼，百战不殆"的指导规律。借鉴到项目设计上，我们在深挖项目需求后，不仅要广泛调研主流技术和解决方案，而且要进行严格的设计评审后再开发。

本章学习目标

★ 熟悉观察者模式工作原理和开发步骤
★ 精通点对点的多种即时通信原理和方法
★ 掌握多线程转发的灵活配置和线程同步技巧

3.1 核心需求分析和优秀解决方案

在第 2 章中的高并发采集服务中，物联网大数据通过高并发 Mina 框架接收到了 UnitA 公司的智能终端运动数据包，并把智能终端运动包存储到了 sports 表，完成了物联网大数据接收和存储。接下来，我们要按照业务需求查询这些智能终端运动数据发送到公司的不同应用上。此时的转发服务要考虑灵活性和可扩展性，同时要兼顾系统的高性能。系统灵活性是指灵活配置和修改，当业务需求不断更新时，需要对 SQL 和系统配置进行修改，如果能在配置文件或者数据库中设置并修改，可以大大减少维护成本；系统可扩展是指转发不同的数据包给不同的公司应用系统，随着业务的不断发展，需要按照需求增加转发对象和数据。为了解决这一难题，可以采用观察者设计模式增加不同的发送对象，同时不会影响在线业务；系统的高性能是指转发效率要高，可以通过自主开发多线程实现高性能转发。接下来就是围绕转发服务的灵活性、可扩展性和高性能进行设计和实现。

3.2 服务引擎的技术架构设计

（1）大数据灵活转发服务引擎包括 5 个核心模块，如图 3-1 所示，每一个模块要考虑灵活性、可扩展性和高性能等关键因素，设计说明如下。

① 核心模块一：构建 Spring MVC 工程，主要包括 web.xml、pom.xml 等文件的配置；构建 Spring Boot 工程，主要包括 pom.xml、application.properties 等文件的配置。

② 核心模块二：考虑到业务的灵活性，在配置文件中配置业务的处理方式以及开关。

③ 核心模块三：一条数据被多个 App 共享，故采用观察者模式来解决 1 对多的应用场景，通过创建主题与观察者来实现数据发送。

第 3 章 大数据灵活转发微服务引擎

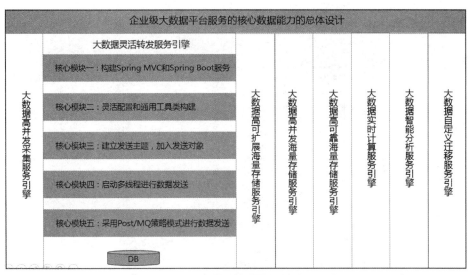

图 3-1　大数据灵活转发服务模块化设计

④ 核心模块四：为了提升发送效率，采用自主研发多线程来发送数据内容，针对一个发送对象启动多个线程，可以提升发送效率。

⑤ 核心模块五：采用 Post 方式发送数据内容，需要调用主流的中间件 Apache HttpClient 来提升发送效率；采用 Active MQ 方式发送数据内容，需要调用 MQ 的发送接口发送消息到消息队列中，然后 MQ 推送给已经订阅了的应用系统。

（2）构建 Spring MVC 版本的 BD_DispatchServer_Maven 服务工程框架，如图 3-2 所示。

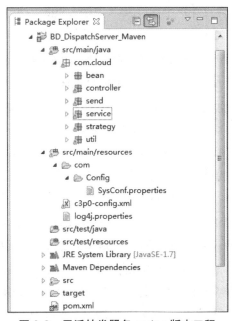

图 3-2　灵活转发服务 spring 版本工程

（3）构建 Spring Boot 版本的 BD_DispatchServer_Boot 服务工程框架，如图 3-3 所示。

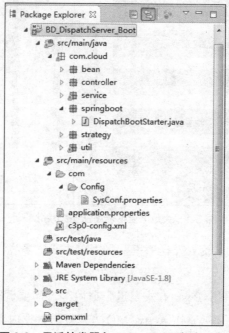

图 3-3　灵活转发服务 Spring Boot 版本工程

（4）针对如上模块设计，编写 UML 代码架构图，完成核心模块和类关系的构建。架构设计如下。

① 核心模块之一架构设计：是建立发送主题 Subject 抽象类和 DataSaveSubject 具体类，存放观察者对象 Observer 接口。同时主题和观察者的创建由发送主题工厂 SbujectFactory 来完成，这样为后续增加信息的发送主题保证了可扩展性。

② 核心模块之二架构设计：初始化发送信息并保存到 AppInfoContext 类的内存中，包括发送目标应用名称 appType、发送开关 appSendFlag、发送标志 sendFlag、发送地址 sendPath 和发送队列名称 appQueueName 等，这些信息保存在数据库 product 表中，发送之前加载到内存 List<HashMap<String, String>> list 中，可以加快发送效率。

③ 核心模块之三架构设计：发送对象是所有接收大数据平台的应用系统，主要通过 post 接口方式和 MQ 消息队列方式进行数据发送。因此通过在系统配置文件 SysConf.properties 中可以灵活配置发送方式 sendWayList=sendWay_1, sendWay_2, 以及对应的发送开关 sendWay_1_toggle=on、sendWay_2_toggle=off。利用配置开关可以实现灵活选择发送方式和启闭，为后续的业务扩展和需求变更提供了便利。

④ 核心模块之四架构设计：为了提升发送效率，我们采用多线程 CommonThread 来发送数据内容。同时为了避免相同的业务处理开启重复的线程，采用了线程"threadKey=

appType + "_" + datatype + "_" + sendWay；"在开启线程前判断是否已经有同样线程 key 的线程在启动。若有，则不启动新线程；否则，开启新线程。这样可以保证线程数目和效率兼顾的平衡。

⑤ 核心模块之五架构设计：采用 PostStrategy 发送策略，调用主流的中间件 Apache HttpClient，创建一个 PostMethod 对象，加载发送路径 url 后，封装发送数据并进行数据转发；采用 MQ 方式发送数据内容，先连接 activeMq 服务，封装发送数据参数，指定消息队列名称并进行数据转发。

⑥ 灵活转发服务整体架构图如图 3-4 所示。

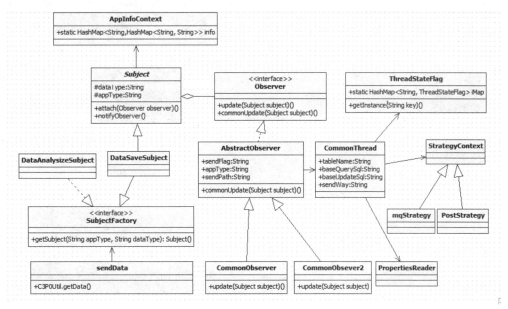

图 3-4　大数据灵活转发服务整体设计

3.3　核心技术讲解及模块化实现

3.3.1　Spring MVC Web 服务构建

（1）配置项目所需要的 jar 包依赖，主要包括 spring、Spring MVC、日志、MySQL 数据库、ActiveMQ 等。在 pom.xml 中添加依赖如下：

```
<project xmlns="http://maven.apache.org/POM/4.0.0" xmlns:xsi="http://
    www.w3.org/2001/XMLSchema-instance"
xsi:schemaLocation="http://maven.apache.org/POM/4.0.0 http://maven.apache.
    org/xsd/maven-4.0.0.xsd">
<modelVersion>4.0.0</modelVersion>
<groupId>com.cloud.send</groupId>
```

```xml
<artifactId>BD_DispatchServer_Maven</artifactId>
<version>0.0.1-SNAPSHOT</version>
<packaging>war</packaging>
<properties>
    <spring.version>4.3.7.RELEASE</spring.version>
</properties>
<build>
    <plugins>
        <plugin>
            <groupId>org.apache.maven.plugins</groupId>
            <artifactId>maven-compiler-plugin</artifactId>
            <version>3.3</version>
            <configuration>
                <source>1.7</source>
                <target>1.7</target>
            </configuration>
        </plugin>
    </plugins>
</build>
<dependencies>
<!-- servlet -->
    <dependency>
        <groupId>javax.servlet</groupId>
        <artifactId>servlet-api</artifactId>
        <version>2.5</version>
    </dependency>
    <dependency>
        <groupId>javax.servlet</groupId>
        <artifactId>jstl</artifactId>
        <version>1.2</version>
    </dependency>
    <!-- spring 常用配置 -->
    <dependency>
        <groupId>org.springframework</groupId>
        <artifactId>spring-core</artifactId>
        <version>${spring.version}</version>
    </dependency>
    <dependency>
        <groupId>org.springframework</groupId>
        <artifactId>spring-beans</artifactId>
        <version>${spring.version}</version>
    </dependency>
    <dependency>
        <groupId>org.springframework</groupId>
        <artifactId>spring-context</artifactId>
```

```xml
        <version>${spring.version}</version>
</dependency>
<dependency>
        <groupId>org.springframework</groupId>
        <artifactId>spring-tx</artifactId>
        <version>${spring.version}</version>
</dependency>
<dependency>
        <groupId>org.springframework</groupId>
        <artifactId>spring-web</artifactId>
        <version>${spring.version}</version>
</dependency>
<dependency>
        <groupId>org.springframework</groupId>
        <artifactId>spring-webmvc</artifactId>
        <version>${spring.version}</version>
</dependency>
<!-- aop -->
<dependency>
        <groupId>org.aspectj</groupId>
        <artifactId>aspectjweaver</artifactId>
        <version>1.7.4</version>
</dependency>
<!-- https://mvnrepository.com/artifact/commons-logging/commons-logging -->
<dependency>
        <groupId>commons-logging</groupId>
        <artifactId>commons-logging</artifactId>
        <version>1.1.3</version>
</dependency>
<dependency>
        <groupId>org.springframework</groupId>
        <artifactId>spring-jms</artifactId>
        <version>4.3.7.RELEASE</version>
</dependency>
<dependency>
        <groupId>org.slf4j</groupId>
        <artifactId>slf4j-log4j12</artifactId>
        <version>1.6.6</version>
</dependency>
<dependency>
        <groupId>log4j</groupId>
        <artifactId>log4j</artifactId>
        <version>1.2.17</version>
</dependency>
```

```xml
<!-- activeMQ -->
<dependency>
    <groupId>org.apache.activemq</groupId>
    <artifactId>activemq-all</artifactId>
    <version>5.9.1</version>
</dependency>
<!-- https://mvnrepository.com/artifact/org.apache.httpcomponents/
   httpcore -->
<dependency>
    <groupId>org.apache.httpcomponents</groupId>
    <artifactId>httpcore</artifactId>
    <version>4.4.1</version>
</dependency>
<dependency>
    <groupId>commons-httpclient</groupId>
    <artifactId>commons-httpclient</artifactId>
    <version>3.1</version>
</dependency>
<!-- commons-codec -->
<dependency>
    <groupId>commons-codec</groupId>
    <artifactId>commons-codec</artifactId>
    <version>1.9</version>
</dependency>
<!-- c3p0 -->
<dependency>
    <groupId>c3p0</groupId>
    <artifactId>c3p0</artifactId>
    <version>0.9.1.2</version>
</dependency>
<!-- https://mvnrepository.com/artifact/commons-logging/commons-
   logging -->
<dependency>
    <groupId>commons-logging</groupId>
    <artifactId>commons-logging</artifactId>
    <version>1.1.3</version>
</dependency>
<!-- 添加数据库驱动 -->
<dependency>
    <groupId>mysql</groupId>
    <artifactId>mysql-connector-java</artifactId>
    <version>5.1.30</version>
</dependency>
<!-- activeMq -->
<dependency>
```

```xml
            <groupId>org.apache.activemq</groupId>
            <artifactId>activemq-core</artifactId>
            <version>5.3.0</version>
        </dependency>
        <!-- json配置 -->
        <dependency>
            <groupId>net.sf.json-lib</groupId>
            <artifactId>json-lib</artifactId>
            <version>2.3</version>
            <classifier>jdk15</classifier>
        </dependency>
    </dependencies>
</project>
```

（2）Web 项目的入口 web.xml 文件，主要用于 Spring MVC、日志记录等配置加载。配置如下：

```xml
<?xml version="1.0" encoding="UTF-8"?>
<web-app xmlns:xsi="http://www.w3.org/2001/XMLSchema-instance"
    xmlns="http://java.sun.com/xml/ns/javaee" xmlns:web="http://
    java.sun.com/xml/ns/javaee/web-app_2_5.xsd"
    xsi:schemaLocation="http://java.sun.com/xml/ns/javaee http://
    java.sun.com/xml/ns/javaee/web-app_2_5.xsd"id="WebApp_ID"
    version="2.5">
<display-name>dispatch</display-name>
<context-param>
    <param-name>webAppRootKey</param-name>
    <param-value>webapp.dispatch.root</param-value>
</context-param>
<context-param>
    <param-name>log4jRefreshInterval</param-name>
    <param-value>60000</param-value>
</context-param>
<context-param>
    <param-name>contextConfigLocation</param-name>
    <param-value>WEB-INF/spring/app-config.xml</param-value>
</context-param>
<filter>
    <filter-name>CharacterEncodingFilter</filter-name>
    <filter-class>org.springframework.web.filter.CharacterEncodingFilter
        </filter-class>
    <init-param>
        <param-name>encoding</param-name>
        <param-value>UTF-8</param-value>
    </init-param>
    <init-param>
```

```xml
        <param-name>forceEncoding</param-name>
        <param-value>true</param-value>
    </init-param>
</filter>
<filter-mapping>
    <filter-name>CharacterEncodingFilter</filter-name>
    <url-pattern>/*</url-pattern>
</filter-mapping>
<listener>

    <listener-class>org.springframework.web.context.ContextLoader Listener
        </listener-class>
</listener>
<listener>
    <listener-class>org.springframework.web.util.Log4jConfigListener
        </listener-class>
</listener>
<servlet>
    <servlet-name>SpringMVC</servlet-name>
    <servlet-class>org.springframework.web.servlet.DispatcherServlet
        </servlet-class>
    <init-param>
        <param-name>contextConfigLocation</param-name>
        <param-value>WEB-INF/spring/mvc-config.xml</param-value>
    </init-param>
    <load-on-startup>1</load-on-startup>
</servlet>
<servlet-mapping>
    <servlet-name>SpringMVC</servlet-name>
    <url-pattern>*.json</url-pattern>
    <url-pattern>/service/*</url-pattern>
    <url-pattern>/</url-pattern>
</servlet-mapping>

</web-app>
```

（3）Spring MVC 的核心配置文件 app-config.xml，主要作用是扫描指定包下的注解，将其注册到 Spring 容器中。配置如下：

```xml
<?xml version="1.0" encoding="UTF-8"?>
<beans xmlns="http://www.springframework.org/schema/beans"
xmlns:xsi="http://www.w3.org/2001/XMLSchema-instance"
xmlns:context="http://www.springframework.org/schema/context"
xmlns:tx="http://www.springframework.org/schema/tx"
xmlns:aop="http://www.springframework.org/schema/aop"
xsi:schemaLocation="
    http://www.springframework.org/schema/beans
```

```
            http://www.springframework.org/schema/beans/spring-beans-3.2.xsd
            http://www.springframework.org/schema/context
            http://www.springframework.org/schema/context/spring-context-3.2.xsd
            http://www.springframework.org/schema/tx
              http://www.springframework.org/schema/tx/spring-tx-3.2.xsd
              http://www.springframework.org/schema/aop
              http://www.springframework.org/schema/aop/spring-aop-3.2.xsd">
    <context:annotation-config />
    <context:component-scan base-package="com.cloud">
        <context:exclude-filter type="annotation" expression=
            "org.springframe work.stereotype.Controller"/>
    </context:component-scan>
</beans>
```

（4）Spring MVC 的核心配置文件 mvc-config.xml，用于扫描 controller 层注解、视图解析器、配置异常拦截器等。配置如下：

```
<?xml version="1.0" encoding="UTF-8"?>
<beans xmlns="http://www.springframework.org/schema/beans"
    xmlns:xsi="http://www.w3.org/2001/XMLSchema-instance" xmlns:mvc=
      "http://www.springframework.org/schema/mvc"
    xmlns:context="http://www.springframework.org/schema/context"
    xsi:schemaLocation="
        http://www.springframework.org/schema/beans http://www.
            springframework.org/schema/beans/spring-beans-3.2.xsd
        http://www.springframework.org/schema/mvc http://www.
            springframework.org/schema/mvc/spring-mvc-3.2.xsd
        http://www.springframework.org/schema/context http://www.
            springframework.org/schema/context/spring-context-3.2.xsd">
    <!-- 1. scan all package -->
    <context:component-scan base-package="com.cloud">
        <context:include-filter type="annotation"
            expression="org.springframework.stereotype.Controller" />
    </context:component-scan>
    <!-- 2. import servlet_view_resolver -->
    <bean
        class="org.springframework.web.servlet.view.
          InternalResourceView Resolver">
        <property name="order" value="2" />
        <property name="viewClass"
            value="org.springframework.web.servlet.view.JstlView" />
        <property name="prefix" value="/jsp/" />
        <property name="suffix" value=".jsp" />
    </bean>
    <!-- 3. support file_upload MultipartResolver -->
    <mvc:default-servlet-handler />
```

```xml
<bean
    class="org.springframework.web.servlet.handler.
      SimpleMapping ExceptionResolver">
    <property name="exceptionMappings">
       <props>
            <prop key="org.springframework.dao.DataAccessException">
                dataAccessFailure</prop>
            <prop key="org.springframework.transaction.
                Transaction Exception">dataAccessFailure</prop>
       </props>
    </property>
</bean>
<!-- 4. support annotation -->
<mvc:annotation-driven />
</beans>
```

(5) Log4j 日志配置文件，主要设置日志输出级别、格式、以及文件位置等。配置如下：

```
log4j.rootCategory = INFO,out,file,LF5
#log4j.appender.LF5=org.apache.log4j.lf5.LF5Appender   //是否弹出日志框
#控制台输出日志
log4j.appender.out=org.apache.log4j.ConsoleAppender
log4j.appender.out.layout=org.apache.log4j.EnhancedPatternLayout
log4j.appender.out.layout.ConversionPattern=[%t]%d{yyyy-MM-dd HH:mm:
    ss.SSS}|%p|%X{userId}%m%n
#文件写入日志
log4j.appender.file = org.apache.log4j.DailyRollingFileAppender
log4j.appender.file.layout = org.apache.log4j.PatternLayout
log4j.appender.file.layout.ConversionPattern = %d %p %L [%t] - %m%n
# logger for apache
log4j.logger.org.apache = ERROR
```

3.3.2 Spring Boot 微服务构建

（1）本服务采用 JDK1.8 版本、Spring Boot 1.5.2.Release 版本、C3P0 0.9.0.4 版本、MySQL 数据库驱动 5.1.6 版本、ActiveMQ5.3.0 版本等。在 pom.xml 文件中添加如下依赖：

```xml
<?xml version="1.0" encoding="UTF-8"?>
<project xmlns="http://maven.apache.org/POM/4.0.0" xmlns:xsi="http://
    www.w3.org/2001/XMLSchema-instance"
 xsi:schemaLocation="http://maven.apache.org/POM/4.0.0 http://maven.
    apache.org/xsd/maven-4.0.0.xsd">
<modelVersion>4.0.0</modelVersion>
<groupId>com.cloud.bigdata</groupId>
<artifactId>BD_DispatchServer_B</artifactId>
```

```xml
<version>0.0.1-SNAPSHOT</version>
<packaging>jar</packaging>
<name>service-hi</name>
<description>Demo project for Spring Boot</description>
<parent>
    <groupId>org.springframework.boot</groupId>
    <artifactId>spring-boot-starter-parent</artifactId>
    <version>1.5.2.RELEASE</version>
    <relativePath/><!-- lookup parent from repository -->
</parent>
<properties>
    <project.build.sourceEncoding>UTF-8</project.build.sourceEncoding>
    <project.reporting.outputEncoding>UTF-8</project.reporting.
       output Encoding>
    <java.version>1.8</java.version>
</properties>
<dependencies>
    <dependency>
        <groupId>org.springframework.cloud</groupId>
        <artifactId>spring-cloud-starter-eureka</artifactId>
    </dependency>
    <dependency>
        <groupId>org.springframework.boot</groupId>
        <artifactId>spring-boot-starter-web</artifactId>
    </dependency>

    <dependency>
        <groupId>org.springframework.boot</groupId>
        <artifactId>spring-boot-starter-test</artifactId>
        <scope>test</scope>
    </dependency>
    <!-- json 配置 -->
    <dependency>
        <groupId>net.sf.json-lib</groupId>
        <artifactId>json-lib</artifactId>
        <version>2.3</version>
        <classifier>jdk15</classifier>
    </dependency>
    <dependency>
        <groupId>org.apache.commons</groupId>
        <artifactId>commons-lang3</artifactId>
        <version>3.0.1</version>
    </dependency>
    <dependency>
        <groupId>commons-httpclient</groupId>
```

```xml
        <artifactId>commons-httpclient</artifactId>
        <version>3.1</version>
    </dependency>
    <!-- activeMq -->
    <dependency>
        <groupId>org.apache.activemq</groupId>
        <artifactId>activemq-core</artifactId>
        <version>5.3.0</version>
    </dependency>
    <!-- jms -->
    <dependency>
        <groupId>javax.jms</groupId>
        <artifactId>jms</artifactId>
        <version>1.1</version>
    </dependency>
    <!--spring-jms -->
    <dependency>
        <groupId>org.springframework</groupId>
        <artifactId>spring-jms</artifactId>
        <version>4.2.1.RELEASE</version>
    </dependency>
    <!-- c3p0 -->
    <dependency>
        <groupId>c3p0</groupId>
        <artifactId>c3p0</artifactId>
        <version>0.9.1.2</version>
    </dependency>
    <!-- 添加数据库驱动 -->
    <dependency>
        <groupId>mysql</groupId>
        <artifactId>mysql-connector-java</artifactId>
    </dependency>
    <dependency>
        <groupId>org.apache.httpcomponents</groupId>
        <artifactId>httpclient</artifactId>
        <version>4.5.5</version>
    </dependency>
</dependencies>
<dependencyManagement>
    <dependencies>
        <dependency>
            <groupId>org.springframework.cloud</groupId>
            <artifactId>spring-cloud-dependencies</artifactId>
            <version>Dalston.RC1</version>
            <type>pom</type>
```

```xml
            <scope>import</scope>
        </dependency>
    </dependencies>
</dependencyManagement>
<build>
    <plugins>
        <plugin>
            <groupId>org.springframework.boot</groupId>
            <artifactId>spring-boot-maven-plugin</artifactId>
        </plugin>
    </plugins>
</build>
<repositories>
    <repository>
        <id>spring-milestones</id>
        <name>Spring Milestones</name>
        <url>https://repo.spring.io/milestone</url>
        <snapshots>
            <enabled>false</enabled>
        </snapshots>
    </repository>
</repositories>
</project>
```

（2）Spring Boot 的核心配置文件 application.properties，用于指定项目应用名称、端口、ip 以及注册中心 eureka 服务地址。配置如下：

```
server.address=127.0.0.1
server.port=8081
spring.application.name=boot-dispatch
#注册中心服务
eureka.client.serviceUrl.defaultZone=http://localhost:8761/eureka/
```

（3）Spring Boot 的启动加载类 DispatchBootStarter，主要功能是扫描指定包下的相关注解并初始化到 spring 容器中，同时启动该服务。代码实现如下：

```
package com.cloud.Spring Boot;
import org.springframework.boot.SpringApplication;
import org.springframework.boot.autoconfigure.Spring BootApplication;
import org.springframework.cloud.netflix.eureka.EnableEurekaClient;
import org.springframework.context.annotation.ComponentScan;
/**
 * spring boot 启动加载类
 *
 * @author changyaobin
 *
```

```
*/
@Spring BootApplication
@ComponentScan(basePackages = { "com.cloud" })
@EnableEurekaClient
public class DispatchBootStarter {
 public static void main(String[] args){
     SpringApplication.run(DispatchBootStarter.class,args);
 }
}
```

3.3.3 灵活配置和通用工具类构建

（1）系统应用配置文件 SysConf.properties 主要功能是：配置发送方式以及开关 sendWay_1_toggle 和 sendWay_2_toggle，相应的发送方式分别为 PostStrategy 和 mqStrategy；同时配置不同发送方式下的数据查询 SQL，实现了真正意义上的可灵活配置；同时配置 mq 服务器地址和消息队列名称。具体配置如下：

```
#mq
#######################
jms.url=tcp://localhost:61616
jms.cachSessionNum=50
jms.queue.AppA=QueueSport

##sendWay
sendWayList=sendWay_1,sendWay_2
#sendWay_1 的开关
sendWay_1_toggle=on
sendWay_1=PostStrategy
#sendWay_1 的查询数据 sql 语句
baseQuerySql_sendWay_1=SELECT s.*,u.company,u.teamName,u.pname FROM %s
  s,usertbl u WHERE s.%s='0' AND s.appType LIKE '%s' AND s.phone=u.
  patientID LIMIT 500;
baseUpdateSql_sendWay_1=update %s set %s\='1',sendTime\=NOW()where id\=%s
#sendWay_2 的开关
sendWay_2_toggle=off
sendWay_2=MqStrategy
baseQuerySql_sendWay_2=select * from %s where %s\='0' and appType like
  '%s' limit 500;
baseUpdateSql_sendWay_2=update %s set %s\='2',sendTime\=NOW()where id\=%s
```

（2）MySQL 数据库操作工具类 C3P0Util，请参考第 2 章数据采集服务的 C3P0Util 类，其默认读取 classpath 下的 c3p0-config.xml 文件。

（3）日期操作工具类 DateUtil，请参考第 2 章数据采集服务的 DateUtil 类。

（4）配置文件读取工具类 PropertiesReader，用于将项目中的 properties 文件加载到流中，根据属性读取其配置值，相关配置如下：

```java
package com.cloud.util;
import java.io.InputStream;
import java.util.Properties;
import org.springframework.core.io.ClassPathResource;
public class PropertiesReader {
    private static Properties prop = new Properties();
    static
    {
        try {
            InputStream SystemIn = new ClassPathResource("com/Config/
                SysConf.properties").getInputStream();
            prop.load(SystemIn);
        } catch(Exception e){
            e.printStackTrace();
        }
    }
    public static String getProp(String name)
    {
        if(prop!=null)
        {
            return prop.get(name).toString();
        }
        return null;
    }
}
```

（5）线程状态标识类 ThreadStateFlag，主要通过一个静态 map 来存放 key 与线程标识对象的关系，并实现一个 key 对应一个实例，不同的 key 有不同的实例。代码实现如下：

```java
package com.cloud.util;
import java.util.HashMap;
/**
 * 线程状态标识,对同一 Key 值只有一个实例,不同的 Key 有不同的实例,习惯称为伪单例
 */
public class ThreadStateFlag {
    private static HashMap<String,ThreadStateFlag> iMap = new HashMap<
        String,ThreadStateFlag>();
    private ThreadStateFlag(){
    }
    public static synchronized ThreadStateFlag getInstance(String key){
        if(iMap.containsKey(key)){
            return iMap.get(key);
        } else {
```

```
            iMap.put(key,new ThreadStateFlag());
            return iMap.get(key);
        }
    }
}
```

（6）Http 客户端工具类 HttpClientUtil，用于通过 http 协议，给指定路径的服务发送业务参数。代码实现如下：

```
package com.cloud.util;
import java.util.Arrays;
import org.apache.commons.httpclient.HttpClient;
import org.apache.commons.httpclient.NameValuePair;
import org.apache.commons.httpclient.methods.PostMethod;
import org.apache.commons.httpclient.params.HttpMethodParams;
/**
 * http 客户端工具类
 * @author changyaobin
 *
 */
public class HttpClientUtil {
    public static boolean sendHttpData(String className,String url,
        NameValuePair[] parameter){
        HttpClient client = new HttpClient();
        PostMethod post = new PostMethod(url);
        boolean isSuccess=true;
        try {
            post.getParams().setParameter(HttpMethodParams.HTTP_
                CONTENT_CHARSET,"UTF-8");
            post.setRequestBody(parameter);
            Log.info(className+" send data:"+Arrays.deepToString
                (parameter));
            //设置连接超时时间
            client.getHttpConnectionManager().getParams().
                setConnectionTimeout(3000);
            //设置响应超时时间
            client.getHttpConnectionManager().getParams().
                setSoTimeout(15000);
            int returnFlag = client.executeMethod(post);
            if(returnFlag!=200){
                isSuccess=false;
            }
            Log.info(className+" success receive form post:" + post.
                getStatusLine().toString()+",returnFlag="+returnFlag);
```

```
        }catch(Exception e){
            e.printStackTrace();
            isSuccess=false;
            Log.info(className+" fail receive form post:" +e.getMessage());
        }finally{
            if(post!=null){
                post.releaseConnection();
                Log.info(className+" post.releaseConnection()" + "is coming");
            }
        }
        return isSuccess;
    }
}
```

（7）日志记录类 Log，用于日志信息写入文件中。代码实现如下：

```
package com.cloud.util;
public class Log {
    static org.apache.log4j.Logger logger = null;
    static{
        logger = org.apache.log4j.Logger.getLogger(Log.class);
    }
    /**
     * @param log
     */
    public static void debug(String log){
        logger.debug(log);
    }
    /**
     * @param log
     */
    public static void info(String log){
        logger.info(log);
    }
    /**
     * @param log
     */
    public static void error(String log){
        logger.error(log);
    }
}
```

3.3.4 创建发送数据主题，注册观察者对象

本服务中一条智能终端运动数据会被多个 App 共享，为了解决这种一对多的应用场景，可以采用观察者模式，同时使用工厂模式来创建数据主题，相关实现过程如下。

（1）数据主题抽象类 Subject，定义观察者的集合和业务相关的参数。代码实现如下：

```java
package com.cloud.service;
import java.util.ArrayList;
import java.util.List;
public abstract class Subject {
    public List<Observer> observers = new ArrayList<Observer>();
    //数据包类型
    protected String dataType;
    //数据包 App 类型
    protected String appType;
    public void add(Observer observer){
        this.observers.add(observer);
    };
    public void del(Observer observer){
    };
    public void notifyObservers(){
        for(Observer observer:observers){
            observer.update(this);
        }
    }
    public String getDataType(){
        return dataType;
    }
    public void setDataType(String dataType){
        this.dataType = dataType;
    }
    public String getAppType(){
        return appType;
    }
    public void setAppType(String appType){
        this.appType = appType;
    };
}
```

（2）数据保存主题实体类 DataSaveSubject。该类没有任何私有属性，只是考虑了以后业务的扩展性创建的该实体类。代码实现如下：

```java
package com.cloud.service;
/**
 * 数据保存主题
 * @author changyaobin
 *
 */
public class DataSaveSubject extends Subject{
}
```

（3）观察者接口 Observer，定义数据更新方法，具体实现由其子类完成。代码实现如下：

```java
package com.cloud.service;
/**
 * 观察者接口
 * @author changyaobin
 *
 */
public interface Observer {
    /**
     * 数据更新方法
     * @param subject    被观察的主题
     */
    public void update(Subject subject);
}
```

（4）app 的信息缓存类 AppInfoContext，主要使用静态 Map 存储 app 的相关信息。当系统启动时即从数据库读取 app 信息到内存，以减少数据库访问次数，若信息有更新，则需要及时刷新此上下文。当然，当数据量较小时，可以采取这种方式。当数据大的时候，最好还是采用 redis 等缓存数据库。app 是指转发的目标应用系统。代码实现如下：

```java
package com.cloud.util;
import java.util.HashMap;
import java.util.List;
import java.util.Map.Entry;
import java.util.Set;
/**
 * 应用系统上下文信息,
 * 保存各个 app 的 appType、sendFlag、sendPath、appToggle 等相关信息
 * <br>系统启动时即从数据库读取到内存,以减少数据库访问次数,若信息有更新,则需要及时
     刷新此上下文
 *
 */
public class AppInfoContext {
    public static HashMap<String,HashMap<String,String>> info = new
        HashMap<String,HashMap<String,String>>();
    static {
        initAppInfo();
    }
    public static void initAppInfo(){
        info.clear();
        String sql = "SELECT appName as appType,appSendFlag as sendFlag,
          appUrl as sendPath,appToggle,appQueueName FROM product";
        List<HashMap<String,String>> list = C3P0Util.getData(sql);
```

```
        if(list != null && list.size()> 0){
            for(HashMap<String,String> map:list){
                info.put(map.get("appType"),map);
            }
        }
    }
    public static String getPropertyByApp(String appType,String prop){
        initAppInfo();
        HashMap<String,String> appInfo = info.get(appType);
        if(appInfo != null){
            return appInfo.get(prop);
        } else {
            return null;
        }
    }
}
```

（5）数据接收 controller 层 SubjectController，用于接收数据采集服务发送的数据，然后构建主题工厂，通知观察者进行更新操作。代码实现如下：

```
package com.cloud.controller;
import org.apache.commons.lang.StringUtils;
import org.springframework.web.bind.annotation.RequestMapping;
import org.springframework.web.bind.annotation.RestController;
import com.cloud.bean.Message;
import com.cloud.service.SubjectFactory;
import com.cloud.util.Log;
/**
 * 转发服务的Controller层,用于接收数据采集发送来的数据
 *
 * @author changyaobin
 *
 */
@RestController
public class SubjectController {
    @RequestMapping("/sendData")
    public Message sendData(String appType,String dataType){
        Message message = new Message();
        if(StringUtils.isBlank(appType)|| StringUtils.isBlank(dataType)){
            message.setCode(1000);
            message.setMessage("参数非法");
            return message;
        }
        try {
            // 构建观察者,并通知观察者进行更新
            SubjectFactory.getSubject(appType,dataType).notifyObservers();
```

```
            message.setCode(1001);
            message.setMessage("数据发送成功");
        } catch(Exception e){
            Log.error("数据处理,异常信息为" + e.getMessage());
            message.setCode(1002);
            message.setMessage("数据处理错误");
        }
        return message;
    }
}
```

（6）主题对象工厂类 SubjectFactory，主要功能是创建 DataSaveSubject 主题，并根据 app 来注册多个观察者 CommonObserver。代码实现如下：

```
package com.cloud.service;
import com.cloud.util.AppInfoContext;
public class SubjectFactory {
    public static Subject getSubject(String appTypes,String dataType){
        DataSaveSubject subject = new DataSaveSubject();
        subject.setAppType(appTypes);
        subject.setDataType(dataType);
        if(appTypes != null && !"".equals(appTypes)){
            String[] app = appTypes.split(";");
            for(String appType:app){
                // 从库中查询 app 的发送标识以及发送路径
                Observer obs =(Observer)new CommonObserver(
                        AppInfoContext.getPropertyByApp(appType,
                          "sendFlag"),
                        appType,AppInfoContext.getPropertyByApp
                          (appType,"sendPath"));
                subject.add(obs);
            }
        }
        return subject;
    }
}
```

（7）AbstractObserver 观察者抽象类，主要功能是提供观察者子类通用的方法。在方法实现中，定义了线程标识 "threadKey = appType + "_" + datatype + "_" + sendWay"。当不存在数据转发的线程（或线程已死亡）时，重新启动线程 CommonThread 进行数据转发。当线程存在时，不需要重新启动线程。这样避免了不必要的性能消耗。代码实现如下：

```
package com.cloud.service;
import java.util.HashMap;
```

```java
import java.util.Random;
import com.cloud.util.PropertiesReader;
import com.cloud.util.ThreadStateFlag;
public abstract class AbstractObserver implements Observer {
    protected void commonUpdate(Subject subject,String appType,
            String sendFlag,String sendPath,
            HashMap<String,CommonThread> threadMap){
        String datatype = subject.getDataType();
        String sendWayList = PropertiesReader.getProp("sendWayList");
        String[] sendWayArray = sendWayList.split(",");
        if(sendWayList == null){
            return;
        }
        for(String sendWay:sendWayArray){
            // 读取配置文件,判断该发送方式是否开启
            if(!"on".equals(PropertiesReader.getProp(sendWay+"_toggle"))){
                continue;
            }
            // 定义线程的唯一性
            String threadKey = appType + "_" + datatype + "_" + sendWay;
            if(threadMap.containsKey(threadKey)){
                CommonThread thread = threadMap.get(threadKey);
                if(thread.isAlive()){
                    continue;
                }
            }
            CommonThread thread = new CommonThread(sendWay,
              ThreadStateFlag.getInstance(threadKey),datatype,
              sendFlag, appType,sendPath);
            thread.setName(threadKey);
            thread.start();
            threadMap.put(threadKey,thread);
        }
    }
}
```

（8）定义观察者具体类CommonObserver，主要功能是接收业务参数，并调用父类的更新方法来实现数据转发。代码实现如下：

```java
package com.cloud.service;
import java.util.HashMap;
public class CommonObserver extends AbstractObserver {
    private String sendFlag = null;
    private String sendPath = null;
    private String appType = null;
```

```
        private static HashMap<String,CommonThread> threadMap = new HashMap
          <String,CommonThread>();
        public CommonObserver(String sendFlag,String appType,String sendPath){
            this.sendFlag = sendFlag;
            this.sendPath = sendPath;
            this.appType = appType;
        }
        @Override
        public void update(Subject subject){
            commonUpdate(subject,appType,sendFlag,sendPath,threadMap);
        }
}
```

3.3.5 启动多线程进行数据发送

设置线程唯一标识"threadKey = appType + "_" + datatype + "_" + sendWay"。若线程需要启动，利用这个线程 key 可以生产一个唯一的单例对象 ThreadStateFlag，将该对象作为线程同步锁，这样可以避免多线程的并发问题。同时在线程中根据不同的发送方式来调用不同的发送策略进行数据发送。

（1）数据发送线程类 CommonThread，主要功能是通过发送标识查询未发送的数据，然后根据发送方式来调用相应的发送策略来进行数据的转发，并更新发送标识。代码实现如下：

```
package com.cloud.service;
import java.util.HashMap;
import java.util.List;
import net.sf.json.JSONObject;
import com.cloud.strategy.SendStrategy;
import com.cloud.strategy.StrategyContext;
import com.cloud.util.C3P0Util;
import com.cloud.util.Log;
import com.cloud.util.PropertiesReader;
import com.cloud.util.ThreadStateFlag;
public class CommonThread extends Thread {
    private String sendWay;
    private ThreadStateFlag threadStateFlag;
    private String datatype;
    private String sendFlag;
    private String appType;
    private String sendPath;
    private String baseQuerySql;
    private String baseUpdateSql;
```

```java
        public CommonThread(String sendWay,ThreadStateFlag threadStateFlag,
            String datatype,String sendFlag,
                String appType,String sendPath){
            super();
            this.threadStateFlag = threadStateFlag;
            this.datatype = datatype;
            this.sendFlag = sendFlag;
            this.appType = appType;
            this.sendPath = sendPath;
            this.sendWay = PropertiesReader.getProp(sendWay);
            this.baseQuerySql = PropertiesReader.getProp("baseQuerySql_" +
                sendWay);
            this.baseUpdateSql = PropertiesReader.getProp("baseUpdateSql_" +
                sendWay);
        }
        public void run(){
            synchronized(threadStateFlag){
                List<HashMap<String,String>> unSendList = null;
                JSONObject data = null;
                String tableName = "sports";
                String querySql = String.format(baseQuerySql,tableName,
                    sendFlag,"%" + appType + "%");
                unSendList = C3P0Util.getData(querySql);
                if(unSendList.size()> 0){
                    for(HashMap<String,String> map:unSendList){
                        data = new JSONObject();
                        data.put("appType",appType);
                        data.put("dataType",map.get("dataType"));
                        data.put("collectDate",map.get("receiveTime"));
                        data.put("phone",map.get("phone"));
                        data.put("dataValue",map.get("dataValue"));
                        data.put("deviceID",map.get("deviceID"));
                        data.put("company",map.get("company"));
                        data.put("pname",map.get("pname"));
                        data.put("teamName",map.get("teamName"));

                        if(StrategyContext.sendData(data.toString(), sendPath,
                            appType,sendWay)){
                        // 数据发送成功,进行发送标识的更新
                            String updateSql = String.format(baseUpdateSql,
                                tableName,sendFlag,map.get("id"));
                            boolean updateSuccess = C3P0Util.executeUpdate
                                (updateSql);
                            System.out.println(updateSql);
                            if(!updateSuccess){
```

```
                    Log.error("send data success,update data 
                        fail, sql is "+updateSql);
                }
            }
        }
    }
    threadStateFlag.notify();
    }
  }
}
```

（2）数据发送策略接口 SendStrategy，定义数据转发方法以及参数，具体实现由其子类完成。代码实现如下：

```
package com.cloud.strategy;
public interface SendStrategy {
    public boolean send(String data,String url,String appType);
}
```

（3）策略发送工厂类 StrategyContext，主要功能是通过不同的发送方式 sendWay 来实例化不同的发送策略。发送策略包括 mqStrategy 和 postStrategy 两种方式。代码实现如下：

```
package com.cloud.strategy;
import com.cloud.util.Log;
public class StrategyContext {
    private static SendStrategy mqStrategy = null;
    private static SendStrategy postStrategy = null;
    static {
        mqStrategy = new MqStrategy();
        postStrategy = new PostStrategy();
    }
    /**
     * 发送接口
     *
     * @param data
     *           //发送的数据
     * @param url
     *           //用户 Post 方式发送的目的地址
     * @param appType
     *           //用于 MQ 方式发送时,查找 queueName
     * @param sendWay
     *           //用于指定发送策略:Post | MQ
     * @return
     */
    public static boolean sendData(String data,String url,String appType,
        String sendWay){
```

```java
            if("MqStrategy".equals(sendWay)){
                Log.debug("MqStrategy 方式发送数据");
                return mqStrategy.send(data,url,appType);
            } else if("PostStrategy".equals(sendWay)){
                Log.debug("PostStrategy 方式发送数据");
                return postStrategy.send(data,url,appType);
            } else {
                Log.info("没有指定发送方式!!!sendWay:" + sendWay);
                return false;
            }
        }
    }
```

3.3.6 采用 Post 策略模式进行数据发送

采用 Post 方式发送数据内容，主要功能是调用主流的中间件 Apache HttpClient，并设置网络传输超时时间、网络响应超时时间等参数，向指定服务路径发送 json 字符串数据。代码实现如下：

```java
package com.cloud.strategy;
import org.apache.commons.httpclient.HttpClient;
import org.apache.commons.httpclient.methods.PostMethod;
import org.apache.commons.httpclient.params.HttpMethodParams;
import com.cloud.util.Log;
public class PostStrategy implements SendStrategy {
    @Override
    public boolean send(String data,String url,String appType){
        HttpClient client = new HttpClient();
        PostMethod post = new PostMethod(url);
        boolean isSuccess = false;
        try {
            System.out.println(url);
            post.getParams().setParameter(
                    HttpMethodParams.HTTP_CONTENT_CHARSET,"UTF-8");
            post.addParameter("data",data);
            Log.debug(this.getClass().getSimpleName()+ " begin send
                data...");
            Log.debug(this.getClass().getSimpleName()+ " send data:" +
                data);
            // 设置连接超时时间
            client.getHttpConnectionManager().getParams().
                setConnectionTimeout(10000);
            // 设置响应超时时间
            client.getHttpConnectionManager().getParams().
                setSoTimeout(15000);
            int returnFlag = client.executeMethod(post);
```

```
                if(returnFlag == 200){
                    isSuccess = true;
                }
                Log.info(this.getClass().getSimpleName()
                        + " success receive form post:"
                        + post.getStatusLine().toString()+ ",returnFlag="
                        + returnFlag);
        } catch(Exception e){
            isSuccess = false;
            Log.error(this.getClass().getSimpleName()
                    + " fail receive form post:" + e.getMessage());
        } finally {
            if(post != null){
                post.releaseConnection();
                Log.info(this.getClass().getSimpleName()
                        + " post.releaseConnection()" + "is coming");
            }
        }
        return isSuccess;
    }
}
```

3.3.7 采用 ActiveMQ 策略模式进行数据发送

采用 ActiveMQ 方式发送数据内容，调用 MQ 的发送接口发送消息给指定队列。ActiveMq 的配置以及实现过程如下：

（1）在 SysConf.properties 中配置 activemq 发送消息队列。

```
jms.url=tcp://localhost:61616
jms.cachSessionNum=50
jms.queue.AppA=QueueSport
```

（2）导入 jar 包 activemq-all-5.9.1.jar。

（3）在网上下载 apache-activemq-5.10.0 服务端压缩包到本地磁盘 D:\apache-activemq-5.10.0 下面。

（4）在本地的系统中配置 activemq 的环境变量。

- 在系统变量中添加：变量名"ACTIVEMQ_HOME"，变量值"D:\apache-activemq-5.10.0\apache-activemq-5.10.0"。
- 在系统变量中添加：变量名"PATH"，变量值"%ACTIVEMQ_HOME%\bin"。
- 此时 activemq 就可以加载 Java 的环境变量 JAVA_HOME。

（5）在 CMD 命令中启动 activemq，输入 activemq start，若有"ActiveMQ WebConsole available at http://127.0.0.1:8161/"提示，则说明 activemq 服务启动成功。

（6）编写发送数据包类的步骤如下：

① 新创建一个连接 activemq 服务的工厂类 ActiveMQConnectionFactory，并设置 mq 的服务地址。

② 新建工厂类 CachingConnectionFactory，将①新建的 ActiveMQConnectionFactory 类传入。

③ 新建 JmsTemplate 类，调用 org.springframework.jms.core.JmsTemplate 来设置发送方式是 P2P 的队列模式还是订阅分发模式的消息主题。

④ 设置发送队列的名称 queueName=QueueSport。

⑤ 调用 jmsTemplate.send 发送数据到 queueName 中。

（7）MQ 消息策略类，主要功能是调用封装的消息发布接口，发送业务参数到消息队列中。代码实现如下：

```java
package com.cloud.strategy;
import com.cloud.util.PropertiesReader;
public class MqStrategy implements SendStrategy {
    static ObsDataMsgPublisher publisher = null;
    static {
        publisher = ObsDataMsgPublisher.getInstance();
    }
    @Override
    public boolean send(String data, String url, String appType) {
        if (publisher == null) {
            return false;
        }
        String queueName = PropertiesReader.getProp("jms.queue." +
            appType);
        System.out.println("queueName="+queueName);
        if (queueName == null || "".equals(queueName)) {
            return false;
        }
        return publisher.sendByQuene(data, queueName);
    }
}
```

（8）MQ 消息发布接口 MessagePublisher，定义发送消息到消息队列的方法。代码实现如下：

```java
package com.cloud.strategy;
import org.apache.log4j.Logger;
public interface MessagePublisher {
    public static Logger log = Logger.getLogger(MessagePublisher.class);
    public boolean sendByQuene(String msg,String queneName);
}
```

（9）MQ 消息发送实现类 ObsDataMsgPublisher，主要功能是连接 activeMq 服务器，发送消息到指定的消息队列。代码实现如下：

```java
package com.cloud.strategy;
import java.net.URLEncoder;
import javax.jms.DeliveryMode;
import javax.jms.JMSException;
import javax.jms.Message;
import javax.jms.Session;
import org.apache.activemq.ActiveMQConnectionFactory;
import org.springframework.jms.JmsException;
import org.springframework.jms.connection.CachingConnectionFactory;
import org.springframework.jms.core.JmsTemplate;
import org.springframework.jms.core.MessageCreator;
import com.cloud.util.PropertiesReader;
public class ObsDataMsgPublisher implements MessagePublisher {
    private static ObsDataMsgPublisher _instance = null;
    private static JmsTemplate jmsTemplate;
    private ObsDataMsgPublisher(){
    };
    public static synchronized ObsDataMsgPublisher getInstance(){
        if(_instance == null){
            _instance = new ObsDataMsgPublisher();
            ActiveMQConnectionFactory mqFactory = new
                ActiveMQConnectionFactory();
            mqFactory.setBrokerURL(PropertiesReader.getProp
                ("jms.url"));
            CachingConnectionFactory cachFactory = new
                CachingConnectionFactory(mqFactory);
            cachFactory.setSessionCacheSize(Integer.parseInt
                (PropertiesReader.getProp("jms.cachSessionNum")));
            jmsTemplate = new JmsTemplate(cachFactory);
            jmsTemplate.setPubSubDomain(false);         // p2p 方式
            jmsTemplate.setDeliveryMode(DeliveryMode.PERSISTENT);
                                                        // 采用持久化方式
        }
        return _instance;
    }
    @Override
    public synchronized boolean sendByQuene(final String msg,final
      String queueName){
        boolean ret = false;
        try {
            log.info(this.getClass().getSimpleName()+ "--准备发送 JMS 消
                息,queueName:" + queueName);
            log.info(this.getClass().getSimpleName()+ "--msg:" + msg);
            jmsTemplate.setDefaultDestinationName(queueName);
```

```java
            jmsTemplate.send(new MessageCreator(){
                @Override
                public Message createMessage(Session session)
                    throws JMSException {
                    try {
                        return session.createTextMessage(URLEncoder.
                            encode(msg,"utf-8"));
                    } catch(Exception e){
                        e.printStackTrace();
                    }
                    return null;
                }
            });
            ret = true;
            log.info(this.getClass().getSimpleName()+ "--JMS 消息发送成功");
        } catch(JmsException e){
            e.printStackTrace();
        }
        return ret;
    }
}
```

（10）测试验证发送数据的结果如图 3-5～图 3-7 所示。

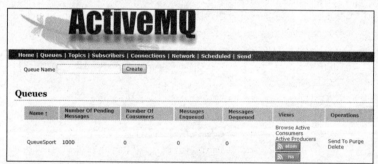

图 3-5　MQ 消息队列列表

图 3-6　消息数据基本信息

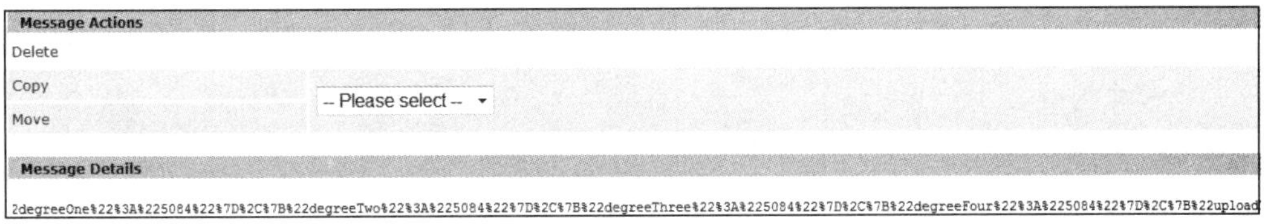

图 3-7　消息数据详情

3.4　项目小结

（1）深入了解观察者模式的架构设计和使用场景：观察者模式将观察对象与主题（也叫作被观察者对象）进行了分离，但是主题必须加入观察对象的集合。当主题内部状态发生变化时，就可以及时地通知观察者。这样，每个对象都做自己的事情，职责分明，提高了程序的复用性和可维护性。结合本章内容，由于一条智能终端运动数据被多个 app 进行共享，那些智能终端运动数据就是主题，一个 app 对应的就是一个观察者。当一条数据接收时，就会创建相应的主题，注册观察者并通知观察者来进行数据转发。

（2）多线程同步锁的设计方法：在本章中数据发送时，采用了多线程的方式。为了避免不必要的线程重复启动，我们使用了一个静态的 map 来存放同步锁 key 与线程的映射关系。至于同步锁 key 的设计，则根据业务设计成"appType + "_" + datatype + "_" + sendWay"。换言之，同一个 app 类型、同一个数据类型、同一种发送方式为一种业务。在启动线程前，根据同步锁 key 去 map 中取出对应的线程对象。若线程对象不存在或者已死，则开启新线程进行数据处理；反之，则不会开启新的线程。这样就保证了一种业务对应一个线程，同样的业务不会启动多个线程，避免了程序的性能消耗。

第 4 章

大数据高可扩展海量存储微服务引擎

架构之道分享之四：孙子兵法的《行军篇》论述了在山地和平原等地区作战要领和观察判断敌情的方法。项目规划不仅要考虑团队的技术实力，更要考察开发周期和应用场景，否则后续很多规划很难实施。

本章学习目标

★ 掌握数据库的弹性设计和通用接口开发
★ 掌握基于 Spring MVC 和 Spring Boot 的存储微服务构建方法
★ 掌握状态模式、策略模式、责任链模式的技术原理和实战方法

4.1 核心需求分析和优秀解决方案

高可扩展海量存储服务引擎是为建立高可扩展的物联网数据资源池而构建，包括物联网的移动设备和体检设备的数据接入，以及和专业机构对接的物联网大数据服务等。随着民众服务需求的变化，物联网设备的广泛采纳，新的物联网服务模式引入，以及体检和医疗生态产业链的不断加入，必须建立统一和标准化的海量存储和并行计算服务，才可以实现不同体检机构、服务机构和物联网公司之间数据互联互通。在大数据时代，物联网数据的统一接入、海量存储、兼容不同数据的开放能力已经是大势所趋。海量存储和并行计算服务通过汇聚不同设备和服务机构的数据，跟踪和预测健康变化，并基于医学知识库和机器学习的智能分析，实现辅助决策支持，对患者进行个性化的慢病防控和干预，为专业医疗机构服务和为患者提供智能分析辅助。

如果构建一个物联网大数据平台的海量存储和并行计算服务，需要对来自多源异构（时间序列）数据进行高效处理。其主要特征包括平台的开放性、通用性、模块化、灵活性和可扩展性等。需要实现多模态、不同时间颗粒度的物联网大数据的统一高效和海量存储，并提供易于扩展的物联网大数据的离线计算和批处理架构。本章介绍基于 MySQL 的高可扩展大数据存储与数据管理方案，适用于构建百万级用户和 TB 级别数据量的平台和应用。在第 5 章介绍基于 MongoDB 的高并发采集大数据处理，适用于构建千万级用户和 PB 级别数据量的平台和应用。在第 6 章介绍基于 Hadoop 集群的高可靠大数据处理，适用于构建亿级用户和 EB 级别以上数据量的平台和应用。

4.2 服务引擎的技术架构设计

（1）海量存储和并行计算服务引擎包括 5 个核心模块，如图 4-1 所示，每一个模块要考虑可扩展性和高性能两个关键因素，具体说明如下。

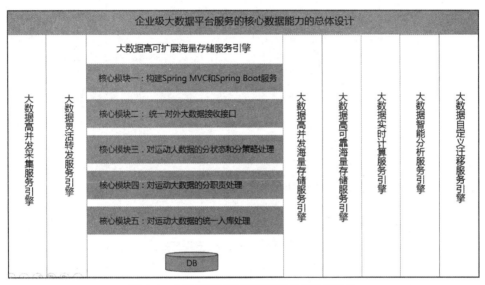

图 4-1 大数据高可扩展存储服务模块化设计

① 核心模块一：基于 Spring MVC 和 Spring Boot 微服务两种方式，构建采集服务框架，借助 maven 强大仓库集成不同中间件。

② 核心模块二：通过统一的对外数据接收接口，接收物联网智能终端运动数据等，实现对物联网移动设备大数据的统一、安全、可靠接入。

③ 核心模块三：提供对物联网智能终端运动大数据分状态和分策略处理，分状态是为了实现业务的可扩展，目前包括接收和存储状态，后续可以灵活扩展。分策略是考虑不同厂家采用不同的策略处理方式，后续可以灵活扩展。

④ 核心模块四：提供对物联网智能终端运动大数据分职责处理，分职责是为了实现数据包业务的可扩展，目前智能终端运动数据包括 1 号和 3 号简要包、2 号详细包，后续可以增加其他数据类型数据包。

⑤ 核心模块五：提供对物联网智能终端运动大数据的统一存储，保证代码的可重用和可扩展性，为后续持续演进提供一站式处理。

（2）构建 Spring MVC 版本的 BD_StorageServer_Maven 服务工程框架，如图 4-2 所示。

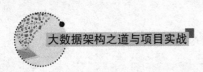

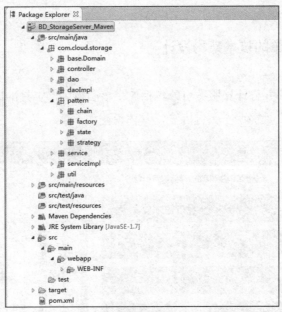

图 4-2　高可扩展存储服务 Spring MVC 版本工程

（3）构建 Spring Boot 版本的 BD_StorageServer_Boot 服务工程框架，如图 4-3 所示。

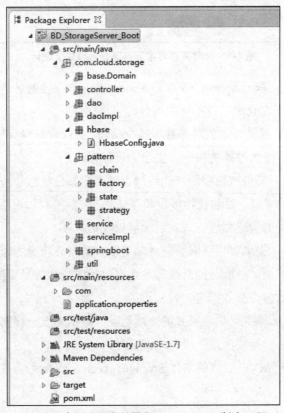

图 4-3　高可扩展存储服务 Spring Boot 版本工程

4.3 核心技术讲解及模块化实现

4.3.1 Spring MVC 的工作原理及执行流程

(1) 在 web.xml 文件中定义前端控制器 DispatcherServlet 来拦截用户请求。Web 应用是基于请求/相应架构的应用，MVC Web 框架需要在 web.xml 中配置核心 Servlet 或 Filter，才可以让该框架介入 Web 应用中。例如，Spring MVC 应用的 web.xml 文件中增加如下配置片段：

```xml
<!--定义Spring MVC 的前端控制器 -->
<servlet>
  <servlet-name>springmvc</servlet-name>
  <servlet-class>
     org.springframework.web.servlet.DispatcherServlet
  </servlet-class>
  <init-param>
     <param-name>contextConfigLocation</param-name>
     <param-value>/WEB-INF/springmvc-config.xml</param-value>
  </init-param>
  <load-on-startup>1</load-on-startup>
</servlet>
<!--让Spring MVC 的前端控制器拦截所有请求-->
<servlet-mapping>
   <servlet-name>springmvc</servlet-name>
   <url-pattern>/</url-pattern>
</servlet-mapping>
```

(2) 定义处理用户请求的 Handle 类，可以实现 Controller 接口或使用@Controller 注释。因为该 DispatcherServlet 就是 MVC 中前端控制器，负责接收请求，并将请求分发给对应的 Handle，即实现 Controller 接口的 Java 类，而该 Java 类负责调用后台业务逻辑代码来处理请求。

(3) 在 Spring MVC 框架中，控制器由两部分组成，即拦截所用用户请求和处理请求的前端控制器 DispatcherServlet，以及实际的业务控制层 Controller。

(4) Handle 映射到 Controller 的方法。在 Spring2.5 之后，推荐使用注解来配置 Handle：

```java
@Controller
public class HelloController{
@RequestMapping(value="/hello")
   public ModelAndView hello(){
     …
```

```
        }
    }
```

（5）创建视图资源。Handle 处理用户请求结束后，通常会返回一个 ModelAndView 对象，该对象中应该包含返回的视图名或视图名和模型，这个视图名就代表需要显示的物理视图资源。如果 Handle 需要把一些数据传给视图资源，则可以通过模型对象。

（6）Spring MVC 执行的具体流程。

① 用户向服务器发送请求，请求被前端控制器 DispatcherServlet 截获。

② DispatcherServlet 对请求 URL 进行解析，得到 URI。调用 HandleMapping 获得该 Handle 配置的所有相关的对象。

③ DispatcherServlet 根据获得的 Handle，选择一个合适的 HandlerAdapter。HandleAdapter 会被用于处理多种 Handle，调用 Handle 实际处理请求的方法。提取请求中的模型数据，开始执行 Handle 对应的 Controller。

④ Handle 执行完成后，向 DispatcherServlet 返回一个 ModelAndView 对象，ModelAndView 对象中应该包含视图名。

⑤ 根据返回的 ModelAndView 对象，选择一个合适的 ViewResolver（视图解析器）返回给 DispatcherServlet。

⑥ ViewResolver 结合 Model 和 View 来渲染视图，将视图渲染结果返回给客户端。

4.3.2 Spring MVC Web 服务构建

（1）配置项目的 pom.xml 文件，用于描述项目信息以及定义本项目中需要用到的依赖，配置如下：

```xml
<project xmlns="http://maven.apache.org/POM/4.0.0" xmlns:xsi="http://
    www.w3.org/2001/XMLSchema-instance"
xsi:schemaLocation="http://maven.apache.org/POM/4.0.0 http://maven.
    apache.org/xsd/maven-4.0.0.xsd">
<modelVersion>4.0.0</modelVersion>
<groupId>com.cloud.storage</groupId>
<artifactId>BD_StorageServer_Maven</artifactId>
<version>0.0.1-SNAPSHOT</version>
<packaging>war</packaging>
<properties>
    <webVersion>3.1</webVersion>
    <project.build.sourceEncoding>UTF-8</project.build.sourceEncoding>
    <spring.version>4.3.7.RELEASE</spring.version>
    <jackson.version>2.5</jackson.version>
    <jdk.version>1.7</jdk.version>
</properties>
<build>
```

```xml
<plugins>
    <plugin>
        <groupId>org.apache.maven.plugins</groupId>
        <artifactId>maven-compiler-plugin</artifactId>
        <version>3.3</version>
        <configuration>
            <source>${jdk.version}</source>
            <target>${jdk.version}</target>
        </configuration>
    </plugin>
</plugins>
</build>
<dependencies>
    <!-- servlet -->
    <dependency>
        <groupId>javax.servlet</groupId>
        <artifactId>servlet-api</artifactId>
        <version>2.5</version>
    </dependency>
    <dependency>
        <groupId>javax.servlet</groupId>
        <artifactId>jstl</artifactId>
        <version>1.2</version>
    </dependency>
    <!-- spring 常用配置 -->
    <dependency>
        <groupId>org.springframework</groupId>
        <artifactId>spring-core</artifactId>
        <version>${spring.version}</version>
    </dependency>
    <dependency>
        <groupId>org.springframework</groupId>
        <artifactId>spring-beans</artifactId>
        <version>${spring.version}</version>
    </dependency>
    <dependency>
        <groupId>org.springframework</groupId>
        <artifactId>spring-context</artifactId>
        <version>${spring.version}</version>
    </dependency>
    <dependency>
        <groupId>org.springframework</groupId>
        <artifactId>spring-tx</artifactId>
        <version>${spring.version}</version>
    </dependency>
```

```xml
<dependency>
    <groupId>org.springframework</groupId>
    <artifactId>spring-web</artifactId>
    <version>${spring.version}</version>
</dependency>
<dependency>
    <groupId>org.springframework</groupId>
    <artifactId>spring-webmvc</artifactId>
    <version>${spring.version}</version>
</dependency>
<!-- 文件上传 -->
<dependency>
    <groupId>commons-fileupload</groupId>
    <artifactId>commons-fileupload</artifactId>
    <version>1.3.2</version>
</dependency>
<!-- aop -->
<dependency>
    <groupId>org.aspectj</groupId>
    <artifactId>aspectjweaver</artifactId>
    <version>1.7.4</version>
</dependency>
<dependency>
    <groupId>commons-logging</groupId>
    <artifactId>commons-logging</artifactId>
    <version>1.1.3</version>
</dependency>
<dependency>
    <groupId>c3p0</groupId>
    <artifactId>c3p0</artifactId>
    <version>0.9.0.4</version>
</dependency>
<dependency>
    <groupId>org.springframework</groupId>
    <artifactId>spring-jdbc</artifactId>
    <version>${spring.version}</version>
</dependency>
<!-- 添加数据库驱动 -->
<dependency>
    <groupId>mysql</groupId>
    <artifactId>mysql-connector-java</artifactId>
    <version>5.1.6</version>
</dependency>
<!-- json 配置 -->
<dependency>
```

```xml
        <groupId>net.sf.json-lib</groupId>
        <artifactId>json-lib</artifactId>
        <version>2.3</version>
        <classifier>jdk15</classifier>
    </dependency>
    <dependency>
        <groupId>com.alibaba</groupId>
        <artifactId>fastjson</artifactId>
        <version>1.2.17</version>
    </dependency>
    <dependency>
        <groupId>org.slf4j</groupId>
        <artifactId>slf4j-log4j12</artifactId>
        <version>1.6.6</version>
    </dependency>
    <dependency>
        <groupId>log4j</groupId>
        <artifactId>log4j</artifactId>
        <version>1.2.17</version>
    </dependency>
</dependencies>
</project>
```

（2）Web 项目核心配置文件 web.xml，用于 Spring 容器、Spring MVC 前端控制器、编码过滤器的加载，它是中央控制器。配置如下：

```xml
<?xml version="1.0" encoding="UTF-8"?>
<web-app xmlns:xsi="http://www.w3.org/2001/XMLSchema-instance" xmlns="
    http://java.sun.com/xml/ns/javaee"xmlns:web="http://java.sun.com/
    xml/ns/javaee/web-app_2_5.xsd"xsi:schemaLocation="http://java.sun.
    com/xml/ns/javaee http://java.sun.com/xml/ns/javaee/web-app_2_5.xsd"
    id="WebApp_ID" version= "2.5">
<display-name>storage</display-name>
<context-param>
  <param-name>webAppRootKey</param-name>
  <param-value>webapp.storage.root</param-value>
</context-param>
<context-param>
  <param-name>log4jConfigLocation</param-name>
  <param-value>classpath:com/cloud/storage/Log/log4j.properties</param-value>
</context-param>
<context-param>
  <param-name>log4jRefreshInterval</param-name>
  <param-value>60000</param-value>
</context-param>
```

```xml
<context-param>
  <param-name>contextConfigLocation</param-name>
  <param-value>WEB-INF/spring/app-config.xml</param-value>
</context-param>
<filter>
  <filter-name>CharacterEncodingFilter</filter-name>
  <filter-class>org.springframework.web.filter.CharacterEncodingFilter
    </filter-class>
  <init-param>
    <param-name>encoding</param-name>
    <param-value>UTF-8</param-value>
  </init-param>
  <init-param>
    <param-name>forceEncoding</param-name>
    <param-value>true</param-value>
  </init-param>
</filter>
<filter-mapping>
  <filter-name>CharacterEncodingFilter</filter-name>
  <url-pattern>/*</url-pattern>
</filter-mapping>
<listener>
<listener-class>org.springframework.web.context.ContextLoaderListener
    </listener-class>
</listener>
<listener>
  <listener-class>org.springframework.web.util.Log4jConfigListener
    </listener-class>
</listener>
<servlet>
  <servlet-name>springmvc</servlet-name>
  <servlet-class>org.springframework.web.servlet.DispatcherServlet
    </servlet-class>
  <init-param>
    <param-name>contextConfigLocation</param-name>
    <param-value>WEB-INF/spring/mvc-config.xml</param-value>
  </init-param>
  <load-on-startup>1</load-on-startup>
</servlet>
<servlet-mapping>
  <servlet-name>springmvc</servlet-name>
  <url-pattern>*.json</url-pattern>
  <url-pattern>/service/*</url-pattern>
  <url-pattern>/</url-pattern>
</servlet-mapping>
</web-app>
```

（3）Spring 的核心配置文件 app-config.xml，主要用于扫描项目注解、初始化 jdbcTemplate，加载业务配置文件、加载 mongodb 配置 xml 文件（第 5 章用到）、加载 hbase 配置 xml 文件（第 6 章用到）等。配置如下：

```xml
<?xml version="1.0" encoding="UTF-8"?>
<beans xmlns="http://www.springframework.org/schema/beans"
xmlns:xsi="http://www.w3.org/2001/XMLSchema-instance"
xmlns:context="http://www.springframework.org/schema/context"
xmlns:tx="http://www.springframework.org/schema/tx"
xmlns:aop="http://www.springframework.org/schema/aop"
xsi:schemaLocation="
    http://www.springframework.org/schema/beans
    http://www.springframework.org/schema/beans/spring-beans-3.2.xsd
    http://www.springframework.org/schema/context
    http://www.springframework.org/schema/context/spring-context- 3.2.xsd
    http://www.springframework.org/schema/tx
    http://www.springframework.org/schema/tx/spring-tx-3.2.xsd
    http://www.springframework.org/schema/aop
    http://www.springframework.org/schema/aop/spring-aop-3.2.xsd">
    <context:annotation-config />
    <context:component-scan base-package="com.cloud.storage">
        <context:exclude-filter type="annotation" expression="org.
            springframe work.stereotype.Controller"/>
    </context:component-scan>
    <bean id="propertyConfigurer" class="org.springframework.beans.
        factory.config.PropertyPlaceholderConfigurer">
        <property name="locations">
            <list>
                <value>classpath:com/cloud/storage/Config/database.
                    properties</value>
                <value>classpath:com/cloud/storage/Config/mongodb.
                    properties</value>
                <value>classpath:com/cloud/storage/Config/hbase.
                    properties</value>
            </list>
        </property>
    </bean>
<bean id="jdbcTemplate" class="org.springframework.jdbc.core.JdbcTemplate">
</bean>
 <import resource="../mongodb.xml" />
 <!-- hbase 配置文件 -->
 <import resource="../hbase.xml" />
</beans>
```

（4）Spring MVC 的配置文件 mvc-config.xml，主要实现 controller 层扫描、视图解析器、异常拦截器、文件上传等功能。配置如下：

```xml
<?xml version="1.0" encoding="UTF-8"?>
<beans xmlns="http://www.springframework.org/schema/beans"
 xmlns:xsi="http://www.w3.org/2001/XMLSchema-instance" xmlns:mvc=
    "http://www.springframework.org/schema/mvc"
 xmlns:context="http://www.springframework.org/schema/context"
 xsi:schemaLocation="
    http://www.springframework.org/schema/beans http://www.
       springframework.org/schema/beans/spring-beans-3.2.xsd
    http://www.springframework.org/schema/mvc http://www.
       springframework.org/schema/mvc/spring-mvc-3.2.xsd
    http://www.springframework.org/schema/context http://www.
       springframework.org/schema/context/spring-context-3.2.xsd">
<!-- 1. scan all package -->
<context:component-scan base-package="com.cloud.storage">
    <context:include-filter type="annotation"
        expression="org.springframework.stereotype.Controller" />
    <context:exclude-filter type="annotation"
        expression="org.springframework.stereotype.Service" />
</context:component-scan>
<!-- 2. import servlet_view_resolver -->
<bean
    class="org.springframework.web.servlet.view.
       InternalResourceViewResolver">
    <property name="order" value="2" />
    <property name="viewClass"
        value="org.springframework.web.servlet.view.JstlView" />
    <property name="prefix" value="/jsp/" />
    <property name="suffix" value=".jsp" />
</bean>
<!-- 3. support file_upload MultipartResolver -->
<mvc:default-servlet-handler />
<bean id="multipartResolver"
    class="org.springframework.web.multipart.commons.
       CommonsMultipartResolver" />
<bean
    class="org.springframework.web.servlet.handler.
       SimpleMappingExceptionResolver">
    <property name="exceptionMappings">
        <props>
            <prop key="org.springframework.dao.DataAccessException">
                dataAccessFailure</prop>
```

```xml
                <prop key="org.springframework.transaction.TransactionException">
                    dataAccessFailure</prop>
            </props>
        </property>
</bean>
<!-- 4. support annotation -->
<mvc:annotation-driven />
</beans>
```

(5) MySQL 数据库配置文件,用于指定数据库路径、账号、连接池相关参数等信息。配置如下:

```
jdbc.driverClassName=com.mysql.jdbc.Driver
jdbc.url=jdbc\:mysql\://localhost\:3306/storage?useUnicode\=
   true&characterEncoding\=UTF-8
jdbc.username=root
jdbc.password=root
jdbc.maxPoolSize=100
jdbc.minPoolSize=40
jdbc.initialPoolSize=20
jdbc.maxIdleTime=600
jdbc.checkoutTimeout=200
jdbc.acquireIncrement=3
jdbc.maxStatements=500
jdbc.automaticTestTable=Test
jdbc.idleConnectionTestPeriod=120
```

4.3.3 Spring Boot Web 微服务构建

(1) 项目的依赖管理文件 pom.xml,用于指定 Spring Boot 的版本以及项目中需要用到的相关依赖。配置如下:

```xml
<project xmlns="http://maven.apache.org/POM/4.0.0" xmlns:xsi="http://
   www.w3.org/2001/XMLSchema-instance"
xsi:schemaLocation="http://maven.apache.org/POM/4.0.0 http://maven.
   apache.org/xsd/maven-4.0.0.xsd">
<modelVersion>4.0.0</modelVersion>
<groupId>com.cloud.bigdata</groupId>
<artifactId>BD_StorageServer_B</artifactId>
<version>0.0.1-SNAPSHOT</version>
<packaging>war</packaging>
<properties>
    <webVersion>3.1</webVersion>
</properties>
<build>
    <plugins>
```

```xml
        <plugin>
            <groupId>org.apache.maven.plugins</groupId>
            <artifactId>maven-compiler-plugin</artifactId>
            <version>3.3</version>
            <configuration>
                <source>1.7</source>
                <target>1.7</target>
            </configuration>
        </plugin>
    </plugins>
</build>
<!--①继承Spring Boot默认配置 -->
<parent>
    <groupId>org.springframework.boot</groupId>
    <artifactId>spring-boot-starter-parent</artifactId>
    <version>1.5.2.RELEASE</version>
</parent>
<dependencies>
    <!--②添加第一个Boot Web启动器 -->
    <dependency>
        <groupId>org.springframework.boot</groupId>
        <artifactId>spring-boot-starter-web</artifactId>
    </dependency>
    <!-- json配置 -->
    <dependency>
        <groupId>net.sf.json-lib</groupId>
        <artifactId>json-lib</artifactId>
        <version>2.3</version>
        <classifier>jdk15</classifier>
    </dependency>
    <!-- https://mvnrepository.com/artifact/com.alibaba/fastjson -->
    <dependency>
        <groupId>com.alibaba</groupId>
        <artifactId>fastjson</artifactId>
        <version>1.2.17</version>
    </dependency>
    <dependency>
        <groupId>org.springframework.boot</groupId>
        <artifactId>spring-boot-configuration-processor</artifactId>
        <optional>true</optional>
    </dependency>
    <dependency>
        <groupId>org.apache.commons</groupId>
        <artifactId>commons-lang3</artifactId>
        <version>3.0.1</version>
```

```xml
        </dependency>
        <dependency>
            <groupId>commons-httpclient</groupId>
            <artifactId>commons-httpclient</artifactId>
            <version>3.1</version>
        </dependency>
        <!-- c3p0 -->
        <dependency>
            <groupId>c3p0</groupId>
            <artifactId>c3p0</artifactId>
            <version>0.9.1.2</version>
        </dependency>
        <dependency>
            <groupId>org.apache.httpcomponents</groupId>
            <artifactId>httpclient</artifactId>
            <version>4.5.5</version>
        </dependency>
        <!-- 添加数据库驱动 -->
        <dependency>
            <groupId>mysql</groupId>
            <artifactId>mysql-connector-java</artifactId>
        </dependency>
        <!-- Spring Boot 集成 JdbcTemplate -->
        <dependency>
            <groupId>org.springframework.boot</groupId>
            <artifactId>spring-boot-starter-jdbc</artifactId>
        </dependency>
        <!-- Spring Boot 集成 mongodb -->
        <dependency>
            <groupId>org.springframework.boot</groupId>
            <artifactId>spring-boot-starter-data-mongodb</artifactId>
        </dependency>
        <dependency>
            <groupId>org.springframework.data</groupId>
            <artifactId>spring-data-mongodb</artifactId>
            <version>1.10.6.RELEASE</version>
        </dependency>
        <!-- hadoop 相关 jar 包 -->
<!-- hbase 相关 -->
        <dependency>
            <groupId>org.apache.hbase</groupId>
            <artifactId>hbase-client</artifactId>
            <version>0.98.13-hadoop2</version>
        </dependency>
        <dependency>
```

```xml
        <groupId>org.springframework.data</groupId>
        <artifactId>spring-data-hadoop</artifactId>
        <version>2.0.2.RELEASE</version>
    </dependency>
</dependencies>
</project>
```

（2）配置 Spring Boot 的核心配置文件 application.properties，主要用于指定项目的端口、项目应用名称、mysql 数据源配置等。配置如下：

```
server.address=127.0.0.1
server.port=8083
server.context-path:/boot_storageServer
###### 设置数据源 ######
spring.datasource.url=jdbc:mysql://localhost:3306/storage?autoReconnect=
    true&useUnicode=true&characterEncoding=utf-8
spring.datasource.username=root
spring.datasource.password=root
spring.datasource.driver-class-name=com.mysql.jdbc.Driver
# 配置初始化大小、最小、最大
spring.datasource.initialSize=5
spring.datasource.minIdle=5
spring.datasource.maxActive=20
# 配置获取连接等待超时的时间
spring.datasource.maxWait=6000
```

（3）Spring Boot 服务启动 StorageBootStarter 类，主要用于扫描项目中的所有注解到 spring 容器中并启动 spring boot 服务。代码实现如下：

```java
package com.cloud.storage.Spring Boot;
import org.springframework.boot.SpringApplication;
import org.springframework.boot.autoconfigure.Spring BootApplication;
import org.springframework.context.annotation.ComponentScan;
/**
 * Spring Boot 的启动类
 *
 * @author changyaobin
 *
 */
@Spring BootApplication
@ComponentScan(basePackages = { "com.cloud.storage" })
public class StorageBootStarter {
 public static void main(String[] args){
     SpringApplication.run(StorageBootStarter.class,args);
 }
}
```

4.3.4 统一对外数据接收接口及通用类

（1）CommonRestfulController 类的 businessDataReceive 接口。

数据接收接口 businessDataReceive 对接 BD_DispatchServer 转发服务数据发送接口的数据，收到的 jsonData 字符串数据需要转换为 JSONObject 对象，之后再进行数据格式校验，当校验通过之后可以进行 mongodb、mysql、hbase 3 种存储方式的选择。考虑到服务的通用型和可扩展性，以上 3 种存储方式的开启与关闭通过配置文件进行了设置，且同时只能一种方式设置为 true。这样，可以满足不同存储类型的项目需求。代码实现如下：

```java
package com.cloud.storage.controller;
import java.io.UnsupportedEncodingException;
import java.util.HashMap;
import java.util.Map;
import javax.servlet.http.HttpServletRequest;
import javax.servlet.http.HttpServletResponse;
import org.apache.log4j.Logger;
import org.springframework.beans.factory.annotation.Autowired;
import org.springframework.stereotype.Controller;
import org.springframework.ui.ModelMap;
import org.springframework.web.bind.annotation.PathVariable;
import org.springframework.web.bind.annotation.RequestMapping;
import org.springframework.web.bind.annotation.RequestMethod;
import com.cloud.storage.base.Domain.SportsData;
import com.cloud.storage.pattern.state.Context;
import com.cloud.storage.service.SportsDataHbaseService;
import com.cloud.storage.service.ObservationService;
import com.cloud.storage.service.PatientService;
import com.cloud.storage.service.SportsDataService;
import com.cloud.storage.util.DateUtil;
import com.cloud.storage.util.JsonUtil;
import com.cloud.storage.util.PropertiesReader;
import com.cloud.storage.util.ResponseUtil;
import com.cloud.storage.util.ValidateUtil;
import net.sf.json.JSONObject;
/**
 * 数据接收接口,与 DispatchServer 转发服务进行数据对接
 *
 * @author changyaobin
 *
 */
@Controller
public class CommonRestfulController {
    @Autowired
```

```java
    private ObservationService observationService;
    @Autowired
    private SportsDataService sportsDataService;
    @Autowired
    private SportsDataHbaseService sportsDataHbaseService;
    @Autowired
    private PatientService patientService;
    private static Logger log = Logger.getLogger(CommonRestfulController.
      class);
    /**
     * 数据采集接口
     *
     * @param request
     * @param response
     * @throws Exception
     */
    @SuppressWarnings({ "rawtypes","unchecked" })
    @RequestMapping(value = "/businessDataReceive")
    public void businessDataReceive(HttpServletRequest request,
      HttpServlet Response response)throws Exception {
        log.info("the start of businessDataReceive ");
        Map result = new HashMap();
        log.info("收到网关 DispatchServer 发来数据*_*... \r\n");
        String jsonData = "";
        try {
            jsonData = new String((request.getParameter("data").getBytes
              ("iso-8859-1")),"UTF-8");
        } catch(UnsupportedEncodingException e){
            e.printStackTrace();
            log.error("receive data occur exception:" + e.getMessage());
        }
        JSONObject jo = JSONObject.fromObject(jsonData);
        // 数据参数校验
        String validateInfo = "" + ValidateUtil.checkAppType(JsonUtil.
          getJsonParamterString(jo,"appType"))+(ValidateUtil.isValid
          (JsonUtil.getJsonParamterString (jo,"dataType"))== true ? "":
          "false")+ ValidateUtil.checkDateTime(JsonUtil.
          getJsonParamterString(jo,"collectDate"))+(ValidateUtil.isValid
          (JsonUtil.getJsonParamterString (jo,"phone"))== true ? "":
          "false");
        // 校验通过
        if("".equals(validateInfo)){
            String isMongo = PropertiesReader.getProp("mongodb");
            String isMysql = PropertiesReader.getProp("mysql");
```

```
        String isHbase = PropertiesReader.getProp("hbase");
        Map<String,Class>classMap=new HashMap<>();
        classMap.put("dataValue",HashMap.class);
        // 入库 mongodbJSONObject
        SportsData sportsData =(SportsData)JSONObject.toBean(JSONObject.
          fromObject(jsonData), SportsData.class,classMap);
        if("true".equals(isMongo)){
            // 入库 mongodb
            sportsDataService.saveSportsData(sportsData);
        } else if("true".equals(isMysql)){
            // 入库 mysql
            new Context(request,response,observationService,
               patientService).request();
        } else if("true".equals(isHbase)){
            // 入 Hbase 库
            sportsDataHbaseService.saveData(sportsData);
        }
    } else {
        response.setStatus(412);
        result.put("status","数据验证失败!" + validateInfo);
        log.info("the end of businessDataReceive has invalidate param
           include " + validateInfo);
    }
    response.setStatus(200);
    ResponseUtil.writeInfo(response,JSONObject.fromObject(result).
       toString());
}
```

（2）系统应用配置文件 SysConf.properties，主要用于配置该服务支持的 app 应用类型、入库方式等。配置如下：

```
apptype=AppA
datatype=bloodpressure
sendData=false
#保存到 mysql 的标识
mysql=true
#保存到 mongodb 的标识
mongodb=false
#保存到 hbase 的标识
hbase=false
#apptype
APPTYPE_AppA=AppA
APPTYPE_AppB=AppB
APPTYPE_AppC=AppC
APPTYPE_AppD=AppD
```

```
#Datatype
DATATYPE_STEPCOUNT=stepCount
DATATYPE_STEPDETIAL=stepDetail
DATATYPE_STEPEFFECTIVE=stepEffective
```

（3）定义智能终端运动数据实体类SportsData，供mysql、mongodb入库（第5章用到）使用。代码实现如下：

```java
package com.cloud.storage.base.Domain;
import java.io.Serializable;
import java.util.List;
import java.util.Map;
import org.springframework.data.mongodb.core.mapping.Document;
/**
 * 智能终端运动信息数据javaBean
 *
 * @author changyaobin
 *
 */
@Document(collection = "SportsData")
public class SportsData implements Serializable {
    private static final long serialVersionUID = 1L;
    private String appType;
    private String dataType;
    private String phone;
    private String collectDate;
    private List<Map<String,String>>dataValue;
    private String deviceID;
    private String sendFlag;
    private String company;
    private String teamName;
    private String pname;
    public String getAppType(){
        return appType;
    }
    public void setAppType(String appType){
        this.appType = appType;
    }
    public String getDataType(){
        return dataType;
    }
    public void setDataType(String dataType){
        this.dataType = dataType;
    }
    public String getPhone(){
        return phone;
```

```java
    }
    public void setPhone(String phone){
        this.phone = phone;
    }
    public String getCollectDate(){
        return collectDate;
    }
    public void setCollectDate(String collectDate){
        this.collectDate = collectDate;
    }
    public List<Map<String,String>> getDataValue(){
        return dataValue;
    }
    public void setDataValue(List<Map<String,String>> dataValue){
        this.dataValue = dataValue;
    }
    public String getSendFlag(){
        return sendFlag;
    }
    public void setSendFlag(String sendFlag){
        this.sendFlag = sendFlag;
    }
    public String getDeviceID(){
        return deviceID;
    }
    public void setDeviceID(String deviceID){
        this.deviceID = deviceID;
    }
    public static long getSerialversionuid(){
        return serialVersionUID;
    }
    public String getCompany(){
        return company;
    }
    public void setCompany(String company){
        this.company = company;
    }
    public String getTeamName(){
        return teamName;
    }
    public void setTeamName(String teamName){
        this.teamName = teamName;
    }
    public String getPname(){
        return pname;
    }
```

```
        public void setPname(String pname){
            this.pname = pname;
        }
}
```

（4）定义微服务消息交互类 Message，包含操作状态码、提示信息、返回数据等参数。代码实现如下：

```
package com.cloud.storage.base.Domain;
/**
 * 服务对接消息 JavaBean
 *
 * @author changyaobin
 *
 */
public class Message {
    // 操作状态码
    private int code;
    // 操作后提示信息
    private String message;
    // 操作返回数据
    private Object data;
    public int getCode(){
        return code;
    }
    public void setCode(int code){
        this.code = code;
    }
    public String getMessage(){
        return message;
    }
    public void setMessage(String message){
        this.message = message;
    }
    public Object getData(){
        return data;
    }
    public void setData(Object data){
        this.data = data;
    }
}
```

（5）日期操作工具类 DateUtil，其主要封装了对日期操作的常用方法，代码实现参考数据采集服务的 DateUtil 类。具体代码参考第 2 章的相关 DateUtil 类。

（6）Json 操作工具类 JsonUtil，封装了对 JSON 数据格式的常用方法，代码实现参考数据采集服务的 JsonUtil 类。具体代码参考第 2 章的相关 Jsonutil 类。

（7）配置文件读取工具类 PropertiesReader，封装了对 properties 格式文件的读取，代码实现参考数据采集服务的 PropertiesReader 类，需要注意的是，读取配置文件的路径要与本服务配置文件的路径吻合。具体代码参考第 2 章的相关 PropertiesReader 类。

（8）数据响应给客户端的工具类 ResponseUtil，封装了 httpResponse 返回给客户端相关的方法。代码实现如下：

```java
package com.cloud.storage.util;
import java.io.IOException;
import javax.servlet.http.HttpServletResponse;
public class ResponseUtil {
    /**
     * write to the client with the value
     *
     * @param response
     */
    public static void writeInfo(HttpServletResponse response,String 
       value){
        try {
            response.setContentType("text/html;charset=utf-8");
            response.getWriter().write(value);
            response.getWriter().flush();
            response.getWriter().close();
        } catch(IOException e){
            e.printStackTrace();
        }
    }

    /**
     * write to the client success
     *
     * @param response
     */
    public static void writeSuccess(HttpServletResponse response){
        try {
            response.setContentType("text/html;charset=utf-8");
            response.getWriter().write("{\"status\":\"success\"}");
            response.getWriter().flush();
            response.getWriter().close();
        } catch(IOException e){
            e.printStackTrace();
        }
    }
    /**
```

```java
 * write to the client success with the value
 *
 * @param response
 */
public static void writeSuccess(HttpServletResponse response,String
   value){
    try {
        System.out.println("{\"status\":\"success\",\"info\":\"" +
           value + "\"}");
        response.setContentType("text/html;charset=utf-8");
        response.getWriter().write("{\"status\":\"success\",
           \"info\":" + value + "}");
        response.getWriter().flush();
        response.getWriter().close();
    } catch(IOException e){
        e.printStackTrace();
    }
}
/**
 * write to the client failure
 *
 * @param response
 */
public static void writeFailture(HttpServletResponse response){
    try {
        response.setContentType("text/html;charset=utf-8");
        response.getWriter().write("{\"status\":\"failure\"}");
        response.getWriter().flush();
        response.getWriter().close();
    } catch(IOException e){
        e.printStackTrace();
    }
}
/**
 * write to the client failure with the value
 *
 * @param response
 */
public static void writeFailture(HttpServletResponse response,
   String value){
    try {
        response.setContentType("text/html;charset=utf-8");
        response.getWriter().write("{\"status\":\"failure\",
           \"error\":\"" + value + "\"}");
        response.getWriter().flush();
```

```
            response.getWriter().close();
        } catch(IOException e){
            e.printStackTrace();
        }
    }
}
```

（9）参数校验工具类 ValidateUtil，其主要封装了对 Java 常用数据类型的校验。代码实现如下：

```
package com.cloud.storage.util;
import java.util.regex.Pattern;
/**
 * 参数校验工具类
 *
 * @author changyaobin
 *
 */
public class ValidateUtil {
    /**
     * @param string
     * @return Boolean
     */
    public static Boolean isValid(String string){
        if(string == null || "".equals(string.trim()))
            return false;
        else
            return true;
    }
    /**
     * @param pageno
     * @param total
     * @return pageno
     */
    @SuppressWarnings("finally")
    public static int isLegalPagenoUtil(int pageno,int total){
        try {
            if(pageno < 1)
                pageno = 1;
            else if(pageno > total)
                pageno = total;
        } catch(Exception e){
            pageno = 1;
        } finally {
            return pageno;
        }
    }
```

```java
}
/**
 * false,
 *
 * @param param
 * @return
 */
public static Boolean paramCheck(Object... param){
    for(Object o:param){
        if(o instanceof String){
            if(!ValidateUtil.isValid((String)o)){
                return false;
            }
        } else {
            if(o == null){
                return false;
            }
        }
    }
    return true;
}
public static boolean isNumeric(String str){
    Pattern pattern = Pattern.compile("[0-9]*");
    return pattern.matcher(str).matches();
}
/***
 * 2013-12-09 or 17:57:45
 *
 * @param dateTime
 * @return
 */
public static String checkDateTime(String dateTime){
    String returnstr = "";
    int length = dateTime == null ? 0:dateTime.length();
    if(length == 14 && isPositiveInteger(dateTime)){
        returnstr = "";
    } else if(length == 10 && dateTime.indexOf('-')!= -1){
        returnstr = "";
    } else if(length == 8 && dateTime.indexOf(':')!= -1){
        returnstr = "";
    } else if(length == 19 && dateTime.indexOf(':')!= -1){
        returnstr = "";
    } else {
        returnstr = "无效的时间 !";
    }
```

```java
        return returnstr;
}
/***
 * validate the idcard
 *
 * @param idcard
 * @return
 */
public static String checkIdCard(String idcard){
    String isIdcard = "";
    if(idcard == null ||(idcard.length()!= 15 && idcard.length()!=
       18)){
        isIdcard = "无效的身份证号!";
    }
    return isIdcard;
}
/***
 * validate the number
 *
 * @param number
 * @param filedname
 * @return
 */
public static String checkNumber(String number,String filedname){
    return isNonNegativeInteger(number)?"":filedname + "字段值无效! ";
}
/***
 * validate the number
 *
 * @param number
 * @param filedname
 * @return
 */
public static String checkNumber(int number,String filedname){
    return number >= 0 ? "":filedname + "字段值无效! ";
}

/***
 * validate the decimal
 *
 * @param number
 * @param filedname
 * @return
 */
public static String checkDecimal(String number,String filedname){
```

```java
        return isPositiveDouble(number)? "":filedname + "字段值无效！";
    }
    /***
     * validate the apptype
     *
     * @param appType
     * @return
     */
    public static String checkAppType(String ape){
        boolean app_boolean = false;
        String apptype = PropertiesReader.getProp("apptype");
        if(apptype.indexOf(ape)!= -1)
            app_boolean = true;
        return app_boolean == true ? "":"系统类型无效!";
    }
    /***
     * validate the positive double/float number
     *
     * @param number
     * @return
     */
    public static boolean isPositiveDouble(String number){
        boolean isNumber = false;
        isNumber = number.matches("^(0|([1-9]+[0-9]*))(\\.[0-9]+)?$");
        return isNumber;
    }
    /***
     * validate the positive int number
     *
     * @param number
     * @return
     */
    public static boolean isPositiveInteger(String number){
        boolean isNumber = false;
        isNumber = number.matches("^([1-9]+[0-9]*)$");
        return isNumber;
    }
    /***
     * validate the NonNegative int number
     *
     * @param number
     * @return
     */
    public static boolean isNonNegativeInteger(String number){
        boolean isNumber = false;
```

```
        isNumber = number.matches("^0|([1-9]+[0-9]*)$");
        return isNumber;
    }
}
```

4.3.5　MySQL 对智能终端运动数据的分状态和分策略处理

数据接收包括两个状态，即接收和存储。考虑到后续可能会增加转发或分析状态，可以使用状态模式来实现高扩展性。首先定义状态模式的抽象类 State 和环境类 Context，其次定义 ReceiveState 接收状态，继承 State 的 handle 方法，实现数据接收和对象转换。最后定义 SaveState 方法，继承 State 的 handle 方法，实现存储数据。

（1）状态模式之抽象类，定义数据处理 handle 方法，其功能实现由接收以及保存状态类去实现。代码实现如下：

```
package com.cloud.storage.pattern.state;
/**
 * 状态抽象类(状态模式)
 *
 * @author Changyaobin
 *
 */
public abstract class State {
    // 数据处理方法,Context 为参数封装类
    public abstract void handle(Context context);
}
```

（2）状态模式之状态的 Context，其主要作用是封装多个状态处理类需要用到的参数以及定义数据处理方法 request。需要注意的是，当初始化一个 Context 类，其状态类会默认初始化为接收状态。代码实现如下：

```
package com.cloud.storage.pattern.state;
import javax.servlet.http.HttpServletRequest;
import javax.servlet.http.HttpServletResponse;
import com.cloud.storage.base.Domain.SportsData;
import com.cloud.storage.service.ObservationService;
import com.cloud.storage.service.PatientService;
/**
 * 多个参数的封装类
 *
 * @author changyaobin
 *
 */
public class Context {
    private State state;
    private SportsData data;
```

```java
private HttpServletRequest request;
private HttpServletResponse response;
private ObservationService observationService;
private PatientService patientService;
public Context(){
}
/**
 *
 * @param request
 * @param response
 * @param transportService
 * @param observationService
 */
public Context(HttpServletRequest request,HttpServletResponse response,
    ObservationService observationService,PatientService patientService){
    this.request = request;
    this.response = response;
    this.observationService = observationService;
    this.patientService=patientService;
    this.state = new ReceiveState();// 数据刚接收进来,初始化为接收状态
}
// 数据处理
public void request(){
    state.handle(this);
}
/**
 * @return the state
 */
public State getState(){
    return state;
}
/**
 * @param state
 *        the state to set
 */
public void setState(State state){
    this.state = state;
}
/**
 * @return the request
 */
public HttpServletRequest getRequest(){
    return request;
}
public SportsData getData(){
```

```java
        return data;
    }
    public void setData(SportsData data){
        this.data = data;
    }
    public void setObservationService(ObservationService
      observation Service){
        this.observationService = observationService;
    }
    /**
     * @param request
     *            the request to set
     */
    public void setRequest(HttpServletRequest request){
        this.request = request;
    }
    /**
     * @return the response
     */
    public HttpServletResponse getResponse(){
        return response;
    }
    /**
     * @param response
     *            the response to set
     */
    public void setResponse(HttpServletResponse response){
        this.response = response;
    }
    /**
     * @return the observationService
     */
    public ObservationService getObservationService(){
        return observationService;
    }
    public PatientService getPatientService(){
        return patientService;
    }
    public void setPatientService(PatientService patientService){
        this.patientService = patientService;
    }
}
```

（3）状态模式之 receiveState，重写 State 的 handle 方法，主要作用是将 JSON 字符串数据转换成 SportsData 对象，同时要切换到保存状态，调用当前状态（已转换为保存状态）进行数据处理。代码实现如下：

```java
package com.cloud.storage.pattern.state;
import java.util.HashMap;
import com.cloud.storage.base.Domain.SportsData;
import com.cloud.storage.util.JsonUtil;
import com.cloud.storage.util.Log;
import net.sf.json.JSONArray;
import net.sf.json.JSONObject;
/**
 * 接收状态(状态模式)
 *
 * @author changyaobin
 *
 */
public class ReceiveState extends State {
    @Override
    public void handle(Context context){
        Log.getLogger().i("now is in receive state");
        String params = context.getRequest().getParameter("data");
        JSONObject jo = JSONObject.fromObject(params);
        String phone = JsonUtil.getJsonParamterString(jo,"phone");
        String appType = JsonUtil.getJsonParamterString(jo,"appType");
        String dataType = JsonUtil.getJsonParamterString(jo,"dataType");
        String pname = JsonUtil.getJsonParamterString(jo,"pname");
        String teamName = JsonUtil.getJsonParamterString(jo,"teamName");
        String company = JsonUtil.getJsonParamterString(jo,"company");
        SportsData data = new SportsData();
        data.setAppType(appType);
        data.setDataType(dataType);
        data.setPhone(phone);
        data.setPname(pname);
        data.setCompany(company);
        data.setTeamName(teamName);
        data.setCollectDate(JsonUtil.getJsonParamterString(jo,
            "collectDate"));
        data.setDataValue(JSONArray.toList(JSONArray.fromObject
            (jo.get("dataValue")),HashMap.class));
        Log.getLogger().i("receive data:" + data.toString());
        context.setData(data);
        context.setState(new SaveState());
        context.request();
    }
}
```

(4)状态模式之 SaveState,重写 State 的 handle 方法,其主要功能是将患者的信息入库后,通过策略工厂 StrategyFactory 来根据不同的 App 类型初始化出不同的策略处理类,进行智能终端运动数据入库操作。在这里采用的工厂以及策略模式,主要考虑到不同厂家采用不同的策略处理方式,如 AppA 业务采用 AppAStrategy 策略模式,AppB 业务采用 AppBStrategy 策略模式,这样保证了以后业务的高度扩展性。代码实现如下:

```
package com.cloud.storage.pattern.state;
import com.cloud.storage.base.Domain.Patient;
import com.cloud.storage.base.Domain.SportsData;
import com.cloud.storage.pattern.factory.StrategyFactory;
import com.cloud.storage.pattern.strategy.Strategy;
import com.cloud.storage.util.Log;
/**
 * 保存状态(状态模式)
 *
 * @author changyaobin
 *
 */
public class SaveState extends State {
    @Override
    public void handle(Context context){
        SportsData data = context.getData();
        Patient patient=new Patient();
        patient.setPhone(data.getPhone());
        patient.setDeviceId(data.getDeviceID());
        patient.setName(data.getPname());
        patient.setAppType(data.getAppType());
        boolean savePatient2MysqlSuccess = context.getPatientService().
            savePatient2Mysql(patient);
        if(savePatient2MysqlSuccess){
            Log.getLogger().i("now begin to save data!");
            Strategy st = new StrategyFactory(context.getRequest().
                getSession().getServletContext(),context.getData().
                getAppType()).getInstance();
            st.dealData(context);
        }else{
            Log.getLogger().error("save patient data fail");
        }
    }
}
```

(5)策略工厂类 StrategyFactory,用于依据不同的 AppType 从 Spring 容器获得不同的策略处理类。需要注意的是,由于是从 Spring 容器中获取策略类,但前提条件是已经在 Spring 容器中注入了 AppAtrategy 等策略类,否则无法从容器中获得这些策略类,就必须将各个策略类注册到 Spring 容器中。代码实现如下：

```java
package com.cloud.storage.pattern.factory;

import javax.servlet.ServletContext;
import org.springframework.context.ApplicationContext;
import org.springframework.web.context.support.
    WebApplicationContextUtils;
import com.cloud.storage.pattern.strategy.Strategy;
/**
 * 策略工厂,根据不同的AppType来生产对应的策略处理类
 *
 * @author changyaobin
 *
 */
public class StrategyFactory {
    private Strategy strategy;
    public StrategyFactory(ServletContext sc,String appType){
        ApplicationContext ctx = WebApplicationContextUtils.
          getRequiredWebApplicationContext(sc);
        strategy =(Strategy)ctx.getBean(appType + "databean");
    }
    public Strategy getInstance(){
        return strategy;
    }
}
```

（6）策略模式之抽象策略类 Strategy，定义数据处理方法规范，具体实现由其子类去完成。代码实现如下：

```java
package com.cloud.storage.pattern.strategy;
import com.cloud.storage.pattern.state.Context;
/**
 * 策略的抽象类(策略模式)
 *
 * @author changyaobin
 *
 */
public abstract class Strategy {
    // 策略的处理方法,context为多参数的封装类
    public abstract boolean dealData(Context context);
}
```

（7）unitA 公司的 AppA 业务数据处理策略类，包括对智能终端运动的 3 种数据包处理。一是简要包 StepCountChainHandler 的处理；二是详细包 StepDetailChainHandler

的处理；三是有效包 StepEffectiveChainHandler 的处理。在这里，使用了设计模式中的责任链模式，将 3 个处理类首尾拼接组成链状，若后续有新的数据包进来，则只要将其责任处理类加入责任链中即可。需要注意的是，进行数据处理时调用的第一个责任处理类必须是责任链的头部处理类。@Component（"AppAdatabean"）表示注入实体类到 Spring 容器中，为 Spring 容器中获取这个对象做好准备。代码实现如下：

```java
package com.cloud.storage.pattern.strategy;
import org.springframework.stereotype.Component;
import com.cloud.storage.pattern.chain.Handler;
import com.cloud.storage.pattern.chain.StepCountChainHandler;
import com.cloud.storage.pattern.chain.StepDetailChainHandler;
import com.cloud.storage.pattern.state.Context;
import com.cloud.storage.util.Log;
/**
 * unitA 公司的 AppA 业务数据处理策略类
 *
 * @author changyaobin
 *
 */
@Component("AppAdatabean")
public class AppAtrategy extends Strategy {
    @Override
    public boolean dealData(Context context){
        Log.getLogger().d("Strategy data start save in db !");
        Handler newStepCountChainHandler = new StepCountChainHandler();
        Handler newStepDetailChainHandler = new StepDetailChainHandler();
        newStepCountChainHandler.setSuccessor(newStepDetailChainHandler);
        return newStepCountChainHandler.HandleRequest(context);
    }
}
```

（8）unitA 公司的 AppB 业务数据处理类，当然，在这里并没有写相关业务代码，只是为了直观地看到策略模式的结构组成。若有真实的业务，只需要在这里组装责任链即可，具体思路与 AppA 的策略处理类一致。代码实现如下：

```java
package com.cloud.storage.pattern.strategy;
import org.springframework.stereotype.Component;
import com.cloud.storage.pattern.state.Context;

/**
 * unitA 公司的 AppB 业务数据处理策略类
 *
 * @author changyaobin
 *
```

```
 */
@Component("AppBdatabean")
public class AppBtrategy extends Strategy {
    @Override
    public boolean dealData(Context context){
        return false;
    }
}
```

4.3.6　MySQL 对智能终端运动数据的分职责处理

（1）责任链的抽象类 Handler。责任链模式（Chain of Responsibility Pattern）为请求创建了一个接收者对象的链。使多个对象都有机会处理请求，从而避免请求的发送者和接收者之间的耦合关系。将这些对象连成一条链，并沿着这条链传递该请求，直到有一个对象处理它为止。责任链模式由以下两个角色组成。

① 抽象处理者角色（Handler）：定义了一个处理请求的接口。

② 具体处理者角色（Concrete Handler）：实现抽象角色中定义的接口，并处理它所负责的请求。如果不能处理则访问它的后继者。

责任链的适用场景主要包括 3 种，一是有多个对象可以处理一个请求，哪个对象处理该请求运行时刻自动确定；二是当在不明确指定接收者的情况下，向多个对象中的一个提交一个请求；三是可处理一个请求的对象集合应被动态指定。责任链的优点是降低了耦合、提高了灵活性。但是责任链模式可能会带来一些额外的性能损耗，因为它每次执行请求都要从链子开头开始遍历。接下来按照设计模式思想，建立 Handler 抽象类，其主要定义了业务数据处理方法规范。代码实现如下：

```
package com.cloud.storage.pattern.chain;
import com.cloud.storage.pattern.state.Context;
/**
 * 责任抽象类(责任链模式)
 *
 * @author changyaobin
 *
 */
public abstract class Handler {
    // 下一个责任类的引用
    protected Handler successor;
    /**
     * @return the successor
     */
    public Handler getSuccessor(){
        return successor;
    }
```

```
    /**
     * @param successor
     *            the successor to set
     */
    public void setSuccessor(Handler successor){
        this.successor = successor;
    }
    /**
     * deal the request
     */
    public abstract boolean HandleRequest(Context context);

}
```

（2）简要包责任类 StepCountChainHandler，处理智能终端运动 1 号和 3 号数据包，调用 ObservationService 的 insertOrUpdateData 方法进行智能终端运动数据存储到 MySQL。如果接收到数据类型不是智能终端运动简要包，需要传递请求到智能终端运动详细包 StepDetailChainHandler，具体方法是 successor.HandleRequest（context）。代码实现如下：

```
package com.cloud.storage.pattern.chain;
import java.util.HashMap;
import java.util.Map;
import com.cloud.storage.base.Domain.SportsData;
import com.cloud.storage.pattern.state.Context;
import com.cloud.storage.util.Log;
import com.cloud.storage.util.PropertiesReader;
import net.sf.json.JSONObject;
/**
 * unitA 1 号数据包和 3 号数据包业务处理责任类
 *
 * @author changyaobin
 *
 */
public class StepCountChainHandler extends Handler {

    @Override
    public boolean HandleRequest(Context context){
        String dataType = context.getData().getDataType();
        // 判断是否是简要步数
        if(PropertiesReader.getProp("DATATYPE_STEPCOUNT").
          equals IgnoreCase(dataType)){
            SportsData data = context.getData();
            Log.getLogger().d("data deal by StepCountChainHandler !");
            if(data.getDataValue()!= null && data.getDataValue().size()!=
              0){
```

```java
            Map<String,Object> map = new HashMap<String,Object>();
            String measureTime = data.getDataValue().get(11).get
                ("measureTime");
            map.put("stepCount",data.getDataValue());
            // insert into database
            boolean bool = context.getObservationService().
                insertOrUpdateData(data.getPhone(),measureTime,
                PropertiesReader.getProp("DATATYPE_STEPCOUNT"),
                PropertiesReader.getProp("APPTYPE_AppA"),
                    JSONObject.fromObject(map).toString(),data.
                        getCollectDate());
            if(!bool)
                Log.getLogger().e("StepCountChainHandler save data
                    into db error!");
            return bool;
        } else {
            Log.getLogger().e("StepCountChainHandler datavalue is
                null!");
            return false;
        }
    } else {
        if(successor != null)
            return successor.HandleRequest(context);
        else
            return false;
    }
}
```

（3）智能终端运动详细包责任类 StepDetailChainHandler，处理智能终端运动 2 号包，调用 ObservationService 的 insertOrUpdateData 方法进行智能终端运动数据存储到 MySQL。若数据不是 2 号包数据且存在下一个责任类，则调用下一个责任类进行处理。代码实现如下：

```java
package com.cloud.storage.pattern.chain;
import java.util.HashMap;
import java.util.Map;
import com.cloud.storage.base.Domain.SportsData;
import com.cloud.storage.pattern.state.Context;
import com.cloud.storage.util.Log;
import com.cloud.storage.util.PropertiesReader;
import net.sf.json.JSONObject;
/**
 * unitA 2号包业务处理责任类
```

```java
 *
 * @author changyaobin
 *
 */
public class StepDetailChainHandler extends Handler {

    @Override
    public boolean HandleRequest(Context context){
        String dataType = context.getData().getDataType();
        // 判断是否是详细步数
        if(PropertiesReader.getProp("DATATYPE_STEPDETIAL").
            equalsIgnoreCase(dataType)){
            SportsData data = context.getData();
            Log.getLogger().d("data deal by StepDetailChainHandler !");
            if(data.getDataValue()!= null && data.getDataValue().size()!= 0){
                Map<String,Object> map = new HashMap<String,Object>();
                String measureTime = data.getDataValue().get(7).get
                    ("measureTime");
                String hour = data.getDataValue().get(6).get("hour");
                if(Integer.parseInt(hour)< 10 && measureTime.length()== 10){
                    hour = "0" + hour;
                }
                measureTime +=(" " + hour + ":" + "00" + ":" + "00");
                map.put("stepDetail",data.getDataValue());
                // insert into database
                boolean bool = context.getObservationService().
                    insertOr UpdateData(data.getPhone(),measureTime,
                    PropertiesReader.getProp("DATATYPE_ STEPDETIAL"),
                    PropertiesReader.getProp("APPTYPE_AppA"),
                        JSONObject.fromObject(map).toString(),data.
                            getCollectDate());
                if(!bool)
                    Log.getLogger().e("StepDetailChainHandler save data
                        into db error!");
                return bool;
            } else {
                Log.getLogger().e("StepDetailChainHandler datavalue is
                    null!");
                return false;
            }
        } else {
            if(successor != null)
                return successor.HandleRequest(context);
            else
                return false;
```

```
            }
        }
}
```

4.3.7　MySQL 对智能终端运动数据的统一入库处理

（1）智能终端运动数据入库接口 ObservationService，主要定义入库方法 insertOrUpdateData 以及参数。代码实现如下：

```
package com.cloud.storage.service;
/**
 * 数据插入 MySQL 的接口层
 *
 * @author changyaobin
 *
 */
public interface ObservationService {
 /**
  * 通用的数据插入接口
  *
  * @param uniqueField
  *         phone number
  * @param dateTime
  *         measureTime
  * @param businessType
  *         from concept table's conceptDescribe filed
  * @param appType
  *         app type
  * @param param
  *         JSONObject string
  * @param receiveDateTime
  *         when the date be collected
  */
 public boolean insertOrUpdateData(String uniqueField,String dateTime,
    String businessType,String appType,
        String param,String collectDate);
}
```

（2）智能终端运动数据入库处理类 ObservationServiceImpl，实现 ObservationService 接口。在入库时，需要对 patientId 进行检验，如果 patientId 在海量存储服务中的 patient 表没有注册，那么就不能接收该智能终端运动数据。同时需要进行 conceptId（也是 App 对应的 id）的检验，如果 conceptId 在海量存储服务中的 concept 表没有注册，那么就不能接收该智能终端运动数据。当校验通过后，进行入库或者更新操作。代码实现如下：

```java
package com.cloud.storage.serviceImpl;
import java.util.List;
import java.util.Map;
import org.springframework.beans.factory.annotation.Autowired;
import org.springframework.stereotype.Service;
import com.cloud.storage.dao.JdbcDao;
import com.cloud.storage.service.ObservationService;
import com.cloud.storage.util.DateUtil;
import net.sf.json.JSONObject;
@SuppressWarnings({ "all" })
@Service
public class ObservationServiceImpl implements ObservationService {
    private static org.apache.log4j.Logger log = org.apache.log4j.
        Logger.getLogger(ObservationServiceImpl.class);
    @Autowired
    private JdbcDao jdbcDao;
    @Override
    @SuppressWarnings("rawtypes")
    public boolean insertOrUpdateData(String uniqueField,String dateTime,
      String businessType,String appType,
        String param,String collectDate){
      // find patientId
      int patientId = this.queryPatientByPhone(uniqueField,appType);
      if(0 == patientId){
          log.debug("patient not found,param is:" + "uniqueField-" +
            uniqueField + " appType-" + appType);
          return false;
      } else {
          String concept_sql = "SELECT conceptId,conceptName FROM concept
            WHERE conceptDescribe = '" + businessType + "'";
          List list = this.jdbcDao.getData(concept_sql);
          JSONObject jo = JSONObject.fromObject(param);
          String[] sql = new String[list == null ? 0:list.size()];
          String check_sql = "SELECT COUNT(1)FROM observation WHERE
            conceptId = '%s' AND patientId = '%s' AND obsDatetime = '%s' ";
          String insert_sql = "INSERT into 'observation'('obsDatetime',
            'value','conceptId','patientId','collectDate',
            'receiveDateTime')values('%s','%s','%s','%s',
            '%s','%s')";
          String update_sql = "UPDATE observation set value = '%s',
            collectDate = '%s',receiveDateTime = '%s' WHERE conceptId
            = '%s' AND patientId = '%s' AND obsDatetime = '%s' ";
          for(int i = 0;i < list.size();i++){
              Map map =(Map)list.get(i);
```

```java
                    int count = this.jdbcDao.queryForInt(String.format
                        (check_sql,map.get ("conceptId").toString(),
                        patientId,dateTime));
                    sql[i] = String.format(insert_sql,dateTime,jo.getString
                        (map.get("conceptName").toString()),
                            map.get("conceptId").toString(),patientId,
                            collectDate,DateUtil.getCurrentTime());
                }
                if(sql.length > 0){
                    try {
                        int[] batchUpdate = this.jdbcDao.batchUpdate(sql);
                        return true;
                    } catch(Exception e){
                        e.printStackTrace();
                        log.debug("batchUpdate data error !\r\n" +
                            e.getMessage());
                        return false;
                    }
                } else {
                    log.debug("can't find concept data error !");
                    return false;
                }
            }
        }
        /**
         * find the patient by unique field
         *
         * @param uniqueField
         * @param appType
         * @return
         */
        public int queryPatientByPhone(String phone,String appType){
            String query_sql = "select patientId from patient p where 1=1 and
                p.appType = '" + appType + "' " + " and p.phone =
                '" + phone + "' ";
            return this.jdbcDao.queryForInt(query_sql);
        }
    }
```

（3）定义接口 JdbcDao，封装了对 MySQL 数据库通用的增、删、改、查方法。代码实现如下：

```java
package com.cloud.storage.dao;
import java.util.HashMap;
import java.util.List;
import java.util.Map;
import org.springframework.jdbc.support.rowset.SqlRowSet;
```

```java
/**
 * jdbcDao层接口
 *
 * @author changyaobin
 *
 */
public interface JdbcDao {
    public String monitor();
    public void execute(final String sql);
    public int delete(final String sql);
    /**
     * add the sql
     *
     * @param sql
     * @return
     */
    public int add(final String sql);
    /**
     * update the sql
     *
     * @param sql
     * @return
     */
    public int update(final String sql);
    /**
     * batch execute sql
     *
     * @param sql
     * @return
     */
    public int[] batchUpdate(final String sql[]);
    /**
     * query for list by sql
     *
     * @param sql
     * @return
     */
    public List<?> queryForList(String sql);
    /**
     * query for list by sql and elementType
     *
     * @param <T>
     * @param sql
     * @param elementType
     * @return
```

```java
 */
public <T> List<T> queryForList(String sql,Class<T> elementType);
/**
 * query for rowset by sql
 *
 * @param sql
 * @return
 */
public SqlRowSet queryForRowSet(String sql);
/**
 * simple query
 *
 * @param sql
 * @return int value
 */
public int queryForInt(String sql);
/**
 * simple query
 *
 * @param sql
 * @return long value
 */
public long queryForLong(String sql);
/**
 * simple query
 *
 * @param sql
 * @return map object
 */
@SuppressWarnings("rawtypes")
public Map queryForMap(String sql);
/**
 * query for object
 *
 * @param sql
 * @param requiredType
 * @return
 */
public Object queryForObject(String sql,Class<?> requiredType);
/**
 * get count
 *
 * @param sql
 * @return
 */
```

```java
public int getCount(String sql);
/**
 * execute update
 *
 * @param sql
 * @return true if success
 */
public boolean executeUpdate(String sql);
/**
 * execute delete
 *
 * @param sql
 * @return
 */
public int executeDelete(String sql);
/**
 * get map data with key is column'name value is column's value
 * @param sql
 * @return
 * @throws Exception
 */
public List<HashMap<String,String>> getData(String sql);
/**
 * get page data
 *
 * @param sql
 *
 * @param pageno
 *         page number
 * @param pagesize
 *         page size
 * @return
 */
public List<HashMap<String,String>> getScollData(String sql,int
  pageno,int pagesize);
/**
 * get obs data
 *
 * @param appType
 * @param conceptDescribe
 * @param startTime
 * @param endTime
 * @param isMq
 * @return
 */
```

```
    public List<HashMap<String,String>> getObsData(String appType,
      String conceptDescribe,String startTime, String endTime,
      String isMq);
    /**
     * @param sql
     * @param changeSecond
     * @return
     */
    public String getJson(String sql);
}
```

（4）数据库操作实现类 JdbcDaoImpl，实现 JdbcDao 接口。JdbcTemplate 是 SpringJDBC 框架的核心，JdbcTemplate 是为不同类型的 JDBC 操作提供模板方法，每个模板方法都能控制整个过程，并允许覆盖过程中的特定任务。通过模板方法可以在尽可能保留灵活性的情况下，将数据库存取的工作量降到最低。JdbcTemplate 主要提供以下五类方法，一是 execute 方法，用于执行任何 SQL 语句，一般执行 DDL 语句；二是 update 方法，用于执行新增、修改、删除等语句；三是 batchUpdate 方法，用于执行批处理相关语句；四是 query 方法及 queryForXXX 方法，用于执行查询相关语句；五是 call 方法，用于执行存储过程、函数相关语句。代码实现如下：

```
package com.cloud.storage.daoImpl;
import java.sql.Connection;
import java.sql.PreparedStatement;
import java.sql.ResultSet;
import java.sql.ResultSetMetaData;
import java.sql.SQLException;
import java.sql.Statement;
import java.util.ArrayList;
import java.util.Collections;
import java.util.Comparator;
import java.util.HashMap;
import java.util.List;
import java.util.Map;
import javax.sql.DataSource;
import org.springframework.beans.factory.annotation.Autowired;
import org.springframework.dao.DataAccessException;
import org.springframework.dao.EmptyResultDataAccessException;
import org.springframework.dao.IncorrectResultSizeDataAccessException;
import org.springframework.jdbc.core.ConnectionCallback;
import org.springframework.jdbc.core.JdbcTemplate;
import org.springframework.jdbc.core.PreparedStatementCreator;
import org.springframework.jdbc.core.ResultSetExtractor;
import org.springframework.jdbc.core.StatementCallback;
import org.springframework.jdbc.support.GeneratedKeyHolder;
```

```java
import org.springframework.jdbc.support.KeyHolder;
import org.springframework.jdbc.support.rowset.SqlRowSet;
import org.springframework.stereotype.Repository;
import com.cloud.storage.dao.JdbcDao;
import com.mchange.v2.c3p0.PooledDataSource;
/**
 * jdbcdao 实现类
 *
 * @author changyaobin
 *
 */
@Repository
public class JdbcDaoImpl implements JdbcDao {
    @Autowired
    JdbcTemplate jdbcTemplate;
    @Override
    public String monitor(){
        StringBuffer re_value = new StringBuffer();
        re_value.append("monitor c3p0 pool");
        DataSource ds = this.jdbcTemplate.getDataSource();
        if(ds instanceof PooledDataSource){
            try {
                PooledDataSource pds =(PooledDataSource)ds;
                re_value.append("\r\nnum_connections:").append(pds.
                    getNumConnectionsDefaultUser())
                        .append("\r\nnum_busy_connections:").append
                            (pds.getNumBusyConnectionsDefaultUser())
                        .append("\r\nnum_idle_connections:").append
                            (pds.getNumIdleConnectionsDefaultUser());
            } catch(SQLException e){
                re_value.append("\r\n" + e.getMessage());
            }
        } else {
            re_value.append("\r\nNot a c3p0 PooledDataSource!");
        }
        return re_value.toString();
    }
    @Override
    public void execute(final String sql){
        jdbcTemplate.execute(sql);
    }
    @Override
    public int delete(String sql){
        return jdbcTemplate.update(sql);
    }
```

```java
@Override
public int add(String sql){
    final String sqlSave = sql.toString();
    KeyHolder keyHolder = new GeneratedKeyHolder();
    jdbcTemplate.update(new PreparedStatementCreator(){
        @Override
        public PreparedStatement createPreparedStatement(Connection
           con)throws SQLException {
            PreparedStatement ps = con.prepareStatement(sqlSave,
                Statement.RETURN_GENERATED_KEYS);
            return ps;
        }
    },keyHolder);
    int code = keyHolder.getKey().intValue();
    return code;
}
@Override
public int update(String sql){
    return jdbcTemplate.update(sql);
}
@Override
public int[] batchUpdate(String[] sql){
    return jdbcTemplate.batchUpdate(sql);
}
@Override
public List<?> queryForList(String sql){
    return jdbcTemplate.queryForList(sql);
}
@Override
public <T> List<T> queryForList(String sql,Class<T> elementType){
    return jdbcTemplate.queryForList(sql,elementType);
}
@Override
public SqlRowSet queryForRowSet(String sql){
    return jdbcTemplate.queryForRowSet(sql);
}
@Override
public int queryForInt(String sql){
    int re_int = 0;
    try {
        re_int = jdbcTemplate.queryForObject(sql,Integer.class);
    } catch(EmptyResultDataAccessException e){
        re_int = 0;
    } catch(IncorrectResultSizeDataAccessException e){
        re_int = -1;
```

```java
        }
        return re_int;
    }
    @Override
    public long queryForLong(String sql){
        return jdbcTemplate.queryForObject(sql,Long.class);
    }
    @SuppressWarnings("rawtypes")
    @Override
    public Map queryForMap(String sql){
        return jdbcTemplate.queryForMap(sql);
    }
    @Override
    public Object queryForObject(String sql,Class<?> requiredType){
        Object obj = null;
        try {
            obj = jdbcTemplate.queryForObject(sql,requiredType);
        } catch(DataAccessException e){
            obj = null;
        }
        return obj;
    }
    @SuppressWarnings({"unchecked","rawtypes"})
    @Override
    public int getCount(String sql){
        return(int)jdbcTemplate.query(sql,new ResultSetExtractor(){
            @Override
            public Integer extractData(ResultSet rs)throws SQLException,
              DataAccessException {
                if(rs.next()){
                    return Integer.parseInt(rs.getString(1));
                } else {
                    return 0;
                }
            }
        });
    }
    @Deprecated
    @Override
    public boolean executeUpdate(String sql){
        boolean ret = true;
        try {
            jdbcTemplate.execute(sql);
        } catch(Exception e){
            e.printStackTrace();
```

```java
            ret = false;
        }
        return ret;
    }
    @SuppressWarnings({"unchecked","rawtypes"})
    @Override
    public int executeDelete(final String sql){
        Integer result = 0;
        result =(Integer)jdbcTemplate.execute(new StatementCallback(){
            @Override
            public Object doInStatement(Statement statement)throws
              SQLException,DataAccessException {
                return statement.executeUpdate(sql);
            }
        });
        return result;
    }
    @SuppressWarnings({"unchecked","rawtypes"})
    @Override
    public List<HashMap<String,String>> getData(final String sql){
        return(List<HashMap<String,String>>)jdbcTemplate.execute(new
          StatementCallback(){
            @Override
            public List doInStatement(Statement statement)throws
              SQLException,DataAccessException {
                List<HashMap<String,String>> result = new ArrayList
                  <HashMap<String,String>>();
                ResultSet rs = statement.executeQuery(sql);
                ResultSetMetaData rsmd = rs.getMetaData();
                while(rs.next()){
                    HashMap<String,String> map = new HashMap<String,
                      String>();
                    for(int i = 0;i < rsmd.getColumnCount();i++){
                        map.put(rsmd.getColumnLabel(i + 1),rs.getString
                          (i + 1));
                    }
                    result.add(map);
                }
                return result;
            }
        });
    }
    @SuppressWarnings({"unchecked","rawtypes"})
    @Override
```

```java
public List<HashMap<String,String>> getScollData(final String sql,
  final int pageno,final int pagesize){
    return(List<HashMap<String,String>>)jdbcTemplate.execute(new
      ConnectionCallback(){
        @Override
        public List doInConnection(Connection conn)throws SQLException,
          DataAccessException {
            PreparedStatement pstat = conn.prepareStatement(sql,
              ResultSet.TYPE_SCROLL_INSENSITIVE,ResultSet.CONCUR_
              READ_ONLY);
            List<HashMap<String,String>> result = new ArrayList
              <HashMap<String,String>>();
            // 最大查询的记录条数
            pstat.setMaxRows(pageno * pagesize);
            ResultSet rs = pstat.executeQuery();
            // 将游标移动到第一条记录
            rs.first();
            // 游标移动到要输出的第一条记录
            rs.relative((pageno - 1)* pagesize - 1);
            ResultSetMetaData rsmd = rs.getMetaData();
            while(rs.next()){
                HashMap<String,String> map = new HashMap<String,
                  String>();
                for(int i = 0;i < rsmd.getColumnCount();i++){
                    map.put(rsmd.getColumnLabel(i + 1),rs.getString
                      (i + 1));
                }
                result.add(map);
            }
            return result;
        }
    });
}
@Override
public List<HashMap<String,String>> getObsData(String appType,
  String conceptDescribe,String startTime,
    String endTime,String isMq){
    String condition = "";
    if(null != startTime && null != endTime){
        condition = "and date(C.obsDatetime)>= '" + startTime + "'
          and date(C.obsDatetime)<= '" + endTime + "'";
    }
    String sql = "select B.encounterId,A.patientId,A.idcard,
      A.phone,A.email,A.name,C.obsDatetime,C.value,D.conceptName
      from patient A,#enc# B,#obs# C,concept D where A.patientId =
      B.patientId and B.encounterId = C.encounterId and C.conceptId =
      D.conceptId and D.conceptDescribe = '" + conceptDescribe + "'
```

```java
            " + condition + " and C.encounter Type ='" + appType + "'";
        if(null == isMq){
            sql = sql.replace("#enc#","encounter");
            sql = sql.replace("#obs#","observation");
        } else {
            sql = sql.replace("#enc#","mqencounter");
            sql = sql.replace("#obs#","mqobservation");
        }
        List<HashMap<String,String>> list = null;
        try {
            list = this.getData(sql);
        } catch(Exception e){
            e.printStackTrace();
        }
        Map<String,HashMap<String,String>> rowData = new HashMap
            < String,HashMap<String,String>>();
        for(HashMap<String,String> temp:list){
            HashMap<String,String> data = rowData.get(temp.get
                ("encounterId").trim());
            if(null == data){
                data = temp;
                rowData.put(temp.get("encounterId").trim(),data);
            }
            data.put(temp.get("conceptName").trim(),temp.get
                ("value"));
        }
        List<HashMap<String,String>> result = new ArrayList<HashMap
            <String,String>>();
        for(HashMap<String,String> map:rowData.values()){
            result.add(map);
        }
        rowData = null;
        Collections.sort(result,new Comparator<HashMap<String,
            String>>(){
            @Override
            public int compare(HashMap<String,String> runOne,HashMap
                < String,String> runTwo){
                return runTwo.get("obsDatetime").compareTo(runOne.get
                    ("obsDatetime"));
            }
        });
        return result;
    }
    @SuppressWarnings({"unchecked","rawtypes"})
    @Override
    public String getJson(final String sql){
        try {
```

```java
        return(String)jdbcTemplate.execute(new StatementCallback(){
            @Override
            public String doInStatement(Statement statement)throws
                SQLException,DataAccessException {
                StringBuffer result = new StringBuffer("[");
                ResultSet rs = statement.executeQuery(sql);
                while(rs.next()){
                    result.append("[");
                    // difference 8 hours
                    result.append(rs.getTimestamp(1).getTime()+
                        1000 * 60 * 60 * 8);
                    result.append(",");
                    result.append(rs.getInt(2)).append("]").
                        append(",");
                }
                if(result.toString().endsWith(","))
                    result.deleteCharAt(result.lastIndexOf(","));
                result.append("]");
                return result.toString();
            }
        });
    } catch(DataAccessException e){
        return "日常数据查询异常!异常信息:\r\n" + e.getMessage();
    }
}
```

（5）同时启动 BD_AggregateServer_Maven 工程服务、BD_DispatchServer_Maven 工程服务、Eureka 注册中心（若是 Spring Boot 服务则需要开启），然后启动模拟端发送数据，本服务就会接收到 dispatcher 服务转发的智能终端运动数据，并存储到本服务的 observation 表中，测试验证结果如图 4-4 所示。

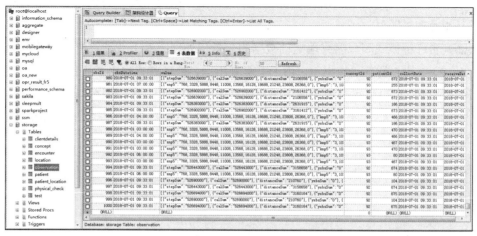

图 4-4　高可扩展存储服务数据入库

4.4 项目小结

(1) 比较 Spring MVC 和 Spring Boot 架构的不同：Spring MVC 主要是通过大量的 xml 文件来完成 Web 服务，Spring、Spring MVC、jdbc 的配置，较为烦琐。而 Spring Boot 则是采用"约定大于俗成"原则，在一个核心配置文件 application.properties 中来完成以上配置，比起 Spring MVC，更为简单和快捷。

(2) 熟悉 Spring 和 Spring MVC 的相关注解的含义。

① @Controller：控制层 controller 注解，用于数据的接收。

② @Autowired：类型自动注入，通过类型自动匹配完成属性注入。

③ @RequestMapping：Springmvc 提供的注解，通过指定名称来完成对外接口路径的定义。

④ @Service：业务层的注解，用于标识该类属于业务层代码。

⑤ @Repository：持久层的直接，用于标识该类属于持久层代码。

(3) 利用 HttpservletReuqest 对象获取 Spring 容器以及注册对象：

```
ApplicationContext ctx =WebApplicationContextUtils.
  getRequiredWebApplicationContext（sc）;
strategy =（Strategy）ctx.getBean（appType + "databean"）;
```

(4) 状态模式和数据的状态处理：接收数据以后，可以分为接收和存储两个内部状态，默认初始化状态为接收状态。当接收状态处理完毕后，内部切换到存储状态。以后若有其他业务，添加相应的状态处理类即可。

(5) 策略模式和智能终端运动数据的分策略处理：依据不同的 app 类型，来调用不同的策略处理类。这样，各个策略类之间相互独立，耦合度低。当有新的应用加入时，开发其相应的类即可。

(6) 责任链模式和智能终端运动数据的分责任处理：对于智能终端运动数据而言，分为 1、2、3 共 3 类运动数据包。由于其类别不同，处理方式也必然不同。考虑到业务以后的扩展性，可以将每个包的业务编写成相应的责任处理类，并将其首尾连接组成链状。当有数据需要处理时，则将其传入链中依次进行匹配处理即可。若有新类别的数据包，则编写责任类，并组装到链的尾部即可。

第 5 章

大数据高并发海量存储微服务引擎

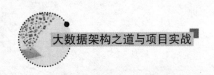

> 架构之道分享之五：孙子兵法的《九地篇》论述了进攻作战中的 9 种地理环境和作战原则，强调了激励士气来提升战斗力的问题。在项目管理中，要考虑到绩效考核和激励制度不让奋斗者吃亏的重要性。

> **本章学习目标**
> ★ 掌握基于 Spring Boot 和 Spring MVC 框架的海量存储服务构建方法
> ★ 掌握 Spring MVC、Spring Boot 对 mongodb 的集成方法，以及 mongoTemplate 操作接口
> ★ 了解 AngularJS 的工作机制和实战技术

5.1 核心需求分析和优秀解决方案

高并发海量存储服务引擎是为建立集中式的物联网数据资源池而构建的，包括对物联网各种数据类型的高并发接入。在实际运营中，我们需要在对业务处理实现升级改造并持续演进。如何应对数据类型和数据格式的灵活变化，是本章要论述的基于 MongoDB 的高并发采集大数据处理服务的主要出发点。MongoDB 不仅是面向文档存储的数据库，更是适合业务扩展的高效处理数据库，对新业务扩展和老业务升级具有一站式处理的能力。目前，MongoDB 支持管道以及 MapReduce，并提供易于扩展的物联网大数据的离线计算和批处理架构，可以满足绝大部分业务需求和个性化服务构建。本章是针对物联网智能终端运动大数据和体检大数据进行海量存储，其中体检大数据实现了基于 AngularJS 架构的可视化展示。

5.2 服务引擎的技术架构设计

大数据高并发海量存储服务引擎包括 5 个核心模块，如图 5-1 所示，每一个模块要考虑通用性和高性能两个关键因素，具体说明如下。

（1）核心模块一：构建主流的服务框架 Spring MVC、Spring Boot 和 MongoDB 集成。

（2）核心模块二：创建统一的 MongoDB 操作数据库 Dao 层，可以适合任务类型业务处理。

（3）核心模块三：提供对物联网智能终端运动大数据的统一处理，包括 Controller 层和 Service 层实现。

（4）核心模块四：提供对物联网体检大数据的统一处理，包括 Controller 层和 Service 层实现。

（5）核心模块五：通过 AngularJS 实现前后端分离，展示体检报告并可以灵活扩展数据内容。

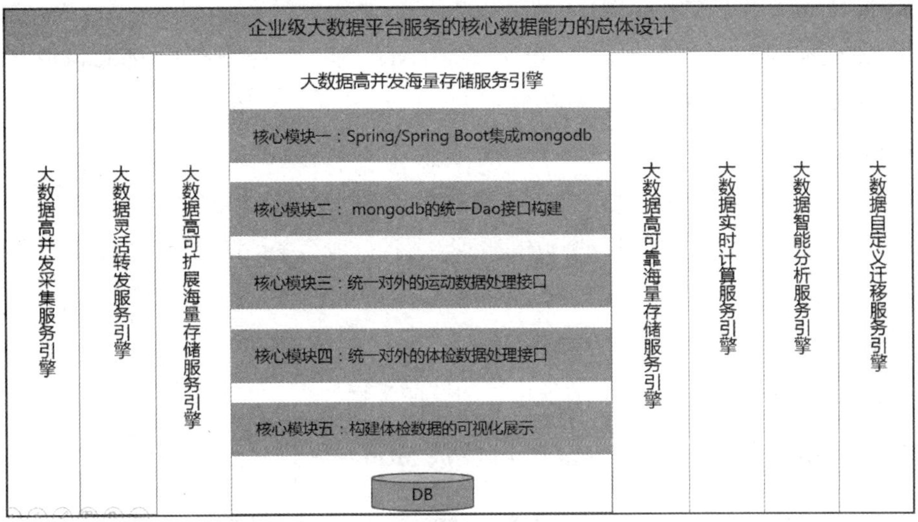

图 5-1　大数据高并发存储服务模块化设计

5.3　核心技术讲解及模块化实现

MongoDB 是由 C++语言编写的基于分布式文件存储的开源数据库系统，其数据存储灵活、查询语句丰富、查询性能快等特点已经越来越受到开发者的青睐。MongoDB 是一个面向文档存储的数据库，所有存储在集合中的数据都是 BSON 格式。BSON 是一种类 json 的二进制形式的存储格式，简称 Binary JSON。MongoDB 字段值可以包含其他文档，数组及文档数组，操作起来比较简单和容易。Mongo 支持丰富的查询表达式，查询指令使用 JSON 形式的标记，可轻易查询文档中内嵌的对象及数组。在安全性方面，MongoDB 可以添加更多的节点实现复制集，保证系统的安全性、容灾性等。在高负载的情况下，可以自定义索引以及搭建分片来提高系统的整体查询性能。同时，MongoDB 支持管道以及并行计算框架 MapReduce，可以通过一些高级语法来完成对数据统计分析等复杂功能。众所周知，传统的 NoSql 数据库对事务的支持不是很理想，但是 2018 年 7 月 MongoDB 宣布推出 4.0 版本，来支持多文档事务。迄今为止已经有接近 3000 多万下载量。MongoDB 是一个可以应用于各种规模的企业、各个行业以及各类应用程序的开源数据库。

本章对体检报告的处理使用了高级查询管道技术。MongoDB 的聚合管道将 MongoDB 文档在一个管道处理完毕后将结果传递给下一个管道处理。管道操作是可以重复的。首先介绍聚合框架中常用的几个操作。

- $project：修改输入文档的结构。可以用来重命名、增加或删除域，也可以用于创建计算结果以及嵌套文档。
- $lookup：用于 MongoDB 中两个 collection 之间关联查询。
- $match：用于过滤数据，只输出符合条件的文档。$match 使用 MongoDB 的标准查询操作。
- $limit：用来限制 MongoDB 聚合管道返回的文档数。
- $skip：在聚合管道中跳过指定数量的文档，并返回余下的文档。
- $unwind：将文档中的某一个数组类型字段拆分成多条，每条包含数组中的一个值。
- $group：将集合中的文档分组，可用于统计结果。
- $sort：将输入文档排序后输出。
- $geoNear：输出接近某一地理位置的有序文档。

5.3.1 Spring MVC 和 Spring Boot 集成 MongoDB

构建 BD_StorageServer_Maven 工程。Java 操作 MongoDB 数据库目前主要有 3 种方式。第一种方式可以通过 MongoDB 底层驱动包提供的 mongoClient 类进行操作，但是较为烦琐且扩展性不好，项目开发中几乎很少使用这种方式。本章主要集中介绍使用 Spring MVC 和 Spring Boot 集成 MongoDB 两种流行的配置方式。配置过程如下：

（1）首先，需要引入 MongoDB 的相关 jar 包依赖，在项目的 pom.xml 文件中添加依赖如下：

① Spring MVC 集成 MongoDB，配置如下：

```xml
<!-- mongodb -->
    <dependency>
        <groupId>org.mongodb</groupId>
        <artifactId>mongo-java-driver</artifactId>
        <version>3.4.0</version>
    </dependency>
    <!-- https://mvnrepository.com/artifact/org.Springframework.data/
       Spring-data-mongodb -->
    <dependency>
        <groupId>org.Springframework.data</groupId>
        <artifactId>Spring-data-mongodb</artifactId>
        <version>1.10.6.RELEASE</version>
    </dependency>
```

② Spring Boot 集成 mongodb，配置如下：

```xml
<!-- Spring Boot 集成 mongodb -->
    <dependency>
```

```
        <groupId>org.Springframework.boot</groupId>
        <artifactId>Spring-boot-starter-data-mongodb</artifactId>
</dependency>
```

（2）MongoDB 的参数配置文件 mongodb.properties（Spring MVC 集成 mongodb 时需用到），用于设置 mongodb 的 ip、端口、连接超时时间、每个主机连接数、线程队列数等。具体配置如下：

```
#主机
mongo.host=localhost
#mongo.host=10.2.44.105
#端口
mongo.port=27017
#主机加端口
mongo.hostport=localhost:27017
#每个主机的连接数
mongo.connectionsPerHost=10
#mongo.connectionsPerHost=2000
#线程队列数
mongo.threadsAllowedToBlockForConnectionMultiplier=20
#连接超时的毫秒。0 是默认和无限
mongo.connectTimeout=3000
#最大等待连接的线程阻塞时间
mongo.maxWaitTime=3000
#mongo.maxWaitTime=4000
mongo.socketKeepAlive=true
#socket 超时。0 是默认和无限
mongo.socketTimeout=3000
#mongo.socketTimeout=6000
```

（3）Spring MVC 集成 mongodb 的配置文件 mongodb.xml，主要指定 mongodb 的 ip 和端口、初始化 MongodbTemplate、进行 mongodb 数据库 Collection 与实体类的映射扫描等。需要注意的是，该配置文件必须在 Spring 核心配置文件 app-config.xml 通过 import 标签引入。具体配置如下：

```
<?xml version="1.0" encoding="UTF-8"?>
<beans xmlns="http://www.Springframework.org/schema/beans"
    xmlns:xsi="http://www.w3.org/2001/XMLSchema-instance" xmlns:
      context="http://www.Springframework.org/schema/context"
    xmlns:mongo="http://www.Springframework.org/schema/data/mongo"
    xsi:schemaLocation="http://www.Springframework.org/schema/context
       http://www.Springframework.org/schema/context/Spring-context-
          3.2.xsd
       http://www.Springframework.org/schema/data/mongo
       http://www.Springframework.org/schema/data/mongo/Spring-
```

```xml
            mongo-1.0.xsd
        http://www.Springframework.org/schema/beans
        http://www.Springframework.org/schema/beans/Spring-beans-
            3.2.xsd">
<!-- 在 Spring 核心配置文件中扫描 mongodb 配置文件 -->
<mongo:mongo host="${mongo.host}" port="${mongo.port}" />
<!-- mongo 的工厂,通过它来取得 mongo 实例,dbname 为 mongodb 的数据库名,没有
    会自动创建 -->
<mongo:db-factory dbname="storage" mongo-ref="mongo" />
<!--mongodb的主要操作对象,所有对mongodb的增、删、改、查的操作都是通过它完成 -->
<bean id="mongoTemplate" class="org.Springframework.data.mongodb.
    core.MongoTemplate">
    <constructor-arg name="mongoDbFactory" ref="mongoDbFactory" />
</bean>
<!-- 映射转换器,扫描 back-package 目录下的文件,根据注释,把它们作为 mongodb
    的一个 collection 的映射 -->
<mongo:mapping-converter base-package="com.cloud.storage.base.
    Domain" />
<context:annotation-config />
</beans>
```

（4）使用 Spring Boot 集成 mongodb，需要在 Spring Boot 的核心配置文件 application.properties 中添加如下参数：

```
#username 和 password 在 mongodb 开启了验证后需要填写,否则只要 ip:port/db 即可
Spring.data.mongodb.uri=mongodb\://localhost\:27017/storage
```

5.3.2 MongoTemplate 核心类实现 Dao 层接口

MongodbDao 自定义封装类，其封装了对 MongoDB 的一些常用操作以及通过泛型来完成对象与 mongodb 的 colletion 之间的转换操作，子类只需要指定要操作哪个实体类即可。代码实现如下：

```
package com.cloud.storage.dao;
import java.util.List;
import org.apache.log4j.Logger;
import org.Springframework.data.mongodb.core.MongoTemplate;
import org.Springframework.data.mongodb.core.query.Query;
import org.Springframework.data.mongodb.core.query.Update;
import com.mongodb.DBCollection;
/**
 * Spring mongodb抽象父类,封装一些常用的方法
 *
 * @author changyaobin
 *
 * @param <T>
```

```java
 */
public abstract class MongodbBaseDao<T> {
    Logger log = Logger.getLogger(this.getClass());
    // Spring mongodb 集成操作类
    protected MongoTemplate mongoTemplate;
    // 链接本地数据库并创建数据表
    public void CreateCollection(String collectionName){
        try {
            log.info("Collection created successfully");
        } catch(Exception e){
            System.err.println(e.getClass().getName()+":"+e.getMessage());
        }
    }
    // 获取数据表
    public DBCollection GetCollection(String collectionName){
        DBCollection coll = null;
        try {
            coll = mongoTemplate.getCollection(collectionName);
            return coll;
        } catch(Exception e){
            System.err.println(e.getClass().getName()+":"+e.getMessage());
        }
        return coll;
    }
    // 通过条件查询实体(集合)
    public List<T> Listfind(Query query){
        return mongoTemplate.find(query,this.getEntityClass());
    }
    // 通过一定的条件查询一个实体
    public T findOne(Query query){
        return(T)mongoTemplate.findOne(query,this.getEntityClass());
    }
    // 通过条件查询更新数据
    public void update(Query query,Update update){
        mongoTemplate.upsert(query,update,this.getEntityClass());
    }
    // 保存一个对象到 mongodb
    public T save(T bean){
        mongoTemplate.save(bean);
        return bean;
    }
    // 通过 ID 获取记录
    public T get(String id){
        return(T)mongoTemplate.findById(id,this.getEntityClass());
    }
```

```
    // 通过 ID 获取记录,并且指定了集合名
    public T get(String id,String collectionName){
        return(T)mongoTemplate.findById(id,this.getEntityClass(),
            collectionName);
    }
    // 获取需要操作的实体类 class
    protected abstract Class getEntityClass();
    // Spring 容器注入 mongodbTemplate
    protected abstract void setMongoTemplate(MongoTemplate mongoTemplate);
}
```

5.3.3 基于 MongoDB 处理智能终端运动数据

（1）先定义一个映射到 mongodb 数据集合的类 SportsData，通过注解标签@Document（collection = "SportsData"）实现。具体代码如下：

```java
package com.cloud.storage.base.Domain;

import java.io.Serializable;
import java.util.List;
import java.util.Map;
import org.Springframework.data.mongodb.core.mapping.Document;
/**
 * 运动信息数据 javaBean
 *
 * @author changyaobin
 *
 */
@Document(collection = "SportsData")
public class SportsData implements Serializable {

    private static final long serialVersionUID = 1L;
    private String appType;
    private String dataType;
    private String phone;
    private String collectDate;
    private List<Map<String,String>>dataValue;
   private String deviceID;
    private String sendFlag;
    private String company;
    private String teamName;
    private String pname;
    public String getAppType(){
        return appType;
    }
    public void setAppType(String appType){
```

```java
        this.appType = appType;
    }
    public String getDataType(){
        return dataType;
    }
    public void setDataType(String dataType){
        this.dataType = dataType;
    }
    public String getPhone(){
        return phone;
    }
    public void setPhone(String phone){
        this.phone = phone;
    }
    public String getCollectDate(){
        return collectDate;
    }
    public void setCollectDate(String collectDate){
        this.collectDate = collectDate;
    }
    public List<Map<String,String>> getDataValue(){
        return dataValue;
    }
    public void setDataValue(List<Map<String,String>> dataValue){
        this.dataValue = dataValue;
    }
    public String getSendFlag(){
        return sendFlag;
    }
    public void setSendFlag(String sendFlag){
        this.sendFlag = sendFlag;
    }
    public String getDeviceID(){
        return deviceID;
    }
    public void setDeviceID(String deviceID){
        this.deviceID = deviceID;
    }
    public static long getSerialversionuid(){
        return serialVersionUID;
    }
    public String getCompany(){
        return company;
    }
    public void setCompany(String company){
```

```java
        this.company = company;
    }
    public String getTeamName(){
        return teamName;
    }
    public void setTeamName(String teamName){
        this.teamName = teamName;
    }
    public String getPname(){
        return pname;
    }
    public void setPname(String pname){
        this.pname = pname;
    }
}
```

（2）智能终端运动数据 controller 层 CommonRestfulController，用于接收第 3 章灵活转发服务发送来的智能终端运动数据。在第 4 章已经提到，数据入库有 mysql、mongodb、hbase 3 种方式，故我们采取了开关的方式。若数据入 mongodb 数据库，只需修改 SysConf.properties 文件的 mongodb 属性值为 true 即可。注意，一定要添加注解 @Controller，以将类依赖注入容器。代码实现如下：

```java
package com.cloud.storage.controller;
import java.io.UnsupportedEncodingException;
import java.util.HashMap;
import java.util.Map;
import javax.servlet.http.HttpServletRequest;
import javax.servlet.http.HttpServletResponse;
import org.apache.log4j.Logger;
import org.Springframework.beans.factory.annotation.Autowired;
import org.Springframework.stereotype.Controller;
import org.Springframework.ui.ModelMap;
import org.Springframework.web.bind.annotation.PathVariable;
import org.Springframework.web.bind.annotation.RequestMapping;
import org.Springframework.web.bind.annotation.RequestMethod;
import com.cloud.storage.base.Domain.SportsData;
import com.cloud.storage.pattern.state.Context;
import com.cloud.storage.service.ObservationService;
import com.cloud.storage.service.PatientService;
import com.cloud.storage.service.SportsDataHbaseService;
import com.cloud.storage.service.SportsDataService;
import com.cloud.storage.util.DateUtil;
import com.cloud.storage.util.JsonUtil;
import com.cloud.storage.util.PropertiesReader;
import com.cloud.storage.util.ResponseUtil;
```

```java
import com.cloud.storage.util.ValidateUtil;
import net.sf.json.JSONObject;
/**
 * 数据接收接口，与DispatchServer转发服务进行数据对接
 *
 * @author Changyaobin
 *
 */
@Controller
public class CommonRestfulController {
    @Autowired
    private ObservationService observationService;
    @Autowired
    private SportsDataService sportsDataService;
    @Autowired
    private SportsDataHbaseService sportsDataHbaseService;
    @Autowired
    private PatientService patientService;
    private static Logger log = Logger.getLogger(CommonRestfulController.
      class);
    /**
     * 数据采集接口
     *
     * @param request
     * @param response
     * @throws Exception
     */
    @SuppressWarnings({"rawtypes","unchecked"})
    @RequestMapping(value = "/businessDataReceive")
    public void businessDataReceive(HttpServletRequest request,
      HttpServlet Response response)throws Exception {
        log.info("the start of businessDataReceive ");
        Map result = new HashMap();
        log.info("收到网关DispatchServer发来数据*_*... \r\n");
        String jsonData = "";
        try {
            jsonData = new String((request.getParameter("data").getBytes
              ("iso-8859-1")),"UTF-8");
        } catch(UnsupportedEncodingException e){
            e.printStackTrace();
            log.error("receive data occur exception:" + e.getMessage());
        }
        JSONObject jo = JSONObject.fromObject(jsonData);
        // 数据参数校验
        String validateInfo = "" + ValidateUtil.checkAppType(JsonUtil.
```

```java
        get JsonParmterString(jo,"appType"))
            +(ValidateUtil.isValid(JsonUtil.getJsonParamterString
                (jo,"dataType"))== true ? "":"false")
            +ValidateUtil.checkDateTime(JsonUtil.getJsonParmterString
                (jo,"collectDate"))
            +(ValidateUtil.isValid(JsonUtil.getJsonParamterString(jo,
                "phone"))== true ? "":"false");
    // 校验通过
    if("".equals(validateInfo)){
        String isMongo = PropertiesReader.getProp("mongodb");
        String isMysql = PropertiesReader.getProp("mysql");
        String isHbase = PropertiesReader.getProp("hbase");
        System.out.println("isMongo------------------------------------" +
            isMongo);
        
        Map<String,Class>classMap=new HashMap<>();
        
        classMap.put("dataValue",HashMap.class);
        // 入库 mongodbJSONObject
        SportsData sportsData =(SportsData)JSONObject.toBean(JSONObject.
            fromObject(jsonData), SportsData.class,classMap);
        if("true".equals(isMongo)){
            // 入库 mongodb
            sportsDataService.saveSportsData(sportsData);
        } else if("true".equals(isMysql)){
            // 入库 mysql
            new Context(request,response,observationService,
                patient Service).request();
        } else if("true".equals(isHbase)){
            // 入 Hbase 库
            sportsDataHbaseService.saveData(sportsData);
        }
    } else {
        response.setStatus(412);
        result.put("status","数据验证失败!" + validateInfo);
        log.info("the end of businessDataReceive has invalidate param
            include " + validateInfo);
    }
    response.setStatus(200);
    ResponseUtil.writeInfo(response,JSONObject.fromObject(result).
        toString());
}
```

（3）智能终端运动数据 service 层接口 SportsDataService，定义智能终端运动数据入 mongodb 的方法，参数为智能终端运动数据 javabean。代码实现如下：

```
package com.cloud.storage.service;
import com.cloud.storage.base.Domain.SportsData;
/**
 * 智能终端运动信息的 service 层
 *
 * @author changyaobin
 *
 */
public interface SportsDataService {
    /**
     * 智能终端运动数据保存入库接口
     *
     * @param sportsData
     */
    public void saveSportsData(SportsData sportsData);
}
```

（4）智能终端运动数据 service 接口实现类 SportsDataServiceImpl，主要功能是调用 dao 层进行入库操作。注意，一定要添加注解@Service，以将类依赖注入容器。代码实现如下：

```
package com.cloud.storage.serviceImpl;
import org.Springframework.beans.factory.annotation.Autowired;
import org.Springframework.stereotype.Service;
import com.cloud.storage.base.Domain.SportsData;
import com.cloud.storage.dao.SportsDataMongoDao;
import com.cloud.storage.service.SportsDataService;
@Service
public class SportsDataServiceImpl implements SportsDataService {
    @Autowired
    private SportsDataMongoDao sportsDataMongoDao;
    @Override
    public void saveSportsData(SportsData sportsData){
        sportsDataMongoDao.saveOne(sportsData);
    }
}
```

（5）智能终端运动数据 DAO 接口 SportsDataMongoDao，定义智能终端运动数据保存入库方法。代码实现如下：

```
package com.cloud.storage.dao;

import com.cloud.storage.base.Domain.SportsData;
/**
 * 智能终端运动数据 dao 层
 *
```

```
 * @author changyaobin
 *
 */
public interface SportsDataMongoDao {
    /**
     * 保存智能终端运动信息数据入Mongodb库
     *
     * @param data
     */
    public void saveOne(SportsData data);
}
```

（6）智能终端运动数据 DAO 接口实现类 SportsDataMongoDaoImpl，主要功能是调用父类方法进行入库操作。注意，一定要添加注解@Repository，以将类依赖注入容器。代码实现如下：

```
package com.cloud.storage.daoImpl;
import org.Springframework.beans.factory.annotation.Autowired;
import org.Springframework.beans.factory.annotation.Qualifier;
import org.Springframework.data.mongodb.core.MongoTemplate;
import org.Springframework.stereotype.Repository;
import com.cloud.storage.base.Domain.SportsData;
import com.cloud.storage.dao.SportsDataMongoDao;
import com.cloud.storage.dao.MongodbBaseDao;
/**
 * 智能终端运动数据 Mongodb 的 dao 层实现类
 *
 * @author changyaobin
 *
 */
@Repository
public class SportsDataMongoDaoImpl extends MongodbBaseDao<SportsData>
    implements SportsDataMongoDao {
    //通知父类该dao层是操作哪个实体类
    @Override
    protected Class getEntityClass(){
        return SportsData.class;
    }
    //从Spring容器中取出mongoTemplate赋值给父类的mongoTemplate属性
    @Autowired
    @Override
    protected void setMongoTemplate(@Qualifier("mongoTemplate")
        Mongo Template mongoTemplate){
        super.mongoTemplate = mongoTemplate;
    }
```

```
    @Override
    public void saveOne(SportsData data){
        this.save(data);
    }
}
```

（7）同时启动高并发采集服务、灵活转发服务、Eureka 注册中心（若是 Spring Boot 服务则需要开启），然后启动智能终端运动模拟端发送数据，本服务就会接收到灵活转发服务转发的数据，并存储到 mongodb 数据库的 SportsData 集合中，测试验证结果如图 5-2 所示。

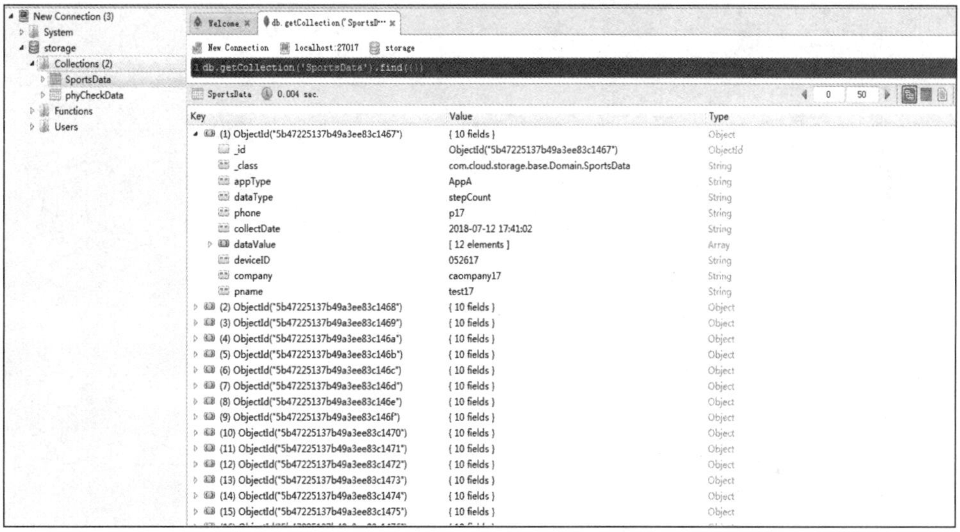

图 5-2　数据入 mongodb 结果

5.3.4　基于 MongoDB 管道技术处理体检数据

其实从上面智能终端运动数据保存可以看出，MongoDB 对单表的操作较为简单。但是对于大部分商用业务来说，单表操作并不能满足复杂的业务需求。MongoDB 不同于传统 NoSql 数据库，其从 3.0 以上版本就开始支持多表关联查询。而在本章中，除了智能终端运动数据，还可以接收用户体检数据，并在页面上完成展示功能，此时页面展示就需要用户信息表和体检信息关联查询。实现过程如下：

（1）用户类 Patient，定义姓名、电话号码、单位部门等基本信息，同时采用 @Document 注解来完成对象与 MongoDB 的 collection 的映射。代码实现如下：

```
package com.cloud.storage.base.Domain;
import org.Springframework.data.annotation.Id;
import org.Springframework.data.mongodb.core.mapping.Document;
/**
 * 用户信息 JavaBean
```

```java
 *
 * @author changyaobin
 *
 */
@Document(collection = "patient")
public class Patient implements java.io.Serializable {
    private static final long serialVersionUID = 2284954391490103232L;
    @Id
    private Integer patientId;
    private String idcard;
    // 姓名
    private String name;
    // 联系电话
    private String phone;
    private String email;
    private String deviceId;
    private String appType;
    // 性别
    private Integer sex;
    private String birth;
    // 年龄
    private String age;
    // 单位
    private String unit;
    // 部门
    private String dept;
    // Constructors
    }
    public Patient(String appType){
        this.appType = appType;
    }
    public Patient(String idcard,String name,String phone,String email,
       String deviceId,String appType,Integer sex,
          String birth){
        this.idcard = idcard;
        this.name = name;
        this.phone = phone;
        this.email = email;
        this.deviceId = deviceId;
        this.appType = appType;
        this.sex = sex;
        this.birth = birth;
    }
    public Integer getPatientId(){
        return this.patientId;
```

```java
    }
    public void setPatientId(Integer patientId){
        this.patientId = patientId;
    }
    public String getIdcard(){
        return this.idcard;
    }
    public void setIdcard(String idcard){
        this.idcard = idcard;
    }
    public String getName(){
        return this.name;
    }
    public void setName(String name){
        this.name = name;
    }
    public String getPhone(){
        return this.phone;
    }
    public void setPhone(String phone){
        this.phone = phone;
    }
    public String getEmail(){
        return this.email;
    }
    public void setEmail(String email){
        this.email = email;
    }
    public String getDeviceId(){
        return this.deviceId;
    }
    public void setDeviceId(String deviceId){
        this.deviceId = deviceId;
    }
    public String getAppType(){
        return this.appType;
    }
    public void setAppType(String appType){
        this.appType = appType;
    }
    public Integer getSex(){
        return this.sex;
    }
    public void setSex(Integer sex){
        this.sex = sex;
```

```
    }
    public String getBirth(){
        return this.birth;
    }
    public void setBirth(String birth){
        this.birth = birth;
    }
    public String getAge(){
        return age;
    }
    public void setAge(String age){
        this.age = age;
    }
    public String getUnit(){
        return unit;
    }
    public void setUnit(String unit){
        this.unit = unit;
    }
    public String getDept(){
        return dept;
    }
    public void setDept(String dept){
        this.dept = dept;
    }
    public static long getSerialversionuid(){
        return serialVersionUID;
    }
}
```

（2）用户 service 层接口定义类 PatientService，定义用户信息入 mongodb 数据库的方法。代码实现如下：

```
package com.cloud.storage.service;
import com.cloud.storage.base.Domain.Patient;
/**
 *
 * @author changyaobin
 *
 */
public interface PatientService {
    public void savePatient2Mongo(Patient patient);
}
```

（3）用户 service 层接口实现类 PatientServiceImpl，主要用于用户信息 mongodb 库的业务实现。代码实现如下：

```
package com.cloud.storage.serviceImpl;
import java.util.HashMap;
import java.util.List;
import org.Springframework.beans.factory.annotation.Autowired;
import org.Springframework.stereotype.Service;
import com.cloud.storage.base.Domain.Patient;
import com.cloud.storage.dao.PatientMongoDao;
import com.cloud.storage.daoImpl.JdbcDaoImpl;
import com.cloud.storage.service.PatientService;
@Service
public class PatientServiceImpl implements PatientService {
    @Autowired
    private PatientMongoDao patientMongoDao;
    @Override
    public void savePatient2Mongo(Patient patient){
        patientMongoDao.savePatient(patient);
    }
}
```

（4）用户信息入 mongodb 库的 dao 层接口，定义入库方法。代码实现如下：

```
package com.cloud.storage.dao;
import com.cloud.storage.base.Domain.Patient;
/**
 * 用户的 Dao 层
 *
 * @author changyaobin
 *
 */
public interface PatientMongoDao {
    void savePatient(Patient patient);
}
```

（5）用户信息入 mongodb 库的 dao 层接口实现类，完成用户信息入库。代码实现如下：

```
package com.cloud.storage.daoImpl;
import org.Springframework.beans.factory.annotation.Autowired;
import org.Springframework.beans.factory.annotation.Qualifier;
import org.Springframework.data.mongodb.core.MongoTemplate;
import org.Springframework.stereotype.Repository;
import com.cloud.storage.base.Domain.Patient;
import com.cloud.storage.dao.MongodbBaseDao;
import com.cloud.storage.dao.PatientMongoDao;
/**
 * 用户信息 dao 层实现类 (用于操作 Mongodb)
```

```java
 *
 * @author changyaobin
 *
 */
@Repository
public class PatientMongoDaoImpl extends MongodbBaseDao<Patient>
    implements PatientMongoDao {
    @Override
    protected Class getEntityClass(){
        return Patient.class;
    }
    @Autowired
    @Override
    protected void setMongoTemplate(@Qualifier(value = "mongoTemplate")
      MongoTemplate mongoTemplate){
        super.mongoTemplate = mongoTemplate;
    }
    @Override
    public void savePatient(Patient patient){
        super.save(patient);
    }
}
```

（6）体检信息类 PhysicalCheckData，定义了体检的一些基本特征属性，并采用 @Document 完成实体与 collection 映射，其中常规检查和内科等属于嵌套类，定义在体检信息类中。代码实现如下：

```java
package com.cloud.storage.base.Domain;
import org.Springframework.data.annotation.Id;
import org.Springframework.data.mongodb.core.index.Indexed;
import org.Springframework.data.mongodb.core.mapping.Document;
/**
 * 体检报告信息 javaBean
 *
 * @author changyaobin
 *
 */
@Document(collection = "phyCheckData")
public class PhysicalCheckData {
    // 监测编号
    @Id
    private String checkNum;
    // 用户 id
    @Indexed
    private Integer patientId;
```

```java
// 常规检查
private GeneralProjecr generalProjecr;
// 内科
private Medical medical;
// 外科
private Surgery surgery;
// 血常规
private RoutineBlood routineBlood;
// 体检日期
private String checkDate;
/**
 * 常规项目检查
 *
 * @author changyaobin
 *
 */
public static class GeneralProjecr {
    // 身高
    private Double height;
    // 体重
    private Double weight;
    // 体重指数
    private Double BMI;
    // 腰围
    private Integer waistline;
    // 收缩压
    private Integer systolicPressure;
    // 舒张压
    private Integer diastolicPressure;
    // 意见
    private String remark;
    public Double getHeight(){
        return height;
    }
    public void setHeight(Double height){
        this.height = height;
    }
    public Double getWeight(){
        return weight;
    }
    public void setWeight(Double weight){
        this.weight = weight;
    }
    public Double getBMI(){
        return BMI;
```

```java
        }
        public void setBMI(Double bMI){
            BMI = bMI;
        }
        public Integer getWaistline(){
            return waistline;
        }
        public void setWaistline(Integer waistline){
            this.waistline = waistline;
        }
        public Integer getSystolicPressure(){
            return systolicPressure;
        }
        public void setSystolicPressure(Integer systolicPressure){
            this.systolicPressure = systolicPressure;
        }
        public Integer getDiastolicPressure(){
            return diastolicPressure;
        }
        public void setDiastolicPressure(Integer diastolicPressure){
            this.diastolicPressure = diastolicPressure;
        }
        public String getRemark(){
            return remark;
        }
        public void setRemark(String remark){
            this.remark = remark;
        }
    }
    // 内科
    public static class Medical {
        // 病史
        private String medicalHistory;
        // 家族史
        private String familyHistory;
        // 心律
        private String heartRhythm;
        // 心音
        private String heartSounds;
        // 肺部听诊
        private String lungsAuscultation;
        // 肝脏听诊
        private String liverAuscultation;
        // 心率
        private Integer heartRate;
```

```java
// 肾脏叩诊
private String kidney;
// 意见
private String remark;
public String getMedicalHistory(){
    return medicalHistory;
}
public void setMedicalHistory(String medicalHistory){
    this.medicalHistory = medicalHistory;
}
public String getFamilyHistory(){
    return familyHistory;
}
public void setFamilyHistory(String familyHistory){
    this.familyHistory = familyHistory;
}
public String getHeartRhythm(){
    return heartRhythm;
}
public void setHeartRhythm(String heartRhythm){
    this.heartRhythm = heartRhythm;
}
public String getHeartSounds(){
    return heartSounds;
}
public void setHeartSounds(String heartSounds){
    this.heartSounds = heartSounds;
}
public String getLungsAuscultation(){
    return lungsAuscultation;
}
public void setLungsAuscultation(String lungsAuscultation){
    this.lungsAuscultation = lungsAuscultation;
}
public String getLiverAuscultation(){
    return liverAuscultation;
}
public void setLiverAuscultation(String liverAuscultation){
    this.liverAuscultation = liverAuscultation;
}
public Integer getHeartRate(){
    return heartRate;
}
public void setHeartRate(Integer heartRate){
    this.heartRate = heartRate;
```

```java
            }
            public String getRemark(){
                return remark;
            }
            public void setRemark(String remark){
                this.remark = remark;
            }
            public String getKidney(){
                return kidney;
            }
            public void setKidney(String kidney){
                this.kidney = kidney;
            }
        }
        // 外科
        public static class Surgery {
            // 淋巴结
            private String lymphGland;
            // 皮肤
            private String skin;
            // 甲状腺
            private String thyroid;
            // 脊柱
            private String spine;
            // 四肢关节
            private String extremitiesJoint;
            // 前列腺
            private String prostate;
            // 肛门
            private String anus;
            // 外科其他
            private String other;
            private String remark;
            public String getLymphGland(){
                return lymphGland;
            }
            public void setLymphGland(String lymphGland){
                this.lymphGland = lymphGland;
            }
            public String getSkin(){
                return skin;
            }
            public void setSkin(String skin){
                this.skin = skin;
            }
```

```java
        public String getThyroid(){
            return thyroid;
        }
        public void setThyroid(String thyroid){
            this.thyroid = thyroid;
        }
        public String getSpine(){
            return spine;
        }
        public void setSpine(String spine){
            this.spine = spine;
        }
        public String getExtremitiesJoint(){
            return extremitiesJoint;
        }
        public void setExtremitiesJoint(String extremitiesJoint){
            this.extremitiesJoint = extremitiesJoint;
        }
        public String getProstate(){
            return prostate;
        }
        public void setProstate(String prostate){
            this.prostate = prostate;
        }
        public String getAnus(){
            return anus;
        }
        public void setAnus(String anus){
            this.anus = anus;
        }
        public String getOther(){
            return other;
        }
        public void setOther(String other){
            this.other = other;
        }
        public String getRemark(){
            return remark;
        }
        public void setRemark(String remark){
            this.remark = remark;
        }
    }
    // 血常规
    public static class RoutineBlood {
```

```java
// 白细胞计数
private Double WBC;
// 淋巴细胞百分比
private Double LYMPH;
// 中间细胞百分比
private Double MON;
// 淋巴细胞绝对值
private Double LYMPHValue;
// 中间细胞绝对值
private Double MONValue;
// 红细胞计数
private Double RBC;
// 血红蛋白
private Double Hb;
// 红细胞压积
private Double HCT;
// 血小板计数
private Double PLT;
private String remark;
public Double getWBC(){
    return WBC;
}
public void setWBC(Double wBC){
    WBC = wBC;
}
public Double getLYMPH(){
    return LYMPH;
}
public void setLYMPH(Double lYMPH){
    LYMPH = lYMPH;
}
public Double getMON(){
    return MON;
}
public void setMON(Double mON){
    MON = mON;
}
public Double getLYMPHValue(){
    return LYMPHValue;
}
public void setLYMPHValue(Double lYMPHValue){
    LYMPHValue = lYMPHValue;
}
public Double getMONValue(){
    return MONValue;
```

```java
    }
    public void setMONValue(Double mONValue){
        MONValue = mONValue;
    }
    public Double getRBC(){
        return RBC;
    }
    public void setRBC(Double rBC){
        RBC = rBC;
    }
    public Double getHb(){
        return Hb;
    }
    public void setHb(Double hb){
        Hb = hb;
    }
    public Double getHCT(){
        return HCT;
    }
    public void setHCT(Double hCT){
        HCT = hCT;
    }
    public Double getPLT(){
        return PLT;
    }
    public void setPLT(Double pLT){
        PLT = pLT;
    }
    public String getRemark(){
        return remark;
    }
    public void setRemark(String remark){
        this.remark = remark;
    }
}
public GeneralProjecr getGeneralProjecr(){
    return generalProjecr;
}
public void setGeneralProjecr(GeneralProjecr generalProjecr){
    this.generalProjecr = generalProjecr;
}
public Medical getMedical(){
    return medical;
}
public void setMedical(Medical medical){
```

```java
        this.medical = medical;
    }
    public Surgery getSurgery(){
        return surgery;
    }
    public void setSurgery(Surgery surgery){
        this.surgery = surgery;
    }
    public RoutineBlood getRoutineBlood(){
        return routineBlood;
    }
    public void setRoutineBlood(RoutineBlood routineBlood){
        this.routineBlood = routineBlood;
    }
    public Integer getPatientId(){
        return patientId;
    }
    public void setPatientId(Integer patientId){
        this.patientId = patientId;
    }
    public String getCheckNum(){
        return checkNum;
    }
    public void setCheckNum(String checkNum){
        this.checkNum = checkNum;
    }
    public String getCheckDate(){
        return checkDate;
    }
    public void setCheckDate(String checkDate){
        this.checkDate = checkDate;
    }
    /**
     * 结果分析
     *
     * @author changyaobin
     *
     */
    public static class ResultAnalysis {
        // BMI 分析
        private String BMIAnalysis;
        // 血压分析
        private String bloodPressureAnalysis;
        // 心率分析
```

```
        private String heartRateAnalysis;
    }
}
```

（7）在体检数据的 controller 层接口类 PhysicalCheckDataController 中，定义了体检信息接收、通过用户 id 查询体检信息、体检信息测试入库等接口。代码实现如下：

```
package com.cloud.storage.controller;
import java.util.HashMap;
import java.util.Map;
import org.apache.commons.lang.StringUtils;
import org.Springframework.beans.factory.annotation.Autowired;
import org.Springframework.web.bind.annotation.RequestMapping;
import org.Springframework.web.bind.annotation.ResponseBody;
import org.Springframework.web.bind.annotation.RestController;
import com.cloud.storage.base.Domain.Message;
import com.cloud.storage.base.Domain.Patient;
import com.cloud.storage.base.Domain.PhysicalCheckData;
import com.cloud.storage.base.Domain.PhysicalCheckData.GeneralProjecr;
import com.cloud.storage.base.Domain.PhysicalCheckData.Medical;
import com.cloud.storage.base.Domain.PhysicalCheckData.RoutineBlood;
import com.cloud.storage.base.Domain.PhysicalCheckData.Surgery;
import com.cloud.storage.service.PatientService;
import com.cloud.storage.service.PhysicalCheckDataService;
import net.sf.json.JSONObject;
/**
 * 体检信息Controller层
 *
 * @author Changyaobin
 *
 */
@RestController
public class PhysicalCheckDataController {
    @Autowired
    private PhysicalCheckDataService physicalCheckDataService;
    @Autowired
    private PatientService patientService;
    @RequestMapping("receivePhyCheckData")
    public Message receivePhyCheckData(String data){
        Message message = new Message();
        if(StringUtils.isNotBlank(data)){
            try {
                PhysicalCheckData physicalCheckData =(com.cloud.storage.
                    base.Domain.PhysicalCheckData)JSONObject.toBean
                    (JSONObject.fromObject(data),Physical CheckData.class);
```

```java
                if(physicalCheckData != null){
                    if(physicalCheckData.getPatientId()!= null
                            || StringUtils.isBlank(physicalCheckData.
                            getCheckNum())){
                        message.setCode(10002);
                        message.setMessage("miss import params");
                    } else {
                        message = physicalCheckDataService.
                            savePhyCheck Data(physicalCheckData);
                        message.setCode(10000);
                        message.setMessage("数据处理成功");
                    }
                }
            } catch(Exception e){
                message.setCode(10001);
                message.setMessage("参数非法");
            }
        } else {
            // 业务状态 10001 非法参数
            message.setCode(10001);
            message.setMessage("参数非法");
        }
        return message;
    }
    @RequestMapping(value="getPhyCheckDataByUserId",produces =
        "application/json;charset=utf-8")
    @ResponseBody
    public String getPhyCheckDataByUserId(Integer userId){
        Map<String,Object> reslut = new HashMap<String,Object>();
        if(userId != null){
            reslut = physicalCheckDataService.getPhyCheckDataByUserId
                (userId);
        }
        return JSONObject.fromObject(reslut).toString();
    }
    @RequestMapping("test")
    public void seceivePhyCheckData(){
        PhysicalCheckData physicalCheckData = new PhysicalCheckData();
        physicalCheckData.setCheckNum("201803204124154");
        physicalCheckData.setPatientId(1000);
        // 常规项目检查
        GeneralProjecr generalProjecr = new GeneralProjecr();
        generalProjecr.setHeight(173.0);
        generalProjecr.setWeight(91.0);
        generalProjecr.setBMI(30.4);
```

```java
        // 收缩压
        generalProjecr.setSystolicPressure(138);
        generalProjecr.setRemark("体重指数增高 腰围增大");
        generalProjecr.setWaistline(80);
        // 内科
        Medical medical = new Medical();
        medical.setMedicalHistory("无");
        medical.setFamilyHistory("无特殊");
        medical.setHeartSounds("正常");
        medical.setHeartRate(72);
        medical.setHeartRhythm("齐");
        // 肾脏叩诊
        medical.setKidney("双肾区无扣痛");
        medical.setRemark("未见明显异常");
        // 血常规
        RoutineBlood routineBlood = new RoutineBlood();
        routineBlood.setHb(154.0);
        routineBlood.setHCT(44.80);
        routineBlood.setLYMPH(32.8);
        routineBlood.setLYMPHValue(2.3);
        routineBlood.setMON(8.4);
        routineBlood.setMONValue(0.6);
        routineBlood.setPLT(291.0);
        routineBlood.setRBC(5.69);
        routineBlood.setWBC(7.0);
        // 外科
        Surgery surgery = new Surgery();
        surgery.setAnus("未见明显异常");
        surgery.setLymphGland("颈部、锁骨上、腋窝及腹股沟未见明显异常");
        surgery.setExtremitiesJoint("未见明显异常");
        surgery.setOther("未见明显异常");
        surgery.setSkin("未见明显异常");
        surgery.setRemark("未见明显异常");
        surgery.setSpine("未见明显异常");
        surgery.setProstate("未见明显异常");
        physicalCheckData.setGeneralProjecr(generalProjecr);
        physicalCheckData.setMedical(medical);
        physicalCheckData.setRoutineBlood(routineBlood);
        physicalCheckData.setSurgery(surgery);
        physicalCheckDataService.savePhyCheckData(physicalCheckData);
}
@RequestMapping("insertPatient")
public void insertPatient(){
    Patient patient = new Patient();
    patient.setAge("37");
```

```
            patient.setName("常耀斌");
            patient.setPatientId(1000);
            patient.setUnit("华为");
            patient.setPhone("131*******120");
            patient.setDept("云平台");
            patientService.savePatient2Mongo(patient);
    }
}
```

（8）在体检信息的 service 层接口类 PhysicalCheckDataService 中，定义了体检信息入库和通过用户 id 查询体检信息的方法。代码实现如下：

```
package com.cloud.storage.service;
import java.util.Map;
import com.cloud.storage.base.Domain.Message;
import com.cloud.storage.base.Domain.PhysicalCheckData;
/**
 * 体检报告的 service 层
 *
 * @author Changyaobin
 *
 */
public interface PhysicalCheckDataService {
    /**
     * 保存体检报告数据
     *
     * @Title:savePhyCheckData
     * @功能描述:TODO
     * @设定文件:@param physicalCheckData
     * @返回类型:void
     * @author:changyaobin
     * @throws:
     */
    public Message savePhyCheckData(PhysicalCheckData physicalCheckData);
    /**
     * 根据用户 id 查询用户基本信息以及体检信息
     *
     * @Title:getPhyCheckDataByUserId
     * @功能描述:TODO
     * @设定文件:@param userId
     * @返回类型:void
     * @author:changyaobin
     * @throws:
     */
    public Map<String,Object> getPhyCheckDataByUserId(Integer userId);
}
```

（9）在体检信息 Service 层接口实现类 PhysicalCheckDataServiceImpl 中，完成体检信息入 mongodb 以及查询等功能实现。代码实现如下：

```
package com.cloud.storage.serviceImpl;
import java.util.HashMap;
import java.util.Map;
import org.Springframework.beans.factory.annotation.Autowired;
import org.Springframework.stereotype.Service;
import com.cloud.storage.base.Domain.Message;
import com.cloud.storage.base.Domain.PhysicalCheckData;
import com.cloud.storage.dao.PhysicalCheckMongoDao;
import com.cloud.storage.service.PhysicalCheckDataService;
@Service
public class PhysicalCheckDataServiceImpl implements
  PhysicalCheckData Service {
    @Autowired
    private PhysicalCheckMongoDao phyCheckDao;
    @Override
    public Message savePhyCheckData(PhysicalCheckData physicalCheck Data){
        Message message = new Message();
        try {
            phyCheckDao.savePhyCheckData(physicalCheckData);
            message.setCode(1000);
            message.setMessage("数据处理成功");
        } catch(Exception e){
            e.printStackTrace();
            message.setCode(10003);
            message.setMessage("数据处理失败");
        }
        return message;
    }
    @Override
    public Map<String,Object> getPhyCheckDataByUserId(Integer userId){
        Map<String,Object> result = new HashMap<String,Object>();
        if(userId != null){
            result = phyCheckDao.getPhyCheckDataByUserId(userId);
        }
        return result;
    }
}
```

（10）体检信息 dao 层接口 PhysicalCheckMongoDao，定义了体检信息入库以及查询方法。代码实现如下：

```java
package com.cloud.storage.dao;
import java.util.Map;
import com.cloud.storage.base.Domain.PhysicalCheckData;
/**
 * 体检信息的dao层接口
 *
 * @author changyaobin
 *
 */
public interface PhysicalCheckMongoDao {
    /**
     * 保存体检信息
     *
     * @param physicalCheckData
     */
    public void savePhyCheckData(PhysicalCheckData physicalCheckData);
    /**
     * 根据用户id查询其体检信息
     *
     * @param userId
     * @return
     */
    public Map<String,Object> getPhyCheckDataByUserId(Integer userId);
}
```

（11）在体检信息dao层接口实现类PhysicalCheckMongoDaoImpl中，通过mongodbTemplate操作管道，完成体检信息入库以及查询功能。代码实现如下：

```java
package com.cloud.storage.daoImpl;
import java.util.ArrayList;
import java.util.List;
import java.util.Map;
import org.Springframework.beans.factory.annotation.Autowired;
import org.Springframework.beans.factory.annotation.Qualifier;
import org.Springframework.data.mongodb.core.MongoTemplate;
import org.Springframework.data.mongodb.core.aggregation.Aggregation;
import org.Springframework.data.mongodb.core.aggregation.
    AggregationOperation;
import org.Springframework.data.mongodb.core.aggregation.AggregationResults;
import org.Springframework.data.mongodb.core.aggregation.TypedAggregation;
import org.Springframework.data.mongodb.core.query.Criteria;
import org.Springframework.stereotype.Repository;
import com.cloud.storage.base.Domain.PhysicalCheckData;
import com.cloud.storage.dao.MongodbBaseDao;
import com.cloud.storage.dao.PhysicalCheckMongoDao;
import com.mongodb.BasicDBObject;
```

```java
/**
 * 体检信息 dao 层接口实现类(用于操作 Mongodb)
 *
 * @author changyaobin
 *
 */
@Repository
public class PhysicalCheckMongoDaoImpl extends MongodbBaseDao
  <PhysicalCheckData> implements PhysicalCheckMongoDao {
    @Override
    protected Class getEntityClass(){
        return PhysicalCheckData.class;
    }
    @Autowired
    @Override
    protected void setMongoTemplate(@Qualifier("mongoTemplate")
      MongoTemplate mongoTemplate){
        super.mongoTemplate = mongoTemplate;
    }
    @Override
    public void savePhyCheckData(PhysicalCheckData physicalCheckData){
        super.save(physicalCheckData);
    }
    @Override
    public Map<String,Object> getPhyCheckDataByUserId(Integer userId){
        // mongodb 管道操作
        List<AggregationOperation> aggOperations = new ArrayList
          <AggregationOperation>();
        // 管道 match 过滤
        aggOperations.add(TypedAggregation.match(Criteria.where
          ("patientId").is(userId)));
        // 表关联查询用户的信息
        aggOperations.add(TypedAggregation.lookup("patient", "patientId",
          "_id","patientInfo"));
        // 切割 lookup 后的用户数组信息
        aggOperations.add(TypedAggregation.unwind("patientInfo", true));

        Aggregation agg = Aggregation.newAggregation(aggOperations);
        // 执行管道查询
        AggregationResults<BasicDBObject> aggregate = mongoTemplate.
          aggregate(agg,PhysicalCheckData.class,BasicDBObject.class);
        List<BasicDBObject> mappedResults = aggregate.getMappedResults();
        if(mappedResults != null && mappedResults.size()> 0){
            return mappedResults.get(0);
```

```
            }
            return null;
        }
}
```

5.3.5 基于 AngularJS 架构可视化体检数据

1. AngularJS 基本原理分析

AngularJS 诞生于 2009 年，由 Misko Hevery 等人创建，现在已经成为 Google 公司的旗舰产品之一。它是一款优秀的前端 JS 框架，有着诸多特性，最为核心的技术是 MVC、模块化、自动化双向数据绑定、语义化标签、依赖注入等。AngularJS 通过指令技术对传统 HTML 实现了自然扩展，通过编译技术实现了数据模型与展现视图的双向自动同步，从而消除了前端开发中烦琐复杂的 DOM 操作，通过模块化设计解决了 JS 代码管理维护和按需加载的问题，提升了前端程序员的高效开发能力。

其中，MVC 模式包含模型 Model、视图 View、控制器 Controller 3 个部分。模型 Model 是用于显示给用户并且与用户互动的数据，视图 View 表示用户看到的 DOM 元素，控制器 Controller 表示技术业务逻辑。背后的业务逻辑用于显示给用户并且与用户互动的数据的优点很多，一是职责清晰：让复杂代码逻辑按不同的职责拆解成 3 个不同的部分，每一部分的职责和功能都独立；二是代码分层模块化：我们知道将一个系统按业务功能竖向划分，就产生了不同的业务模块；三是耦合性低：将项目代码结构拆解到非常小的粒度以后，各个小的模块结构间依赖性降低；四是可重用性高：每个模块独立出来以后，抽象性更高，复用的可能性就更高；五是可维护性高：业务发生变更时影响的范围最小，方便维护；六研发成本低：UI 和逻辑都相对复杂的情况下，MVC 模式能明显地降低研发成本。

2. 体检报告的前端代码结构

体检报告主页面是 report_review.html，依赖 angular.min.js（本章源码文件夹提供）、JQuery-1.10.1.min.js（本章源码文件夹提供）、report.js 等文件，通过 html 和 angular 相关代码完成体检报告展示。代码实现如下：

```
<!DOCTYPE HTML PUBLIC "-//W3C//DTD HTML 4.01 Transitional//EN">
<html>
<head>
<title>健康体检报告</title>
<meta http-equiv="pragma" content="no-cache">
<meta http-equiv="cache-control" content="no-cache">
<meta http-equiv="expires" content="0">
<meta http-equiv="Content-Type" content="text/html;charset=UTF-8">
<meta http-equiv="keywords" content="keyword1,keyword2,keyword3">
<meta http-equiv="description" content="This is my page">
```

```html
<style type="text/css">
body,title,td,th,input {
    font-size:16px;
    text-align:center;
}
</style>
<style type="text/css">
body {
    font-size:16px;
    text-align:left;
}
</style>
<script src="js/angular.min.js"></script>
<script src="js/JQuery-1.10.1.min.js"></script>
<script src="js/report.js"></script>
</head>
<body>
    <div ng-app="myApp" ng-controller="ReportReviewCtrl">
        <h3 id="error"
            style="margin-left:200px;margin-top:100px;display:none;">
            标注操作未完成,所以无法计算得到监测报告。请继续完成标注并上传完整
            的标注结果</h3>
        <div id="report" style="display:none;">
            <h1 align="center">健康体检报告</h1>
            <p align="center"font-family:"微软雅黑";font-weight:bolder;">
                基本信息</p>
            <table cellspacing="0" border="1px" style="border-collapse:
                collapse"
                    bordercolor="#000000" cellspacing="1" cellpadding="7"
                        width="50%"
                    align="center">
                <tr bgcolor=#FFFFFF>
                    <td colspan="2">体检编号:{{result._id}}</td>
                    <td colspan="2">体检日期:{{result.checkDate}}</td>
                </tr>
                <tr bgcolor=#FFFFFF>
                    <td width="220px">姓名:{{result.patientInfo.name}}
                        </td>
                    <td>性别:{{result.patientInfo.sex}}</td>
                    <td colspan="2">年龄:{{result.patientInfo.age}} </td>
                </tr>
                <tr bgcolor=#FFFFFF>
                    <td width="220px">单位:{{result.patientInfo.unit}}
                        </td>
```

```html
            <td>部门:{{result.patientInfo.dept}}</td>
            <td colspan="2">联系电话:{{result.patientInfo.phone}}
                </td>
        </tr>
    </table>
    <p align="center"font-family:"微软雅黑";font-weight:bold;">
        常规项目</p>
    <table cellspacing=0 border="1px" style="border-collapse:
        collapse"
        bordercolor="#000000" cellspacing=1 cellpadding=7
            width=50%
        align="center">
        <tr bgcolor=#FFFFFF>
            <td>身高:{{result.generalProjecr.height}}(cm)</td>
            <td>体重:{{result.generalProjecr.weight}}</td>
            <td>体重指数:{{result.generalProjecr.BMI}}</td>
            <td>腰围:{{result.generalProjecr.waistline}}</td>
        </tr>
        <tr bgcolor=#FFFFFF>
            <td colspan="2">收缩压:{{result.generalProjecr.
                systolicPressure}}</td>
            <td colspan="2">舒张压:{{result.generalProjecr.
                diastolicPressure}}</td>
        </tr>
        <tr bgcolor=#FFFFFF>
            <td colspan="4">初步意见:{{result.generalProjecr.
                remark}}</td>
        </tr>
    </table>
    <p align="center"font-family:"微软雅黑";font-weight:bold;">
        内科</p>
    <table cellspacing=0 border="1px" style="border-collapse:
        collapse"
        bordercolor="#000000" cellspacing=1 cellpadding=7 width=50%
        align="center">
        <tr bgcolor=#FFFFFF>
            <td width="200">病史:{{result.medical.
                medicalHistory}}</td>
            <td width="200">家族史:{{result.medical.
                familyHistory}}</td>
            <td width="200">心律:{{result.medical.
                heartRhythm}}</td>
            <td width="200">心音:{{result.medical.
                heartSounds}}</td>
        </tr>
```

```html
<tr bgcolor=#FFFFFF>
    <td>肺部听诊:{{result.medical.lungsAuscultation}}
        </td>
    <td>肝脏听诊:{{result.medical.liverAuscultation}}
        </td>
    <td>心率:{{result.medical.heartRate}}</td>
    <td>肾脏叩诊:{{result.medical.kidney}}</td>
</tr>
<tr bgcolor=#FFFFFF>
    <td colspan="4">初步意见:{{result.medical.remark}}
        </td>
</tr>
</table>
<p align="center"font-family:"微软雅黑";font-weight:bold;">
外科</p>
<table cellspacing=0 border="1px" style="border-collapse:
    collapse"
    bordercolor="#000000" cellspacing=1 cellpadding=7
        width=50%
    align="center">
    <tr bgcolor=#FFFFFF>
        <td width="200">淋巴结:{{result.surgery.lymphGland}}
            </td>
        <td width="200">皮肤:{{result.surgery.skin}}</td>
        <td width="200">甲状腺:{{result.surgery.thyroid}}
            </td>
        <td width="200">脊柱:{{result.surgery.spine}}</td>
    </tr>
    <tr bgcolor=#FFFFFF>
        <td>四肢关节:{{result.surgery.extremitiesJoint}}
            </td>
        <td>前列腺:{{result.surgery.prostate}}</td>
        <td>肛门:{{result.surgery.anus}}</td>
        <td>外科其他:{{result.surgery.other}}</td>
    </tr>
    <tr bgcolor=#FFFFFF>
        <td colspan="4">初步意见:{{result.surgery.remark}}
            </td>
    </tr>
</table>
<p align="center"font-family:"微软雅黑";font-weight:bold;">
血常规</p>
<table cellspacing=0 border="1px" style="border-collapse:
    collapse"
    bordercolor="#000000" cellspacing=1 cellpadding=7 width=
```

```
                    50%
                align="center">
                <tr bgcolor=#FFFFFF>
                    <td width="200">白细胞计数:{{result.routineBlood.
                        WBC}}</td>
                    <td width="200">淋巴细胞百分比:{{result.routineBlood.
                        LYMPH}}%</td>
                    <td width="200">中间细胞百分比:{{result.routineBlood.
                        MON}}%</td>
                    <td width="200">淋巴细胞绝对值:{{result.routineBlood.
                        LYMPHValue}}</td>
                </tr>
                <tr bgcolor=#FFFFFF>
                    <td>中间细胞绝对值:{{result.routineBlood.
                        MONValue}}</td>
                    <td>红细胞计数:{{result.routineBlood.RBC}}</td>
                    <td>血红蛋白:{{result.routineBlood.Hb}}</td>
                    <td>红细胞压积:{{result.routineBlood.HCT}}</td>
                </tr>
                <tr bgcolor=#FFFFFF>
                    <td colspan="4">初步意见:{{result.routineBlood.
                        remark}}</td>
                </tr>
            </table>
        </div>
    </div>
</body>
</html>
```

3. Angular 核心概念和技术实现

作用域 Scope，用来存储模型 Model 的语境 context。模型放在这个语境中才能被控制器、指令和表达式等访问到。双向数据绑定就是自动同步模型 Model 中的数据和视图 View 表现，Angular 的实现方式允许把应用中的模型看成单一数据源。而视图始终是数据模型的一种展现形式。当模型改变时，视图就能反映这种改变。依赖注入就是负责创建和自动装载对象或函数。体检报告 js 代码 report.js，主要用于发送请求给后台查询出体检信息，并调用作用域 Scope 给页面赋值，完成信息展示。代码实现如下：

```
var app = angular.module('myApp',[]);
app.config(['$locationProvider',function($locationProvider){
        $locationProvider.html5Mode(true);
    }]);
app.controller('ReportReviewCtrl',function($scope,$location,$http,
    $rootScope){
        var param = $location.search();
```

```
          param.userId = 1000;
          $http.post("../BD_StorageServer_Maven/getPhyCheckDataByUserId?
            userId="+param.userId).success(function(response){
          if(response.status){
              //$scope.showError = true;
              $("#error").css("display","block");
          }else{
                $scope.result = eval(response);
                $scope.showResult = true;
                $("#report").css("display","block");
          }
          });
      });
```

4. 体检报告展示效果（如图 5-3 所示）

健康体验报告

基本信息

体检编号：201803204124154		体检日期：	
姓名：常耀斌	性别：	年龄：37	
单位：华为	部门：云平台	联系电话：131*****120	

常规项目

身高：173（cm）	体重：91	体重指数：30.4	腰围：80
收缩压：138		舒张压：	
初步意见：体重指数增高　腰围增大			

内科

病史：无	家族史：无特殊	心律：齐	心音：正常
肺部听诊：	肝脏听诊：	心率：72	肾脏叩诊：双肾区无叩痛
初步意见：未见明显异常			

外科

淋巴结：颈部、锁骨上、腋窝及腹股沟未见明显异常	皮肤：未见明显异常	甲状腺：	脊柱：未见明显异常
四肢关节：未见明显异常	前列腺：未见明显异常	肛门：未见明显异常	外科其他：未见明显异常
初步意见：未见明显异常			

血常规

白细胞计数：7	淋巴细胞百分比：32.8%	中间细胞百分比：8.4%	淋巴细胞绝对值：2.3
中间细胞绝对值：0.6	红细胞计数：5.69	血红蛋白：154	红细胞压积：44.8
初步意见：			

图 5-3　体检报告可视化展示

5.4 项目小结

1. 掌握 Spring MVC 以及 Spring Boot 集成 mongodb 两种配置方式

无论 Spring MVC 还是 Spring Boot 对 mongodb 集成，其实都是等价于对 mongoTemplate 的集成。在 Spring MVC 中，是通过 xml 文件以及 mongodb 的参数来完成集成的。对于 Spring Boot，为了简化开发，只需要在 Spring Boot 的核心配置文件 application.properties 中配置 mongodb 的相关参数即可。

2. 掌握 Spring MVC 集成 mongodb 后常用注解

（1）@Document（collection=" "）用于声明该类与 mongodb 库的 collection 进行映射，括号里的字符串参数即为 collection 名称，例如@Document（collection = "SportsData"）。

（2）@Id 用于声明该字段作为 mongodb 的主键。

（3）@Indexed 用于在 mongodb 数据库中给该字段建立索引。

（4）@RequestMapping（value = "/businessDataReceive"）表请求 URL 注解，调用这个接口的方式就是 http: //localhost: 8080/BD_StorageServer_Maven/businessData Receive。

（5）@Controller 表示 Controller 层接口的注解，说明注入 Spring 中，相当于创建一个类。

（6）@Service 表示 Service 层接口的注解，说明注入 Spring 中，相当于创建一个类。

（7）@Repository 表示 Dao 层接口的注解，说明注入 Spring 中，相当于创建一个类。

3. Mongodb 管道高级查询

Mongodb 的管道主要是为了满足复杂的业务应用而生的。其实，Mongodb 管道语法和 mysql 相似，只需要掌握 lookup、group、project、unwind、skip、limit 等常用管道操作符用法即可。对于 lookup 来说，通过指定关联的字段来完成表关联查询。group 用来做分组聚合使用，一般搭配 sum 等聚合函数使用。project 用于查询指定字段以及字段的重命名等。unwind 用于将数组类型数据切割。skip 用于指定跳过多少条数据，一般与 limit 组合使用完成分页功能。

4. 熟悉 AngularJS 调用后台的接口的方法

"$http.post（url）.success（function（response）{…}）"。

第 6 章

大数据高可靠海量存储微服务引擎

> 架构之道分享之六：孙子兵法的《势篇》提出了优秀将领运用各种谋略和手段创造优势的动态作战思想。优秀的架构师着重关注产品的商用性能指标，包括高可靠、高可扩展、高伸缩等互联网微架构设计，让系统真正实现在线升级和高效演进。

本章学习目标

★ 掌握 Hbase 的工作原理和核心技术
★ 掌握基于 Spring Boot 和 Spring 对 Hbase 集成实现
★ 掌握 Hbase 的模板类实现统一数据操作接口

6.1 核心需求分析和优秀解决方案

随着移动设备和物联网平台的普及，产生了大量的异构多源数据。如何让数据应用满足3个需求：一是离线数据分析；二是实时的数据处理和并行计算；三是高并发、高可靠、低延迟。离线数据分析常用的底层框架就是基于 HDFS 文件系统的 Hadoop MapReduce，上层引擎工具就是 HBase、Hive、Sqoop。HBase 在 2008 年成为 Apache 的开源子项目，2010 年成为 Apache 顶级项目，HBase 凭借列式存储和缓存机制大大提升了数据表的操作效率，成为迄今为止在体系架构和数据模型都非常创新的优秀产品之一。接下来我们熟悉 HBase 产品，并将它应用到对智能终端运动数据的高效处理上。

6.2 服务引擎的技术架构设计

海量存储和并行计算服务引擎包括5个核心模块，如图6-1所示，每一个模块要考虑可扩展性和高性能两个关键因素，具体说明如下。

（1）核心模块一：Spring 框架和 Hbase 集群集成，包括 Spring Boot 微架构的实现方式。

（2）核心模块二：HbaseTemplate 类的配置和实现，核心配置文件 hbase-site.xml 配置实现集群操作方式，HbaseConifg 配置方式实现 HbaseTemplate 方式操作 Hbase 数据库。

（3）核心模块三：提供智能终端运动数据的 Controller 接口，用于接收转发服务转发过来的智能终端运动数据。

（4）核心模块四：提供 Hbase 集群的智能终端运动数据 Service 接口，调用 dao 层接口保存智能终端运动数据入 hbase 库，以及将智能终端运动数据写入日志文件中，供 flume 监控使用。

（5）核心模块五：提供了 Hbase 集群的 dao 层接口，实现了添加智能终端运动数据以及根据 hquery 对象查询智能终端运动数据。

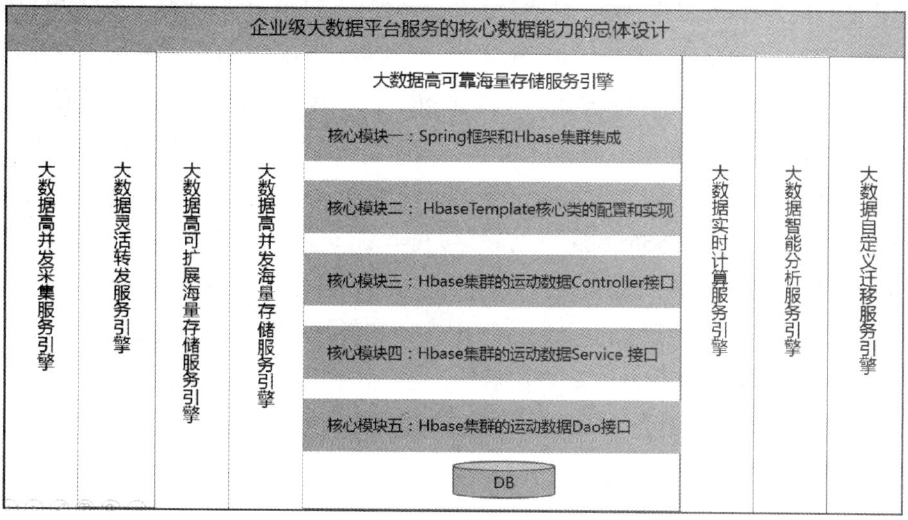

图 6-1　大数据高可靠海量存储服务模块化设计

6.3　核心技术讲解及模块化实现

　　HBase 属于存储层，是一个高可靠性、高性能、面向列、可伸缩的分布式存储系统，可在廉价 PC Server 上搭建起大规模结构化存储集群。Hbase 依托于很多框架和工具。其中，Hadoop HDFS 为 HBase 提供了高可靠性的底层存储支持，Hadoop MapReduce 为 HBase 提供了高性能的计算能力，ZooKeeper 为 HBase 提供了稳定服务和 failover 机制。Pig 和 Hive 还为 HBase 提供了高层语言支持，使得在 HBase 上进行数据统计处理简单快捷。Sqoop 为 HBase 提供了方便的 RDBMS 数据导入功能，使得传统数据库数据向 HBase 中迁移更灵活。

　　HBase 的 Client 客户端借助 HBase 的 RPC 机制与 HMaster 和 HRegionServer 进行通信，ZooKeeper Quorum 中除了存储-ROOT-表的地址和 HMaster 的地址，HRegionServer 也注册到 ZooKeeper 中，使得 HMaster 可以随时感知到各个 HRegionServer 的存活状态。HMaster 解决了单点故障问题，HBase 中可以启动多个 HMaster，通过 ZooKeeper 的 Master Election 机制保证总有一个 Master 运行，HMaster 在功能上主要负责 Table 和 Region 的管理工作，包括管理用户对 Table 的增、删、改、查操作，管理 HRegionServer 的负载均衡，调整 Region 分布，在 Region Split 后负责新 Region 的分配，在 HRegionServer 停机后负责失效 HRegionServer 上的 Regions 迁移。

HRegionServer 职责是负责响应用户 I/O 请求,向 HDFS 文件系统中读写数据,属于 HBase 中最核心的模块。它内部管理了一系列 HRegion 对象,每个 HRegion 对应了 Table 中的一个 Region,HRegion 由多个 HStore 组成。每个 HStore 对应了 Table 中的一个 Column Family 的存储,每个 Column Family 就是一个集中的存储单元,设计师最好将具备共同 IO 特性的 column 放在一个 Column Family 中,一般来说,我们只设置一个 Column Family。HStore 存储是 HBase 存储的核心,其中由两部分组成,一是 MemStore;二是 StoreFiles。MemStore 是 Sorted Memory Buffer,用户写入的数据首先会放入 MemStore,当 MemStore 满了以后会 Flush 成一个 StoreFile(底层实现是 HFile),当 StoreFile 文件数量增长到一定阈值,会触发 Compact 合并操作,将多个 StoreFiles 合并成一个 StoreFile,合并过程中会进行版本合并和数据删除,所以 HBase 其实只有增加数据,所有的更新和删除操作都是在后续的 compact 过程中进行的,这使得用户的写操作只要进入内存中就可以立即返回,保证了 HBase I/O 的高性能。具体过程如图 6-2 所示。

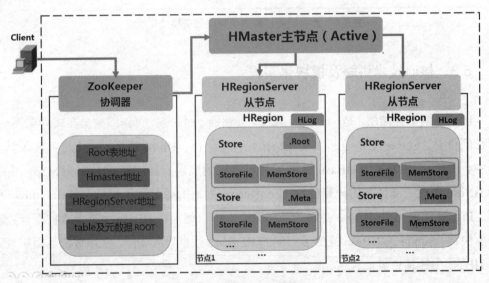

图 6-2 Hbase 工作原理

6.3.1 Hadoop 完全分布式集群构建

1. 配置 Linux 环境

(1)配置好各虚拟机的网络(采用 NAT 联网模式)

(2)通过 Linux 图形界面进行修改(桌面版本 Centos)

进入 Linux 图形界面→右击右上方的两个小电脑→单击 Edit connections→选中当前网络 System eth0→单击 edit 按钮→选择 IPv4→method 选择为 manual→单击 add 按钮→添加 IP:192.168.1.101,子网掩码:255.255.255.0,网关:192.168.1.1→单击 apply 按钮。

(3) 修改配置文件方式

```
vi /etc/sysconfig/network-scripts/ifcfg-eth0
DEVICE="eth0"
BOOTPROTO="static"                ###
HWADDR="00:0C:29:3C:BF:E7"
IPV6INIT="yes"
NM_CONTROLLED="yes"
ONBOOT="yes"
TYPE="Ethernet"
UUID="ce22eeca-ecde-4536-8cc2-ef0dc36d4a8c"
IPADDR="192.168.1.101"            ###
NETMASK="255.255.255.0"           ###
GATEWAY="192.168.1.1"             ###
```

(4) 修改各个虚拟机主机名

```
vi /etc/sysconfig/network
NETWORKING=yes
HOSTNAME=node-1
```

(5) 修改主机名和 IP 的映射关系

```
vi /etc/hosts
192.168.1.101 node-1
192.168.1.102 node-2
192.168.1.103 node-3
```

(6) 关闭防火墙

```
#查看防火墙状态
service iptables status
#关闭防火墙
service iptables stop
#查看防火墙开机启动状态
chkconfig iptables --list
#关闭防火墙开机启动
chkconfig iptables off
```

(7) 配置 ssh 免登录

```
#生成 ssh 免登录密钥
ssh-keygen -t rsa (4 个回车)
```

执行完这个命令后,会生成两个文件 id_rsa(私钥)、id_rsa.pub(公钥)。
将公钥复制到要免密登录的目标机器上:

```
ssh-copy-id node-2
ssh-copy-id node-3
```

（8）同步集群时间

常用的手动进行时间的同步：

```
date -s "2018-03-03 03:03:03"
```

或者网络同步：

```
yum install ntpdate
ntpdate cn.pool.ntp.org
```

2. 安装 JDK 并配置环境变量

（1）上传 jdk

```
rz jdk-8u65-linux-x64.tar.gz
```

（2）解压 jdk

```
tar -zxvf jdk-8u65-linux-x64.tar.gz -C /root/apps
```

（3）将 java 添加到环境变量中

```
vim /etc/profile
#在文件最后添加
export JAVA_HOME=/root/apps/jdk1.8.0_65
export PATH=$PATH:$JAVA_HOME/bin
export CLASSPATH=.:$JAVA_HOME/lib/dt.jar:$JAVA_HOME/lib/tools.jar
#刷新配置
source /etc/profile
```

3. 安装 hadoop2.7.4

（1）上传 hadoop 的安装包到服务器

```
hadoop-2.7.4-with-centos-6.7.tar.gz
```

（2）解压安装包

```
tar zxvf hadoop-2.7.4-with-centos-6.7.tar.gz
```

注意：hadoop2.x 的配置文件目录：$HADOOP_HOME/etc/hadoop。

4. 配置 hadoop 的核心配置文件

（1）配置文件 hadoop-env.sh

```
vi hadoop-env.sh
export JAVA_HOME=/root/apps/jdk1.8.0_65
```

（2）配置文件 core-site.xml

说明：指定 HADOOP 所使用的文件系统 schema（URI），HDFS 的主节点（NameNode）地址。

```
<property>
```

```xml
    <name>fs.defaultFS</name>
    <value>hdfs://node-1:9000</value>
</property>
```

说明：指定 hadoop 运行时产生文件的存储目录，默认/tmp/hadoop-${user.name} -->。

```xml
<property>
    <name>hadoop.tmp.dir</name>
    <value>/home/hadoop/hadoop-2.4.1/tmp</value>
</property>
```

（3）配置文件 hdfs-site.xml

```xml
<!-- 指定 HDFS 副本的数量 -->
<property>
    <name>dfs.replication</name>
    <value>2</value>
</property>
<property>
    <name>dfs.namenode.secondary.http-address</name>
    <value>node-2:50090</value>
</property>
```

（4）配置文件 mapred-site.xml

```
mv mapred-site.xml.template mapred-site.xml
vi mapred-site.xml
<!-- 指定 mr 运行时框架,这里指定在 yarn 上,默认是 local -->
<property>
    <name>mapreduce.framework.name</name>
    <value>yarn</value>
</property>
```

（5）配置文件 yarn-site.xml

```xml
<!-- 指定 YARN 的主节点(ResourceManager)的地址 -->
<property>
    <name>yarn.resourcemanager.hostname</name>
    <value>node-1</value>
</property>
<!-- NodeManager 上运行的附属服务。需配置成 mapreduce_shuffle,才可运行
    MapReduce 程序默认值:"" -->
<property>
    <name>yarn.nodemanager.aux-services</name>
    <value>mapreduce_shuffle</value>
</property>
```

（6）配置文件 slaves，里面写上从节点所在的主机名字

```
vi slaves
node-1
```

```
node-2
node-3
```

5. 将 hadoop 添加到环境变量

```
vim/etc/proflie
export JAVA_HOME=/root/apps/jdk1.8.0_65
export HADOOP_HOME=/root/apps/hadoop-2.7.4
export PATH=$PATH:$JAVA_HOME/bin:$HADOOP_HOME/bin:$HADOOP_HOME/sbin
source /etc/profile
```

6. 格式化 namenode（本质是对 namenode 进行初始化）

```
hdfs namenode -format（hadoop namenode -format）
```

7. 启动 hadoop，验证是否启动成功，如图 6-1 所示。

（1）先启动 HDFS

```
sbin/start-dfs.sh
```

（2）再启动 YARN

```
sbin/start-yarn.sh
```

（3）使用 jps 命令验证

```
27408 NameNode
28218 Jps
27643 SecondaryNameNode（secondarynamenode）
28066 NodeManager
27803 ResourceManager
27512 DataNode
http://192.168.1.101:50070（HDFS 管理界面）
http://192.168.1.101:8088（MR 管理界面）
```

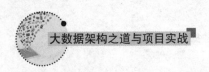

图 6-3　hadoop 集群启动成功

6.3.2 Spring MVC 和 Spring Boot 集成 Hbase

构建 BD_StorageServer_Maven 工程。本服务使用 Hbase 主要来存储用户的基本信息以及智能终端运动数据等，设计了 sport 和 patient 两张表来存放数据。Hbase 是采用集群的方式部署在 VMware 虚拟机的 Linux 系统下，系统镜像可以在本书提供的网盘上下载。Hbase 的相关配置如下。

（1）使用 hbase 之前要引入其相关依赖，需要在项目 pom.xml 文件添加依赖如下：

```
<!-- hbase 相关 -->
    <dependency>
        <groupId>org.apache.hbase</groupId>
        <artifactId>hbase-client</artifactId>
        <version>0.98.13-hadoop2</version>
    </dependency>
    <!-- Spring Hbase -->
    <dependency>
        <groupId>org.Springframework.data</groupId>
        <artifactId>Spring-data-jpa</artifactId>
        <version>1.6.0.RELEASE</version>
    </dependency>
    <dependency>
        <groupId>org.Springframework.data</groupId>
        <artifactId>Spring-data-hadoop</artifactId>
        <version>2.0.2.RELEASE</version>
    </dependency>
```

（2）Hbase 的参数配置文件 hbase.properties，根据业务配置 hbase 的表名、列簇名、列名、flume 的日志文件路径、hbase 在 zookeeper 中的注册名等。具体配置如下：

```
#master
hbase.master=192.168.106.111
#zookeeper ip port
hbase.zookeeper.quorum=hadoop
hbase.zookeeper.property.clientPort=2181
#hbase table
patientTable=patient
sportTable=sport
#hbhbaseTablease family
patientFamily=info
sportFamily=data
#flume 日志文件路径
flumeLogPath=d:/flume/log/data.log
#hbase 启动命令
start hadoop enviroment
start-all.sh
```

```
cd/usr/local/zookeeper/bin
zkServer.sh start
start-bhase.sh
hbase shell
qualifierPhone=phone
qualifierDeviceId=deviceId
qualifierCompany=company
qualifierAppType=appType
qualifierDataType=dataType
qualifierDataValue=dataValue
qualifierReceiveDateTime=receiveDateTime
qualifierPname=pname
qualifierTeamName=teamName
```

6.3.3 HbaseTemplate 核心类实现 Dao 层接口

（1）为了简化代码开发，Spring 一般都会对主流的框架进行集成，并提供封装类，Hbase 也不例外。若使用 Spring 集成 hbase 的方式，需要在 Spring 的核心配置文件中通过 import 标签引入两个配置文件 hbase.xml 和 hbase-site.xml。这两个 xml 文件主要是配置 Hbase 的系统参数，包括系统配置和性能参数。其中 hbase.zookeeper.quorum 名称是 Hadoop（这是我的集群 Master 的配置名称），hbase.zookeeper.property.clientPort 端口号是 2181，这些都是程序连接集群的关键配置参数。

① hbase.xml 文件，在 app-config.xml 中加载<import resource="../hbase.xml" />，配置如下：

```xml
<?xml version="1.0" encoding="UTF-8"?>
<beans xmlns="http://www.Springframework.org/schema/beans"
    xmlns:xsi="http://www.w3.org/2001/XMLSchema-instance"
    xmlns:context="http://www.Springframework.org/schema/context"
    xmlns:hdp="http://www.Springframework.org/schema/hadoop"
    xsi:schemaLocation="http://www.Springframework.org/schema/beans
        http://www.Springframework.org/schema/beans/Spring-beans.xsd
  http://www.Springframework.org/schema/context http://www.
    Springframework.org/schema/context/Spring-context.xsd
  http://www.Springframework.org/schema/hadoop http://www.
    Springframework.org/schema/hadoop/Spring-hadoop.xsd">
<hdp:configuration resources="classpath:/hbase-site.xml"></hdp:
    configuration>
<hdp:hbase-configuration configuration-ref="hadoopConfiguration"/>
<!-- 配置 HbaseTemplate -->
<bean id="hbaseTemplate" class="org.Springframework.data.hadoop.hbase.
    HbaseTemplate">
<property name="configuration" ref="hbaseConfiguration">
</property>
```

```xml
        <property name="encoding" value="UTF-8"></property>
    </bean>
</beans>
```

② hbase-site.xml 文件是配置并加载 Hadoop 集群环境，配置如下：

```xml
<?xml version="1.0" encoding="UTF-8"?>
<configuration>
    <property>
        <name>hbase.client.write.buffer</name>
        <value>62914560</value>
    </property>
    <property>
        <name>hbase.client.pause</name>
        <value>1000</value>
    </property>
    <property>
        <name>hbase.client.retries.number</name>
        <value>10</value>
    </property>
    <property>
        <name>hbase.client.scanner.caching</name>
        <value>1</value>
    </property>
    <property>
        <name>hbase.client.keyvalue.maxsize</name>
        <value>6291456</value>
    </property>
    <property>
        <name>hbase.rpc.timeout</name>
        <value>60000</value>
    </property>
    <property>
        <name>hbase.security.authentication</name>
        <value>simple</value>
    </property>
    <property>
        <name>zookeeper.session.timeout</name>
        <value>60000</value>
    </property>
    <property>
        <name>hbase.zookeeper.quorum</name>
        <value>${hbase.zookeeper.quorum}</value>
    </property>
    <property>
        <name>hbase.zookeeper.property.clientPort</name>
```

```
            <value>${hbase.zookeeper.property.clientPort}</value>
    </property>
</configuration>
```

（2）若使用 Spring Boot 进行开发，则需要通过注解的方式来完成对 hbaseTemplate 的初始化（与 Spring 集成 hbase 的目的一样，只是实现方式不同）。我们可以定义一个 Hbase 的配置类 HbaseConifg 来完成相关功能。代码实现如下：

```
package com.cloud.storage.hbase;
import org.apache.hadoop.hbase.HBaseConfiguration;
import org.Springframework.beans.factory.annotation.Value;
import org.Springframework.context.annotation.Bean;
import org.Springframework.context.annotation.PropertySource;
import org.Springframework.data.hadoop.hbase.HbaseTemplate;
import org.Springframework.stereotype.Component;
/**
 * hbase 的配置类
 *
 * @author changyaobin
 *
 */
@Component
@PropertySource(value = {"classpath:com/cloud/storage/Config/hbase.
    properties"})
public class HbaseConfig {
 @Bean
 public HbaseTemplate hbaseTemplate(@Value("${hbase.zookeeper.quorum}")
    String zookerQuorum,
        @Value("${hbase.zookeeper.property.clientPort}")String port){
    HbaseTemplate hbaseTemplate = new HbaseTemplate();
    org.apache.hadoop.conf.Configuration conf = HBaseConfiguration.
        create();
    conf.set("hbase.zookeeper.quorum",zookerQuorum);
    conf.set("hbase.zookeeper.port",port);
    hbaseTemplate.setConfiguration(conf);
    hbaseTemplate.setAutoFlush(true);
    return hbaseTemplate;
 }
}
```

（3）不管操作什么数据库，其相应的 jar 包都会提供给我们对应的操作类或接口，但是调用其底层的接口或类是相当的烦琐。对于有经验的开发人员，会寻找相应的框架或者自己封装框架，这样可以大大减少一些重复的代码操作，把精力全部投入到业务开发中。遵循这个原则，我们这里对 HBase 一些常用的查询、插入参数进行了封装，这样就可以实现构建查询对象来灵活地构建 hbase 查询语句。下面是自定义区查询类 HbaseConfig，代码实现如下：

```java
package com.cloud.storage.base.Domain;
import java.util.HashMap;
import java.util.List;
import java.util.Map;
import org.apache.hadoop.hbase.client.Scan;
import org.apache.hadoop.hbase.filter.CompareFilter.CompareOp;
import org.apache.hadoop.hbase.filter.Filter;
import org.apache.hadoop.hbase.filter.PageFilter;
import com.google.common.collect.Lists;
public class HQuery {
    private String table;
    private String family;
    private String qualifier;
    // 列值查询,key 为列的值,value 为 hbase 比较类型
    private Map<String,CompareOp>qualifierValues=new HashMap<>();
    private String row;
    private String startRow;
    private String stopRow;
    private Filter filter;
    private PageFilter pageFilter;
    private Scan scan;
    private String searchLimit;
    private List<HBaseColumn> columns = Lists.newArrayList();
    public String getTable(){
        return table;
    }
    public void setTable(String table){
        this.table = table;
    }
    public String getFamily(){
        return family;
    }
    public void setFamily(String family){
        this.family = family;
    }
    public String getQualifier(){
        return qualifier;
    }
    public void setQualifier(String qualifier){
        this.qualifier = qualifier;
    }
    public String getRow(){
```

```java
        return row;
    }
    public void setRow(String row){
        this.row = row;
    }
    public String getStopRow(){
        return stopRow;
    }
    public void setStopRow(String stopRow){
        this.stopRow = stopRow;
    }
    public Filter getFilter(){
        return filter;
    }
    public void setFilter(Filter filter){
        this.filter = filter;
    }
    public PageFilter getPageFilter(){
        return pageFilter;
    }
    public void setPageFilter(PageFilter pageFilter){
        this.pageFilter = pageFilter;
    }
    public Scan getScan(){
        return scan;
    }
    public void setScan(Scan scan){
        this.scan = scan;
    }
    public List<HBaseColumn> getColumns(){
        return columns;
    }
    public void setColumns(List<HBaseColumn> columns){
        this.columns = columns;
    }
    public String getSearchLimit(){
        return searchLimit;
    }
    public void setSearchLimit(String searchLimit){
        this.searchLimit = searchLimit;
    }
    public String getStartRow(){
        return startRow;
    }
    public void setStartRow(String startRow){
```

```
            this.startRow = startRow;
    }
    public Map<String,CompareOp> getQualifierValues(){
        return qualifierValues;
    }
    public void setQualifierValues(Map<String,CompareOp> qualifierValues){
        this.qualifierValues = qualifierValues;
    }
}
```

（5）自定义 hbase 库的列簇类 HBaseColumn，主要用于数据插入 HBase 时构建列簇、列名、列值等参数。代码实现如下：

```
package com.cloud.storage.base.Domain;

public class HBaseColumn {
    private String family;
    private String qualifier;
    private String value;
    public String getFamily(){
        return family;
    }
    public void setFamily(String family){
        this.family = family;
    }
    public String getQualifier(){
        return qualifier;
    }
    public void setQualifier(String qualifier){
        this.qualifier = qualifier;
    }
    public String getValue(){
        return value;
    }
    public void setValue(String value){
        this.value = value;
    }
    public HBaseColumn(String family,String qualifier,String value){
        super();
        this.family = family;
        this.qualifier = qualifier;
        this.value = value;
    }
}
```

6.3.4 Hbase 集群的智能终端运动数据 Controller 接口

智能终端运动数据 Controller 接口，用于接收第 3 章的灵活转发服务的智能终端运动数据，对数据参数进行校验，并调用 Service 层接口保存到 HBase 库。在第 4 章已经提到，入库方式可以通过修改 SysConf.properties 的入库开关来完成，若要入 HBase 库只需将 SysConf.properties 的 HBase 置为 true 即可。文件代码实现如下：

```java
package com.cloud.storage.controller;
import java.io.UnsupportedEncodingException;
import java.util.HashMap;
import java.util.Map;
import javax.servlet.http.HttpServletRequest;
import javax.servlet.http.HttpServletResponse;
import org.apache.log4j.Logger;
import org.Springframework.beans.factory.annotation.Autowired;
import org.Springframework.stereotype.Controller;
import org.Springframework.ui.ModelMap;
import org.Springframework.web.bind.annotation.PathVariable;
import org.Springframework.web.bind.annotation.RequestMapping;
import org.Springframework.web.bind.annotation.RequestMethod;
import com.cloud.storage.base.Domain.SportsData;
import com.cloud.storage.pattern.state.Context;
import com.cloud.storage.service.ObservationService;
import com.cloud.storage.service.PatientService;
import com.cloud.storage.service.SportsDataHbaseService;
import com.cloud.storage.service.SportsDataService;
import com.cloud.storage.util.DateUtil;
import com.cloud.storage.util.JsonUtil;
import com.cloud.storage.util.PropertiesReader;
import com.cloud.storage.util.ResponseUtil;
import com.cloud.storage.util.ValidateUtil;
import net.sf.json.JSONObject;
/**
 * 数据接收接口,与DispatchServer 转发服务进行数据对接
 *
 * @author changyaobin
 *
 */
@Controller
public class CommonRestfulController {
    @Autowired
    private ObservationService observationService;
    @Autowired
    private SportsDataService sportsDataService;
    @Autowired
```

```java
    private SportsDataHbaseService sportsDataHbaseService;
@Autowired
    private PatientService patientService;
    private static Logger log = Logger.getLogger(CommonRestfulController.
      class);
/**
 * 数据采集接口
 *
 * @param request
 * @param response
 * @throws Exception
 */
@SuppressWarnings({"rawtypes","unchecked"})
@RequestMapping(value = "/businessDataReceive")
public void businessDataReceive(HttpServletRequest request,
  Http ServletResponse response)throws Exception {
    log.info("the start of businessDataReceive");
    Map result = new HashMap();
    log.info("收到网关 DispatchServer 发来数据*_*... \r\n");
    String jsonData = "";
    try {
        jsonData = new String((request.getParameter("data").getBytes
          ("iso-8859-1")),"UTF-8");
    } catch(UnsupportedEncodingException e){
        e.printStackTrace();
        log.error("receive data occur exception:" + e.getMessage());
    }
    JSONObject jo = JSONObject.fromObject(jsonData);
    // 数据参数校验
    String validateInfo = "" + ValidateUtil.checkAppType(JsonUtil.
      getJsonParamterString(jo,"appType"))
            +(ValidateUtil.isValid(JsonUtil.getJsonParamterString
              (jo,"dataType"))== true ? "":"false")
            + ValidateUtil.checkDateTime(JsonUtil.
              getJsonParamter String(jo,"collectDate"))
            +(ValidateUtil.isValid(JsonUtil.getJsonParamterString
              (jo,"phone"))== true ? "":"false");
    // 校验通过
    if("".equals(validateInfo)){
        String isMongo = PropertiesReader.getProp("mongodb");
        String isMysql = PropertiesReader.getProp("mysql");
        String isHbase = PropertiesReader.getProp("hbase");
        Map<String,Class>classMap=new HashMap<>();
        classMap.put("dataValue",HashMap.class);
        // 入库 mongodbJSONObject
```

```
                SportsData sportsData =(SportsData)JSONObject.toBean
                  (JSONObject.fromObject(jsonData),SportsData.class,
                  classMap);
                if("true".equals(isMongo)){
                    // 入库mongodb
                    sportsDataService.saveSportsData(sportsData);
                } else if("true".equals(isMysql)){
                    // 入库mysql
                    new Context(request,response,observationService,
                      patient Service).request();
                } else if("true".equals(isHbase)){
                    // 入Hbase库
                    sportsDataHbaseService.saveData(sportsData);
                }
            } else {
                response.setStatus(412);
                result.put("status","数据验证失败!" + validateInfo);
                log.info("the end of businessDataReceive has invalidate
                  param include " + validateInfo);
            }
            response.setStatus(200);
            ResponseUtil.writeInfo(response,JSONObject.fromObject(result).
              toString());
        }
```

6.3.5　Hbase 集群的智能终端运动数据 Service 接口

（1）智能终端运动数据保存接口 SportsDataHbaseService，定义保存智能终端运动数据的方法。具体代码如下：

```
package com.cloud.storage.service;
import com.cloud.storage.base.Domain.SportsData;
public interface SportsDataHbaseService {
    boolean saveData(SportsData data)throws Exception;
}
```

（2）定义接口实现类 SportsDataHbaseServiceImpl，主要调用 Dao 层接口保存智能终端运动数据入 HBase 库，以及将智能终端运动数据写入日志文件中，供 Flume 监测使用。@Value 表示从配置文件 hbase.properties 中读取列名，代码实现如下：

```
package com.cloud.storage.serviceImpl;
import java.io.BufferedWriter;
import java.io.File;
import java.io.FileWriter;
import java.util.List;
import java.util.Map;
```

```java
import org.apache.log4j.Logger;
import org.Springframework.beans.factory.annotation.Autowired;
import org.Springframework.beans.factory.annotation.Value;
import org.Springframework.stereotype.Service;
import com.cloud.storage.base.Domain.HBaseColumn;
import com.cloud.storage.base.Domain.HQuery;
import com.cloud.storage.base.Domain.SportsData;
import com.cloud.storage.dao.SportsDataHbaseDao;
import com.cloud.storage.service.SportsDataHbaseService;
import com.cloud.storage.util.DateUtil;
import com.cloud.storage.util.ValidateUtil;
import net.sf.json.JSONArray;
@Service
public class SportsDataHbaseServiceImpl implements SportsDataHbase Service {
    private static Logger log = Logger.getLogger
        (SportsDataHbaseService Impl.class);
    @Autowired
    private SportsDataHbaseDao sportsDataHbaseDao;
    @Value("${sportTable}")
    private String sportTable;
    @Value("${patientTable}")
    private String patientTable;
    @Value("${patientFamily}")
    private String patientFamily;
    @Value("${sportFamily}")
    private String sportFamily;
    @Value("${qualifierPhone}")
    private String qualifierPhone;
    @Value("${qualifierDeviceId}")
    private String qualifierDeviceId;
    @Value("${qualifierCompany}")
    private String qualifierCompany;
    @Value("${qualifierAppType}")
    private String qualifierAppType;
    @Value("${qualifierDataType}")
    private String qualifierDataType;
    @Value("${qualifierDataValue}")
    private String qualifierDataValue;
    @Value("${qualifierReceiveDateTime}")
    private String qualifierReceiveDateTime;
    @Value("${qualifierPname}")
    private String qualifierPname;
    @Value("${qualifierTeamName}")
    private String qualifierTeamName;
```

```java
@Value("${flumeLogPath}")
private String flumeLogPath;
public boolean saveData(SportsData data)throws Exception {
    boolean saveSuccess = false;
    String dataType = data.getDataType();
    String appType = data.getAppType();
    String collectDate = data.getCollectDate();
    List<Map<String,String>> dataValue = data.getDataValue();
    String phone = data.getPhone();
    String deviceID = data.getDeviceID();
    if(ValidateUtil.paramCheck(appType,deviceID,collectDate,
       dataType,phone)){
        if(ValidateUtil.paramCheck(phone)){
            if("stepCount".equals(dataType)|| "stepDetail".equals
               (dataType)){
                HQuery patienQuery = new HQuery();
                List<HBaseColumn> infoColums = patienQuery.
                   getColumns();
                String patientRowkey = phone + "_" + deviceID + "_" +
                   appType;
                patienQuery.setRow(patientRowkey);
                patienQuery.setTable(patientTable);
                infoColums.add(new HBaseColumn(patientFamily,
                   qualifierPhone,data.getPhone()));
                infoColums.add(new HBaseColumn(patientFamily,
                   qualifierPname,data.getPname()));
                infoColums.add(new HBaseColumn(patientFamily,
                   qualifierPname,data.getPname()));
                infoColums.add(new HBaseColumn(patientFamily,
                   qualifierTeamName,data.getTeamName()));
                infoColums.add(new HBaseColumn(patientFamily,
                   qualifierCompany,data.getCompany()));
                infoColums.add(new HBaseColumn(patientFamily,
                   qualifierAppType,appType));
                sportsDataHbaseDao.addSportsData(patienQuery);
                HQuery hquery = new HQuery();
                String stringDate = DateUtil.dateToString
                   (collectDate);
                long time = DateUtil.getTime(stringDate);
                long Datetime = Long.MAX_VALUE - time;
                String rowkey = phone + "_" + Datetime + "_" + appType +
                   "_" + dataType + DateUtil.getCurrentTime();
                hquery.setRow(rowkey);
                hquery.setTable(sportTable);
                List<HBaseColumn> columns = hquery.getColumns();
```

```java
            columns.add(new HBaseColumn(sportFamily,
                qualifier_Phone,phone));
            columns.add(new HBaseColumn(sportFamily,
                qualifier_DeviceId,deviceID));
            columns.add(new HBaseColumn(sportFamily,
                qualifier_Company,"bigData"));
            columns.add(new HBaseColumn(sportFamily,
                qualifier_AppType,appType));
            columns.add(new HBaseColumn(sportFamily,
                qualifier_DataType,dataType));
            columns.add(new HBaseColumn(sportFamily,
                qualifier_DataValue,JSONArray.fromObject
                (dataValue).toString()));
            columns.add(new HBaseColumn(sportFamily,
                qualifier_ReceiveDateTime,collectDate.toString()));
            columns.add(new HBaseColumn(sportFamily,"id",rowkey));
            sportsDataHbaseDao.addSportsData(hquery);
            saveSuccess = true;
        }
    }
}
// 数据保存到 hbase 后,写入日志文件中,供 flume 读取
if(saveSuccess){
    if("stepCount".equals(dataType)){
        String stepSum = "";
        String distanceSum = "";
        String calSum = "";
        if("stepCount".equals(dataType)){
            if(dataValue != null && dataValue.size()> 0){
                Map<String,String> map = dataValue.get(0);
                for(Map<String,String> dataMap:dataValue){
                    if(dataMap.containsKey("stepSum")){
                        stepSum = dataMap.get("stepSum");
                    } else if(dataMap.containsKey("distanceSum")){
                        distanceSum = dataMap.get("distanceSum");
                    } else if(dataMap.containsKey("calSum")){
                        calSum = dataMap.get("calSum");
                    }
                }
            }
        }
    }
    // 数据日志信息
    String dataLog = data.getCollectDate()+ "\t" + phone +
```

```
                    "SOCKET/1.0\t" + "bigData\t" + deviceID + "\t"
                        + stepSum + "\t" + distanceSum + "\t" + calSum;
                File file = new File(flumeLogPath);
                if(!file.exists()){
                    file.mkdirs();
                }
                BufferedWriter fileWriter = new BufferedWriter(new
                    FileWriter(file,true));
                fileWriter.write(dataLog);
                fileWriter.newLine();
                fileWriter.close();
            }
        }
        return saveSuccess;
    }
}
```

6.3.6 Hbase 集群的智能终端运动数据 Dao 接口

（1）智能终端运动数据 Hbase 库操作 Dao 层接口 SportsDataHbaseDao，定义了添加智能终端运动数据以及根据 Hquery 对象查询智能终端运动数据的方法。代码实现如下：

```
package com.cloud.storage.dao;
import java.util.List;
import com.cloud.storage.base.Domain.HQuery;
import com.cloud.storage.base.Domain.SportsData;
public interface SportsDataHbaseDao {
    boolean addSportsData(HQuery query);
    List<SportsData> selectByQuery(HQuery query);
}
```

（2）智能终端运动数据 Dao 层实现类 SportsDataHbaseDaoImpl，通过自定义 HBaseTemplate 类来操作 HBase 库。代码实现如下：

```
package com.cloud.storage.daoImpl;

import java.util.List;
import org.Springframework.beans.factory.annotation.Autowired;
import org.Springframework.stereotype.Repository;
import com.cloud.storage.base.Domain.HQuery;
import com.cloud.storage.base.Domain.SportsData;
import com.cloud.storage.dao.SportsDataHbaseDao;
import com.cloud.storage.util.HBaseTemplate;
@Repository
public class SportsDataHbaseDaoImpl implements SportsDataHbaseDao{
    @Autowired
```

```java
    private HBaseTemplate hbaseTemplate;
    public boolean addSportsData(HQuery query){
        boolean addSuccess=false;
        try {
            hbaseTemplate.execute(query);
            addSuccess=true;
        } catch(Exception e){
            e.printStackTrace();
        }
        return addSuccess;
    }
    @Override
    public List<SportsData> selectByQuery(HQuery query){
        // TODO Auto-generated method stub
        List<SportsData> datas = hbaseTemplate.find(query,SportsData.
            class);
        return datas;
    }
}
```

（3）自定义 HBase 操作模板工具类，通过操作 Spring-hbase 提供给我们的 HbaseTemplate 模板类，以及自定义封装的 HBase 查询类 HQuery 完成对 HBase 的新增、条件查询等操作。代码实现如下：

```java
package com.cloud.storage.util;
import java.util.Iterator;
import java.util.List;
import java.util.Set;
import org.apache.commons.lang.StringUtils;
import org.apache.hadoop.hbase.Cell;
import org.apache.hadoop.hbase.client.HTableInterface;
import org.apache.hadoop.hbase.client.Put;
import org.apache.hadoop.hbase.client.Result;
import org.apache.hadoop.hbase.client.Scan;
import org.apache.hadoop.hbase.filter.FilterList;
import org.apache.hadoop.hbase.filter.SingleColumnValueFilter;
import org.apache.hadoop.hbase.util.Bytes;
import org.apache.log4j.Logger;
import org.Springframework.beans.factory.annotation.Autowired;
import org.Springframework.data.hadoop.hbase.HbaseTemplate;
import org.Springframework.data.hadoop.hbase.RowMapper;
import org.Springframework.data.hadoop.hbase.TableCallback;
import org.Springframework.stereotype.Component;
import org.Springframework.stereotype.Repository;
import com.alibaba.fastjson.JSON;
```

```java
import com.alibaba.fastjson.JSONArray;
import com.alibaba.fastjson.JSONObject;
import com.cloud.storage.base.Domain.HBaseColumn;
import com.cloud.storage.base.Domain.HQuery;
@Component
public class HBaseTemplate {
    private final Logger logger = Logger.getLogger(this.getClass());
    @Autowired
    private HbaseTemplate htemplate;
    /**
     * 写数据
     *
     * @param tableName
     * @param action
     * @return
     */
    public Object execute(final HQuery query){
        if(StringUtils.isBlank(query.getRow())|| query.getColumns().
          isEmpty()){
            return null;
        }
        synchronized(Object.class){
            return htemplate.execute(query.getTable(),new TableCall back
               <Object>(){
                  @SuppressWarnings("deprecation")
                  @Override
                  public Object doInTable(HTableInterface table)throws
                    Throwable {
                      try {
                          byte[] rowkey = query.getRow().getBytes();
                          Put put = new Put(rowkey);
                          if(query.getColumns()!= null){
                              Iterator<HBaseColumn> iterator = query.
                                 getColumns().iterator();
                              while(iterator.hasNext()){
                                  HBaseColumn col = iterator.next();
                                  if(StringUtils.isNotBlank(col.
                                     getFamily())
                                           && StringUtils.isNotBlank(col.
                                              getQualifier())
                                           && StringUtils.isNotBlank(col.
                                              getValue())){

                                      put.add(Bytes.toBytes(col.getFamily()),
                                          Bytes.toBytes(col.getQualifier()),
```

```java
                                Bytes.toBytes(col.getValue()));
                    }
                }
                table.put(put);
            }
            System.out.println("data into hbase success!");
        } catch(Exception e){
            e.printStackTrace();
            logger.warn("==> hbase get object fail> " + query.
                getRow());
        }
        return null;
    }
});
}
/**
 * 通过表名和 key 获取一行数据
 *
 * @param tableName
 * @param rowName
 * @return
 */
public <T> T get(HQuery query,final Class<T> c){
    if(StringUtils.isBlank(query.getTable())|| StringUtils.isBlank
      (query.getRow())){
        return null;
    }
    return htemplate.get(query.getTable(),query.getRow(),new RowMapper
      <T>(){
        public T mapRow(Result result,int rowNum)throws Exception {
            List<Cell> ceList = result.listCells();
            T item = c.newInstance();
            JSONObject obj = new JSONObject();
            if(ceList != null && ceList.size()> 0){
                for(Cell cell:ceList){
                    obj.put(Bytes.toString(cell.getQualifierArray(),
                        cell.getQualifierOffset(),cell.
                        getQualifierLength()),
                            Bytes.toString(cell.getValueArray(),
                                cell.getValueOffset(),cell.
                                getValueLength()));
                }
```

```java
                } else {
                    return null;
                }
                item = JSON.parseObject(obj.toJSONString(),c);
                return item;
            }
        });
    }
    /**
    * 通过表名、key、列族和列获取一个数据
    *
    * @param tableName
    * @param rowName
    * @param familyName
    * @param qualifier
    * @return
    */
    public String getColumn(HQuery query){
        if(StringUtils.isBlank(query.getTable()) || StringUtils.isBlank
          (query. getRow()) || StringUtils.isBlank(query.getFamily())||
          StringUtils. isBlank(query.getQualifier())){
            return null;
        }
        return htemplate.get(query.getTable(),query.getRow(),query.
          get Family(),query.getQualifier(),
            new RowMapper<String>(){
                public String mapRow(Result result,int rowNum)
                  throws Exception {
                    List<Cell> ceList = result.listCells();
                    String res = "";
                    if(ceList != null && ceList.size()> 0){
                        for(Cell cell:ceList){
                            res = Bytes.toString(cell.getValue
                                Array(),cell.getValueOffset(),cell.
                                getValueLength());
                        }
                    }
                    return res;
                }
            });
    }
    /**
    * 通过表名,开始行键和结束行键、列名、列值获取数据
    *
```

```java
 * @param HQuery
 * @return
 */
public <T> List<T> find(HQuery query,final Class<T> c){
    // 如果未设置scan,设置scan
    if(query.getScan()== null){
        Scan scan = new Scan();
        FilterList localFilterList = new FilterList(FilterList.
          Operator.MUST_PASS_ALL);
        // 起止搜索
        if(StringUtils.isNotBlank(query.getStartRow())&& StringUtils.
          isNotBlank(query.getStopRow())){
            scan.setStartRow(Bytes.toBytes(query.getStartRow()));
            scan.setStartRow(Bytes.toBytes(query.getStopRow()));
        }
        // 列匹配搜索
        if(StringUtils.isNotBlank(query.getFamily())&& StringUtils.
          isNotBlank(query.getQualifier())
                && query.getQualifierValues()!= null){
            Set<String> keySet = query.getQualifierValues().keySet();
            for(String key:keySet){
                SingleColumnValueFilter singleColumnValueFilter =
                    new SingleColumnValueFilter(Bytes.toBytes(query.
                    getFamily()),Bytes. toBytes(query.getQualifier()),
                    query.getQualifierValues().get(key),Bytes.
                    toBytes(key));
                localFilterList.addFilter(singleColumn ValueFilter);
            }
        }
        // 分页搜索
        if(query.getPageFilter()!= null){
            localFilterList.addFilter(query.getPageFilter());
        }
        scan.setFilter(localFilterList);
        query.setScan(scan);
    }
    // 设置缓存
    query.getScan().setCacheBlocks(false);
    query.getScan().setCaching(2000);
    return htemplate.find(query.getTable(),query.getScan(),
      new Row Mapper<T>(){
        @Override
        public T mapRow(Result result,int rowNum)throws Exception {
            List<Cell> ceList = result.listCells();
```

```java
                JSONObject obj = new JSONObject();
                T item = c.newInstance();
                if(ceList != null && ceList.size()> 0){
                    for(Cell cell:ceList){
                        String value = Bytes.toString(cell.getValue
                            Array(),cell.getValueOffset(),cell.
                            getValueLength());
                        String quali = Bytes.toString(cell.getQualifier
                            Array(),cell.getQualifierOffset(),cell.
                            getQualifierLength());
                        if(value.startsWith("[")){
                            obj.put(quali,JSONArray.parseArray (value));
                        } else {
                            obj.put(quali,value);
                        }
                    }
                }
                item = JSON.parseObject(obj.toJSONString(),c);
                return item;
            }
        });
    }
    /**
     * 通过表名、列名、列值、比较运算符获取数据
     *
     * @param HQuery
     * @return
     */
    public <T> List<T> findByCompare(HQuery query,final Class<T> c){
        // 如果未设置 scan,设置 scan
        if(query.getScan()== null){
            Scan scan = new Scan();
            FilterList localFilterList = new FilterList(FilterList.
              Operator.MUST_PASS_ALL);
            // 列匹配搜索
            if(StringUtils.isNotBlank(query.getFamily())&& StringUtils.
              isNotBlank(query.getQualifier())
                    && query.getQualifierValues()!=null){
                Set<String> keySet = query.getQualifierValues().keySet();
                for(String key:keySet){
                    SingleColumnValueFilter singleColumnValueFilter =
                      new SingleColumnValueFilter(Bytes.toBytes(query.
                      getFamily()),Bytes. toBytes(query.getQualifier()),
```

```
                    query.getQualifierValues().get(key),Bytes.
                        toBytes(key));
                localFilterList.addFilter(singleColumnValue Filter);
            }
            // query.setSearchEqualFilter(query.getFamily(),query.
                getQualifier(),query.getQualifierValue());
        }
        scan.setFilter(localFilterList);
        query.setScan(scan);
    }
    // 设置缓存
    query.getScan().setCacheBlocks(false);
    query.getScan().setCaching(2000);
    return htemplate.find(query.getTable(),query.getScan(),
      new Row Mapper<T>(){
        @Override
        public T mapRow(Result result,int rowNum)throws Exception {
            List<Cell> ceList = result.listCells();
            JSONObject obj = new JSONObject();
            T item = c.newInstance();
            if(ceList != null && ceList.size()> 0){
                for(Cell cell:ceList){
                    String value = Bytes.toString(cell.getValueArray(),
                       cell.getValueOffset(),cell.getValueLength());
                    String quali = Bytes.toString(cell.
                      getQualifierArray(),cell.getQualifierOffset(),
                      cell.getQualifierLength());
                    if(value.startsWith("[")){
                        obj.put(quali,JSONArray.parseArray
                           (value));
                    } else {
                        obj.put(quali,value);
                    }
                }
            }
            item = JSON.parseObject(obj.toJSONString(),c);
            return item;
        }
    });
}
```

6.4 项目小结

本章重点介绍了 Hbase 的工作原理以及特性，为后续工程构建和开发奠定了基础。接着讲述了 Spring MVC 和 Spring Boot 对 HbaseTemplate 集成方式。具体如下：

（1）使用 Hbase 主要来存储用户的基本信息以及智能终端运动数据等，设计了 sport 和 patient 两张表来分别存放运动和用户信息数据。

（2）加载 pom.xml 文件，引入开发的相关 JAR 包，包括 Spring 和 Hadoop 集成等。

（3）构建一个集群环境，Hbase 的参数配置文件 hbase.properties（配置了列簇和列名等），并加入 hbase-site.xml 文件（配置了 Hadoop 环境的 hbase.zookeeper.quorum 的名称为 hadoop 和 zookeeper 端口号 hbase.zookeeper.property.clientPort）。

（4）需要在 Spring 的核心配置文件中通过 import 标签引入配置文件 hbase.xml。hbase.xml 依赖注入 HbaseTemplate 类，实现了 Dao 层数据操作接口的封装。

（5）自己封装了类 HQuery（定义了 Hbase 的表名、列簇、列名、rowkey 等）和 HBaseColumn（具体的列名称），实现了利用对象方式统一操作数据库的基础。Spring 读取配置文件属性的注解使用是主流的开发模式，就是 SportsDataHbaseServiceImpl 中添加注解，如@Value ("${sportTable}") 等，来读取配置文件 hbase.properties 的列和列簇等定义信息。

（6）在全面完成了（1）～（5）的集群配置环境之后（Hadoop 集群环境构建，Spring 和 Hadoop 集成，引入 HbaseTemplate 类和 Hbase 构建列和列簇等），可以进行编程实现 DAO 层接口 hbaseTemplate.execute（query），让用户和运动的数据进入 Hbase 的 sport 和 patient 两张表。

（7）常用的数据库查询操作，就是构建一个查询类，然后查出整个对象。具体如下：

```
HQuery query = new HQuery();
query.setTable(sportTable);
query.setFamily("data");
query.setQualifier("phone");
query.setQualifierValue(phone);
List<SportsData> datas = sportsDataHbaseDao.selectByQuery(query);
```

第 7 章

大数据实时计算微服务引擎

架构之道分享之七：孙子兵法的《地形篇》论述了各种地形特点以及军队在不同地形条件下的作战原则。项目中要按照项目规模和数据规模，灵活地选择机器学习的语言和成熟框架，从而得到事半功倍的效果。

本章学习目标

- ★ 掌握分布式采集服务 Flume 部署及数据采集
- ★ 掌握分布式消息服务 Kafka 部署及数据发送
- ★ 掌握 Hbase 数据库设计和 Spark 集群环境构建
- ★ 掌握分布式实时处理引擎 SparkStreaming 原理及数据处理方法
- ★ 掌握微服务实现数据处理
- ★ 掌握微服务实现数据可视化

7.1 核心需求分析和优秀解决方案

大数据实时计算服务引擎是为新型大数据应用实时提供计算和分析的服务。当物联网大数据保存在海量数据存储服务引擎的数据库时，需要通过大数据实时计算服务引擎对数据进行抽取、过滤、统计分析和结果保存等处理。具体过程分两步进行，第一步，借助高性能数据采集服务 Flume，采集物联网数据到分布式消息服务中间件 Kafka 中，作为消费端的 SparkStreaming 从 Kafka 中实时拉取物联网数据，进行分析处理后把结果保存到 HBase 中。第二步，通过大数据可视化工具展示数据处理结果，采用 HBase+SpringBoot+Echarts 框架实现数据查询和展示，用饼状图把每个公司的总步数实时统计并展现在可视化平台上。

7.2 服务引擎的技术架构设计

大数据实时计算服务引擎包括 5 个核心模块，如图 7-1 所示，模块之间都要考虑高内聚耦合的设计理念，尤其要重视高性能和可扩展性等关键因素，具体说明如下。

（1）核心模块一：部署 Flume 服务及相关 conf 配置文件，使用 Flume 采集服务框架将海量存储服务的数据采集到 Kafka 中。

（2）核心模块二：部署 Kafka 服务，SparkStreaming 引擎从 Kafka 中读取数据并且进行数据分析。

（3）核心模块三：把每个公司员工的总步数统计结果实时保存到 HBase 数据库中。

（4）核心模块四：利用 Spring Boot 框架构建大数据实时计算服务引擎。

（5）核心模块五：利用大数据可视化工具 Echarts 进行数据可视化展示。

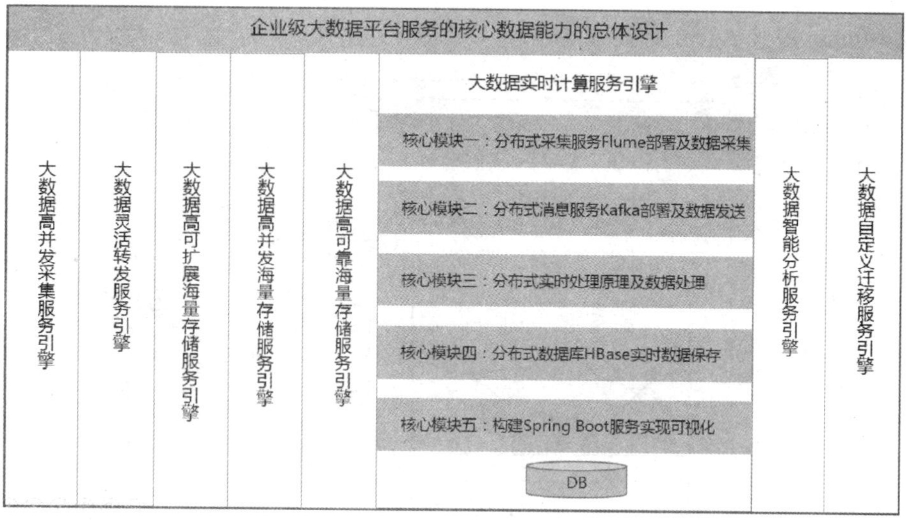

第 7 章 大数据实时计算微服务引擎

图 7-1 大数据实时计算服务引擎的模块化设计

7.3 核心技术讲解及模块化实现

7.3.1 分布式采集服务 Flume 部署及数据采集

Flume 是一个分布式、高可靠和高可用的海量日志采集、聚合和传输的实时系统。Flume 可以采集文件、Socket 数据包等各种形式源数据，又可以将采集到的数据输出到 HDFS、HBase、Hive、Kafka 等诸多外部存储系统中。对绝大多数业务数据采集而言，无须太多代码开发，通过对 Flume 的灵活配置就可以实现。对于复杂的应用场景，Flume 也具备良好的自定义扩展能力。因此，Flume 可以满足目前所有互联网主流业务的日常数据处理需求。Flume 的运行原理是：Flume 的核心角色为 Agent，Flume 分布式系统常常是由很多 Agent 连接而形成的。Agent 内部有三个组件，一是 Source 采集源，用于和数据源对接并获取数据；二是 Channel 通道，属于 Agent 内部的数据传输通道，用于从 Source 将数据传递到 Sink；三是 Sink 目标地，采集数据的传送目的地，用于往下一级 Agent 传递数据或者往外部存储系统传递数据。通常情况下，Flume 的采集模式分为单级和多级采集模式。

1. Flume 单级 Agent 采集数据模式（如图 7-2 所示）

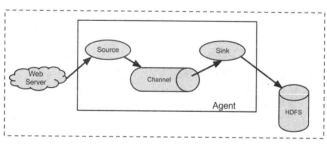

图 7-2 单级 Agent 采集数据过程模型

2. Flume 多级 Agent 采集数据模式（如图 7-3 所示）

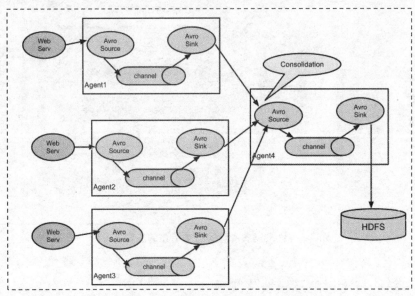

图 7-3　多级 Agent 采集数据过程模型

3. Flume 的安装和服务发布

Flume 的安装和所有的第三方中间件安装步骤类似，包括解压、修改配置文件、启动服务 3 个步骤。具体如下：

（1）上传安装包并解压。前提是在高可靠海量存储服务的 hadoopV2.7.3 环境下，上传安装包到数据源所在节点上。解压 tar -zxvf apache-flume-1.6.0-bin.tar.gz。

（2）进入 Flume 的目录，修改 conf 下的 flume-env.sh，在里面配置 JAVA_HOME。

（3）根据数据采集的需求配置采集方案，描述在配置文件（文件名可任意自定义）中。指定采集方案配置文件，在相应的节点上启动 flume agent。

（4）使用 Flume 将本地文件采集到 Kafka 中，在 Flume 的 conf 目录中创建一个名为 log_Kafka 的 conf 文件，并进行如下配置：

```
# 定义 agent
a1.sources = src1
a1.channels = ch1
a1.sinks = k1
# 定义 sources 监控 root 目录下的名为 log 的文件
a1.sources.src1.type = exec
a1.sources.src1.command=tail -F /root/companylog
a1.sources.src1.channels=ch1
# 定义 sinks
a1.sinks.k1.type = org.apache.flume.sink.Kafka.KafkaSink
a1.sinks.k1.topic = companytopic
```

```
a1.sinks.k1.brokerList =192.168.106.111:9092
a1.sinks.k1.batchSize = 20
a1.sinks.k1.requiredAcks = 1
a1.sinks.k1.channel = ch1
# 定义 channels
a1.channels.ch1.type = memory
a1.channels.ch1.capacity = 1000
```

（5）启动 Flume，加载配置文件 log_kafka.conf 启动。

```
bin/flume-ng agent --conf conf --conf-file conf/log_kafka.conf --name a1
```

7.3.2　分布式消息服务 Kafka 部署及数据发送

Apache Kafka 是一个开源消息系统，由 Scala 语言实现，是 Apache 软件基金会开发的一个开源消息系统项目。Kafka 于 2011 年年初实现了开源。Kafka 服务目标是为处理实时数据提供一个统一、高通量、低延迟的平台。Kafka 是一个分布式消息队列，包括生产者和消费者两个组件。它提供了类似于 JMS 的特性，但是在工作原理上不一样。Kafka 对消息保存根据 Topic 类型进行分组，发送消息者称为 Producer 服务，接收消息者称为 Consumer 服务，此外 Kafka 集群由多个 Kafka 实例组成，每个实例（server）称为一个 broker，producer 和 consumer 都依赖 zookeeper 集群来保存一些 meta 信息，来保证系统高可用性。Kafka 作为一个主流的分布式消息队列，主要作用有 3 个关键特点：解耦、异步和并行。

1. Kafka 的主要组件

（1）Topic：对消息保存根据 Topic 类型进行分组。

（2）Producer：发送消息者。

（3）Consumer：接收消息者。

（4）broker：每个 Kafka 实例（server）。

（5）Zookeeper：依赖集群保存 meta 信息。

2. Kafka 架构图（如图 7-4 所示）

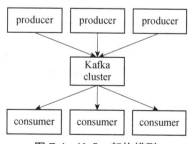

图 7-4　Kafka 架构模型

3. Kafka 的安装和服务发布

（1）从 http://Kafka.apache.org/downloads.html 下载 Kafka_2.11-1.0.1.tgz 安装包。

（2）把安装包解压到指定目录：tar –zxvf /root/ Kafka_2.11-1.0.1 -C / usr/local/。

（3）修改配置文件 vi/usr/local/ Kafka_2.11-1.0.1/config/server.properties，具体如下：

```
#修改 broker 的编号,必须唯一且不能重复
broker.id=0
#处理网络请求的线程数量
num.network.threads=3
#处理磁盘请求的线程数量
num.io.threads=8
#发送套接字的缓冲区大小
socket.send.buffer.bytes=102400
#接受套接字的缓冲区大小
socket.receive.buffer.bytes=102400
#数据存放目录
log.dirs=/tmp/Kafka-logs
#topic 在当前 broker 上的分片数量
num.partitions=1
#segment 文件保存时间,默认 7 天
log.retention.hours=168
#连接 zookeeper 的节点
zookeeper.connect=192.168.16.100:2181
```

（4）分发安装包，命令：scp -r /usr/servers/ Kafka_2.11-1.0.1 hadoop02：/usr/local。

（5）修改配置文件，依次修改各服务器上配置文件的 broker.id，不得重复。

（6）启动集群，在各个节点运行 Kafka 命令：bin/Kafka-server-start.sh 和 config/server.properties。说明一下，运行 Kafka 需要使用 zookeeper，需要先启动 zookeeper。

（7）启动 Kafka，加载配置文件 server.properties 启动服务。

```
bin/kafka-server-start.sh config/server.properties &
```

（8）创建一个 topic，名字是 companytopic。

```
bin/kafka-topics.sh --create --zookeeper hadoop: 2181 --replication-
    factor 1--partitions 1 --topic companytopic
```

```
[root@hadoop kafka_2.10-0.8.2.1]# cd ..
[root@hadoop local]# cd kafka_2.10-0.8.2.1/
[root@hadoop kafka_2.10-0.8.2.1]# bin/kafka-topics.sh --create --zookeeper hadoop:218
1 --replication-factor 1 --partitions 1 --topic companytopic
```

（9）启动一个消费者，并监听 companytopic 主题。

```
bin/kafka-console-consumer.sh --zookeeper hadoop: 2181 --topic companytopic
    --from-beginning
```

```
[root@hadoop local]# cd kafka_2.10-0.8.2.1/
[root@hadoop kafka_2.10-0.8.2.1]# bin/kafka-console-consumer.sh --zookeeper hadoop:21
81 --topic companytopic --from-beginning
```

7.3.3 创建 HBase 数据库和 Spark 环境

（1）创建 HBase 数据库，表名 companySportAnalysis，列簇 info，并保存运算结果。

```
hbase(main):002:0> create "companySportAnalysis","info"
0 row(s) in 4.6390 seconds

=> Hbase::Table - companySportAnalysis
hbase(main):003:0>
```

（2）安装 Spark 2.2.0，构建 Spark 开发环境。

① 下载安装包，下载地址：http://spark.apache.org/downloads.html，选择 spark-2.2.0-bin-hadoop2.7 版本。

② 新建安装目录：/usr/local。

③ 解压 tar -zxvf spark-2.2.0-bin-hadoop2.7.tgz。

④ 重命名 mv spark-2.2.0-bin-hadoop2.7 spark。

⑤ 修改配置文件 vi spark-env.sh（先把 spark-env.sh.template 重命名为 spark-env.sh）。

⑥ 配置 Java 环境变量。

```
export JAVA_HOME=/usr/java
#指定 spark 主服务 Master 的 IP
export SPARK_MASTER_HOST=hadoop
#指定 spark 主服务 Master 的端口
export SPARK_MASTER_PORT=7077
#woker 使用1g 和 1 个核心进行任务处理
export SPARK_WORKER_CORES=1
export SPARK_WORKER_MEMORY=1g
slaves 修改文件
```

⑦ 修改文件 slaves：vi slaves（先把 slaves.template 重命名为 slaves）。

```
hadoop02
hadoop03
```

⑧ 复制到其他节点。通过 scp 命令将 spark 的安装目录复制到其他机器上：

```
scp -r/opt/bigdata/spark hadoop02:/usr/local
scp -r/opt/bigdata/spark hadoop03:/usr/local
```

⑨ 配置 spark 环境变量。

将 spark 添加到环境变量，添加以下内容到 /etc/profile：

```
export SPARK_HOME=/usr/local/spark
export PATH=$PATH:$SPARK_HOME/bin
```

注意，最后 source /etc/profile 刷新配置。

⑩ 启动 spark。

```
#在主节点上启动 spark，/opt/bigdata/spark/sbin/start-all.sh
```

```
[root@hadoop sbin]# ./start-all.sh
starting org.apache.spark.deploy.master.Master, logging to /usr/local/spark-2.2.0-bin
-hadoop2.7/logs/spark-root-org.apache.spark.deploy.master.Master-1-hadoop.out
hadoop: starting org.apache.spark.deploy.worker.Worker, logging to /usr/local/spark-2
.2.0-bin-hadoop2.7/logs/spark-root-org.apache.spark.deploy.worker.Worker-1-hadoop.out
[root@hadoop sbin]#
```

⑪ 上传文件到集群并执行方法。

```
./spark-submit --class com.test.spark.project.CompanyStreamingApp --master
    local --executor-memory 1G --total-executor-cores 1 /root/spark.jar
```

```
[root@hadoop spark-2.2.0-bin-hadoop2.7]# cd bin/
[root@hadoop bin]# ./spark-submit --class com.test.spark.project.CompanyStreamingApp
--master local --executor-memory 1G --total-executor-cores 1 /root/spark.jar
Using Spark's default log4j profile: org/apache/spark/log4j-defaults.properties
18/06/30 12:25:09 INFO SparkContext: Running Spark version 2.2.0
18/06/30 12:25:11 INFO SparkContext: Submitted application: CompanyStepCount
18/06/30 12:25:11 INFO SecurityManager: Changing view acls to: root
18/06/30 12:25:11 INFO SecurityManager: Changing modify acls to: root
18/06/30 12:25:11 INFO SecurityManager: Changing view acls groups to:
18/06/30 12:25:11 INFO SecurityManager: Changing modify acls groups to:
18/06/30 12:25:11 INFO SecurityManager: SecurityManager: authentication disabled; ui
acls disabled; users  with view permissions: Set(root); groups with view permissions:
Set(); users  with modify permissions: Set(root); groups with modify permissions: Se
t()
18/06/30 12:25:13 INFO Utils: Successfully started service 'sparkDriver' on port 4942
9.
18/06/30 12:25:13 INFO SparkEnv: Registering MapOutputTracker
18/06/30 12:25:13 INFO SparkEnv: Registering BlockManagerMaster
18/06/30 12:25:13 INFO BlockManagerMasterEndpoint: Using org.apache.spark.storage.Def
aultTopologyMapper for getting topology information
18/06/30 12:25:13 INFO BlockManagerMasterEndpoint: BlockManagerMasterEndpoint up
18/06/30 12:25:13 INFO DiskBlockManager: Created local directory at /tmp/blockmgr-467
bbcd2-d732-44b5-96b4-9a4489140a36
18/06/30 12:25:13 INFO MemoryStore: MemoryStore started with capacity 366.3 MB
18/06/30 12:25:14 INFO SparkEnv: Registering OutputCommitCoordinator
18/06/30 12:25:14 INFO Utils: Successfully started service 'SparkUI' on port 4040.
```

7.3.4 分布式实时处理引擎 Spark Streaming 原理及数据处理

Spark Streaming 类似于 Apache Storm，用于流式数据的处理。据官方介绍，Spark Streaming 有高吞吐量和容错能力强等特点。Spark Streaming 支持的数据输入源很多，例如 Kafka、Flume、Twitter、ZeroMQ 和简单的 TCP 套接字等。数据输入后可以用 Spark 的高度抽象原语（如 map、reduce、join、window 等）进行运算。而结果也能保存在很多地方，如 HDFS 和数据库等。另外，Spark Streaming 也能和 MLlib（机器学习）以及 Graphx 完美融合。Spark Streaming 的基础抽象是 Discretized Stream，表示持续性的数据流和经过各种 Spark 原语操作后的结果数据流。

（1）在内部实现上，DStream 是一个时间序列的 RDD。每个 RDD 含有一段时间间隔内的数据，如图 7-5 所示。

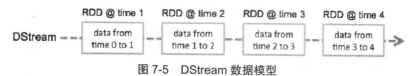

图 7-5　DStream 数据模型

（2）对数据的操作也是以 RDD 为单位来进行，如图 7-6 所示。

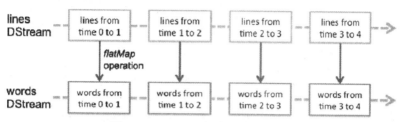

图 7-6　DStream flatMap 操作过程

（3）计算过程是由 Spark 引擎来进行计算的，如图 7-7 所示。

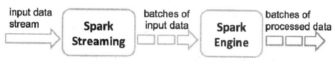

图 7-7　Spark 引擎实现计算

（4）DStream 的常用操作：对 Dstre DStream 的操作算子与 RDD 的类似，分为 Transformations（转换）和 Output Operations（输出）两种，此外转换操作中还有一些比较特殊的原语，如 updateStateByKey()、transform() 以及各种 Window 相关的原语。具体操作如表 7-1 所示。

表 7-1　DStream 操作算子表

Transformation 算子	含　义
map（func）	对 DStream 中的每个元素应用给定函数，返回由各元素组成的 DStream
flatMap（func）	对 DStream 中的每个元素应用给定函数，返回由各元素输出的迭代器组成的 DStream
filter（func）	返回由给定 DStream 中通过筛选的元素组成的 DStream
repartition（numPartitions）	改变 DStream 的分区数
reduceByKey（func，[numTasks]）	将每个批次中键相同的记录归约
groupByKey	将每个批次中记录根据键分组

7.3.5　构建 BD_RTPServer_DP 工程实现数据处理

（1）Sparkstreaming 流式处理引擎从 Kafka 服务中拉取数据并处理，首先构建 BD_RTPServer_DP 工程，实现数据处理，代码架构如图 7-8 所示。

图 7-8　BD_RTPServer_DP 工程代码架构

（2）BD_RTPServer_DP 工程的配置文件 pom，代码如下：

```xml
<project xmlns="http://maven.apache.org/POM/4.0.0" xmlns:xsi="http://
    www.w3.org/2001/XMLSchema-instance" xsi:schemaLocation="http://maven.
    apache.org/POM/4.0.0 http://maven.apache.org/maven-v4_0_0.xsd">
<modelVersion>4.0.0</modelVersion>
<groupId>com.test.spark</groupId>
<artifactId>spark</artifactId>
<version>1.0-SNAPSHOT</version>
<inceptionYear>2008</inceptionYear>
<properties>
  <scala.version>2.11.8</scala.version>
  <kafka.version>0.10.0.0</kafka.version>
  <spark.version>2.2.0</spark.version>
  <hadoop.version>2.7.2</hadoop.version>
  <HBase.version>1.2.0</HBase.version>
</properties>
<dependencies>
  <dependency>
    <groupId>org.apache.kafka</groupId>
    <artifactId> kafka_2.11</artifactId>
    <version>0.10.0.0</version>
  </dependency>
  <dependency>
    <groupId>org.apache.hadoop</groupId>
    <artifactId>hadoop-client</artifactId>
```

```xml
        <version>${hadoop.version}</version>
    </dependency>
    <dependency>
        <groupId>org.apache.HBase</groupId>
        <artifactId>HBase-client</artifactId>
        <version>1.2.0</version>
    </dependency>
    <dependency>
        <groupId>org.apache.HBase</groupId>
        <artifactId>HBase-server</artifactId>
        <version>1.2.0</version>
    </dependency>
    <dependency>
        <groupId>org.apache.spark</groupId>
        <artifactId>spark-streaming_2.11</artifactId>
        <version>2.1.0</version>
    </dependency>
    <dependency>
        <groupId>org.apache.spark</groupId>
        <artifactId>spark-streaming-kafka-0-10_2.11</artifactId>
        <version>2.1.0</version>
    </dependency>
    <dependency>
        <groupId>org.scala-lang</groupId>
        <artifactId>scala-library</artifactId>
        <version>${scala.version}</version>
    </dependency>
</dependencies>
<build>
    <sourceDirectory>src/main/scala</sourceDirectory>
    <testSourceDirectory>src/test/scala</testSourceDirectory>
    <plugins>
        <plugin>
            <groupId>org.scala-tools</groupId>
            <artifactId>maven-scala-plugin</artifactId>
            <executions>
                <execution>
                    <goals>
                        <goal>compile</goal>
                        <goal>testCompile</goal>
                    </goals>
                </execution>
            </executions>
            <configuration>
                <scalaVersion>${scala.version}</scalaVersion>
```

```xml
        <args>
          <arg>-target:jvm-1.5</arg>
        </args>
      </configuration>
    </plugin>
    <plugin>
      <groupId>org.apache.maven.plugins</groupId>
      <artifactId>maven-eclipse-plugin</artifactId>
      <configuration>
        <downloadSources>true</downloadSources>
        <buildcommands>
          <buildcommand>ch.epfl.lamp.sdt.core.scalabuilder</buildcommand>
        </buildcommands>
        <additionalProjectnatures>
          <projectnature>ch.epfl.lamp.sdt.core.scalanature</projectnature>
        </additionalProjectnatures>
        <classpathContainers>
          <classpathContainer>org.eclipse.jdt.launching.JRE_CONTAINER
          </classpathContainer>
          <classpathContainer>ch.epfl.lamp.sdt.launching.SCALA_CONTAINER
          </classpathContainer>
        </classpathContainers>
      </configuration>
    </plugin>
  </plugins>
</build>
<reporting>
  <plugins>
    <plugin>
      <groupId>org.scala-tools</groupId>
      <artifactId>maven-scala-plugin</artifactId>
      <configuration>
        <scalaVersion>${scala.version}</scalaVersion>
      </configuration>
    </plugin>
  </plugins>
</reporting>
</project>
```

（3）Sparkstreaming 从 Kafka 拉取数据，代码实现如下：

```
import org.apache.kafka.clients.consumer.ConsumerRecord
import org.apache. kafka.common.serialization.StringDeserializer
import org.apache.spark.streaming. kafka._
import org.apache.spark.streaming. kafka.LocationStrategies.
  PreferConsistent
```

```
import org.apache.spark.streaming. kafka.ConsumerStrategies.Subscribe
val conf = new SparkConf().setAppName("CompanyStepCount").setMaster
    ("local[2]")
 val sc = new SparkContext(conf)
 // 设置每 3 秒切分一次 RDD
     val ssc = new StreamingContext(sc,Seconds(3))
// 设置 Kafka 参数,从 Kafka 上接收数据
val KafkaParams = Map[String,Object](
    "bootstrap.servers" -> "192.168.106.111:9000",
    "key.deserializer" -> classOf[StringDeserializer],
    "value.deserializer" -> classOf[StringDeserializer],
    "group.id" -> "example",
    "auto.offset.reset" -> "latest",
    "enable.auto.commit" ->(false:java.lang.Boolean)
  )
// 从名为 companytopic 的 topic 中拉取数据
    val topics = Array("companytopic")
    val lines = KafkaUtils.createDirectStream[String,String](
      ssc,
      PreferConsistent,
      Subscribe[String,String](topics,KafkaParams)
    ).map(_.value())
// 测试从 Kafka 接收数据,并进行打印
lines.print()
```

（4）开发操作 HBase 数据库的工具类，代码实现如下：

```
package com.test.spark.project.untils;

import org.apache.hadoop.conf.Configuration;
import org.apache.hadoop.HBase.client.HBaseAdmin;
import org.apache.hadoop.HBase.client.HTable;
import org.apache.hadoop.HBase.client.Put;
import org.apache.hadoop.HBase.util.Bytes;
import java.io.IOException;
/**
 * HBase 操作工具类:Java 工具类建议采用单例模式封装
 */
public class HBaseUtils {
    HBaseAdmin admin = null;
    Configuration configuration = null;
    /**
     * 私有构造方法
     */
    private HBaseUtils(){
// 指定 zookeeper 地址和 HBase 的 root dir
```

```java
        configuration = new Configuration();
        configuration.set("HBase.zookeeper.quorum","192.168.16.100:
            2181");
        configuration.set("HBase.rootdir","hdfs://192.168.16.100:9000/
            HBase");
        try {
            admin = new HBaseAdmin(configration);
        } catch(IOException e){
            e.printStackTrace();
        }
    }
    private static HBaseUtils instance = null;
    public static synchronized HBaseUtils getInstance(){
        if(null == instance){
            instance = new HBaseUtils();
        }
        return instance;
    }
    /**
     * 根据表名获取 Htable 实例
     */
    public HTable getHtable(String tableName){
        HTable table = null;
        try {
            table = new HTable(configuration,tableName);
        } catch(IOException e){
            e.printStackTrace();
        }
        return table;
    }
    /**
     * 添加数据到 HBase 里
     * @param tableName  表名
     * @param rowKey     对应 key 的值
     * @param cf         HBase 列簇
     * @param colum      HBase 对应的列
     * @param value      HBase 对应的值
     */
    public void put(String tableName,String rowKey,String cf,String
      colum,String value){
        HTable table = getHtable(tableName);
        Put put = new Put(Bytes.toBytes(rowKey));
        put.add(Bytes.toBytes(cf),Bytes.toBytes(colum),Bytes.toBytes
            (value));
        try {
```

```
            table.put(put);
        } catch(IOException e){
            e.printStackTrace();
        }
    }
// 测试能否将一条数据添加到HBase数据库中
public static void main(String[] args){
        String tableName = "companySportAnalysis";
        String rowkey = "20171122_1";
        String cf = "sort";
        String colum = "step_count";
        String value = "100";
        HBaseUtils.getInstance().put(tableName,rowkey,cf,colum,value);
    }
}
```

（5）使用 spark-streaming 完成日期转换和数据清洗操作，代码实现如下：

```
package com.test.spark.project.untils

import java.util.Date
import org.apache.commons.lang3.time.FastDateFormat
/**
  * 此工具类是做时间格式转换
  * 2017-11-20 00:39:26 => 20171120
  */
object DateUtils {
  val YYYYMMDDHHMMSS_FORMAT = FastDateFormat.getInstance("yyyy-MM-dd HH:
    mm:ss")
  val TAG_FORMAT = FastDateFormat.getInstance("yyyyMMdd")
  /**
    * 把当前时间转换成时间戳
    * @param time
    * @return
    */
  def getTime(time:String)= {
    YYYYMMDDHHMMSS_FORMAT.parse(time).getTime
  }
  def parseToMin(time:String)= {
    TAG_FORMAT.format(new Date(getTime(time)))
  }
  def main(args:Array[String]):Unit = {
    print(parseToMin("2017-11-20 00:39:26"))
  }
}
  val cleanLog = lines.map(line => {
```

```
    val fields = line.split("\t")
// 取出公司序号
    val company = fields(2)
// 先把公司设成0
    var companyID = 0
// 如果公司以Company开头,用"_"进行分割,取出下画线后面的数据,并且将其变为Int类型
    if(company.startsWith("Company")){
      companyID = company.split("_")(1).toInt
    }
    CompanyLog(DateUtils.parseToMin(fields(0)),fields(1),companyID,
      fields(3),fields(4).toInt,fields(5).toDouble,fields(6).toDouble)
// 把companyID开头为0的,不是以Company开头的脏数据过滤掉
    }).filter(CompanyLog => CompanyLog.companyID != 0)
// 把清洗后的数据进行打印
    .cleanLog.print()
}
```

(6) 智能终端运动数据实体类定义,代码实现如下:

```
package com.test.spark.domain
/**
  * 清洗后的日志信息
  * @param time    发送信息的时间
  * @param user    用户
  * @param companyID  公司名称ID
  * @param refer   发送源
  * @param step    总步数
  * @param stepsize 步长
  * @param calorie 所需要的卡路里
  */
case class CompanyLog(time:String,user:String,companyID:Int,refer:String,
  step:Int,stepsize:Double,calorie:Double)
package com.test.spark.domain

/**
  * 每个公司步数实体类
  * @param CompanyID  公司ID
  * @param stepCount  公司总步数
  */
case class CompanyStepCount(CompanyID:String,stepCount:Int)
```

(7) 智能终端运动数据对象的DAO层的开发,代码实现如下:

```
package com.test.spark.dao

import com.test.spark.domain.CompanyStepCount
import com.test.spark.project.untils.HBaseUtils
```

```
import org.apache.hadoop.HBase.client.Get
import org.apache.hadoop.HBase.util.Bytes
import scala.collection.mutable.ListBuffer
object CompanyStepCountDAO {
  val tableName = "companySportAnalysis"
  val cf = "info"
  val qulifer = "step_count"
  // 保存数据并按rowkey进行累加
  def save(list:ListBuffer[CompanyStepCount]):Unit = {
    val table = HBaseUtils.getInstance().getHtable(tableName)
    for(els <- list){
      table.incrementColumnValue(Bytes.toBytes(els.CompanyID),
        Bytes.toBytes(cf),Bytes.toBytes(qulifer),els.stepCount)
    }
  }
  def Count(day_companyID:String):Long = {
    val table = HBaseUtils.getInstance().getHtable(tableName)
    val get = new Get(Bytes.toBytes(day_companyID))
    val value = table.get(get).getValue(Bytes.toBytes(cf),Bytes.toBytes
      (qulifer))
    if(value == null){
      0L
    } else {
      Bytes.toLong(value)
    }
  }
  // 进行测试
  def main(args:Array[String]):Unit = {
    val list = new ListBuffer[CompanyStepCount]
    list.append(CompanyStepCount("20180401_1",300))
    list.append(CompanyStepCount("20180401_3",900))
    list.append(CompanyStepCount("20180401_8",200))
    // save(list)
    print(Count("20180401_10")+ "------" + Count("20180401_9")+ "-----" +
      Count("20180401_8"))
  }
}
```

（8）计算每个公司的总步数并把数据保存在HBase中，代码实现如下：

```
cleanLog.map(log => {
   (log.time.substring(0,8)+ "_" + log.companyID,log.step)
  }).reduceByKey(_ + _).foreachRDD(rdd => {
    rdd.foreachPartition(partions => {
      val list = new ListBuffer[CompanyStepCount]
      partions.foreach(pair => {
```

```
            list.append(CompanyStepCount(pair._1,pair._2))
       })
// 把数据保存在 HBase 上
   CompanyStepCountDAO.save(list)
      })
   })
```

7.3.6　构建 BD_RTPServer_Boot 服务实现可视化

（1）Spring Boot 构建 Web 项目，实现一个公司内所有人员的总步数统计。在 Idea 中新建 Spring Boot 工程如图 7-9 所示。

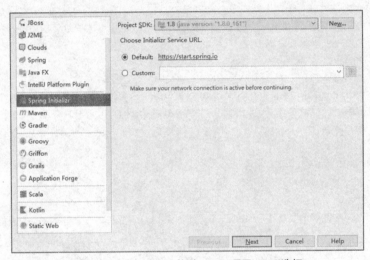

图 7-9　Spring Boot 构建 Web 项目 JDK 选择

（2）构建 Web 项目，选择 JDK1.8 之后，选择依赖，如图 7-10 所示。

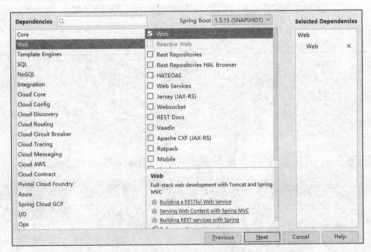

图 7-10　Spring Boot 构建 Web 项目依赖选择

（3）构建 BD_RTPServer_Boot 工程，实现一个公司内所有人员的总步数统计服务。在 Idea 中新建 Spring Boot 工程，架构如图 7-11 所示。

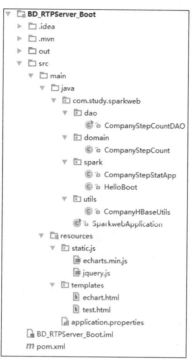

图 7-11　BD_RTPServer_Boot 工程架构

（4）BD_RTPServer_Boot 工程的配置文件 pom.xml，代码实现如下：

```xml
<?xml version="1.0" encoding="UTF-8"?>
<project xmlns="http://maven.apache.org/POM/4.0.0" xmlns:xsi="http://
   www.w3.org/2001/XMLSchema-instance"
xsi:schemaLocation="http://maven.apache.org/POM/4.0.0 http://maven.apache.
   org/xsd/maven-4.0.0.xsd">
<modelVersion>4.0.0</modelVersion>
<groupId>com.study.sparkweb</groupId>
<artifactId>sparkweb</artifactId>
<version>0.0.1-SNAPSHOT</version>
<packaging>jar</packaging>
<name>sparkweb</name>
<description>Demo project for Spring Boot</description>
<parent>
   <groupId>org.Springframework.boot</groupId>
   <artifactId>Spring-boot-starter-parent</artifactId>
   <version>1.5.8.RELEASE</version>
   <relativePath/><!-- lookup parent from repository -->
</parent>
```

```xml
<properties>
    <project.build.sourceEncoding>UTF-8</project.build.sourceEncoding>
    <project.reporting.outputEncoding>UTF-8</project.reporting.outputEncoding>
    <java.version>1.8</java.version>
</properties>
<repositories>
    <repository>
        <id>cloudera</id>
        <url>https://repository.cloudera.com/artifactory/cloudera-repos/</url>
    </repository>
</repositories>
<dependencies>
    <dependency>
        <groupId>org.Springframework.boot</groupId>
        <artifactId>Spring-boot-starter-web</artifactId>
    </dependency>
    <dependency>
        <groupId>org.Springframework.boot</groupId>
        <artifactId>Spring-boot-starter-test</artifactId>
        <scope>test</scope>
    </dependency>
    <dependency>
        <groupId>org.Springframework.boot</groupId>
        <artifactId>Spring-boot-starter-thymeleaf</artifactId>
    </dependency>
    <dependency>
        <groupId>org.apache.HBase</groupId>
        <artifactId>HBase-client</artifactId>
        <version>1.2.0</version>
    </dependency>
    <dependency>
        <groupId>net.sf.json-lib</groupId>
        <artifactId>json-lib</artifactId>
        <version>2.4</version>
        <classifier>jdk15</classifier>
    </dependency>
</dependencies>
<build>
    <plugins>
        <plugin>
            <groupId>org.Springframework.boot</groupId>
            <artifactId>Spring-boot-maven-plugin</artifactId>
        </plugin>
```

```
        </plugins>
    </build>
</project>
```

（5）在 rescources 的 static 中新建一个 js 文件夹，然后把 echarts.min.js 和 JQuery.js 复制到文件夹内。这两个文件可以在官网上下载。

（6）根据天来获取 HBase 表中的类目访问次数，代码实现如下：

```
package com.study.sparkweb.until;

import org.apache.hadoop.conf.Configuration;
import org.apache.hadoop.HBase.client.*;
import org.apache.hadoop.HBase.filter.Filter;
import org.apache.hadoop.HBase.filter.PrefixFilter;
import org.apache.hadoop.HBase.util.Bytes;
import java.io.IOException;
import java.util.HashMap;
import java.util.Map;
/**
 *HBase 操作工具类
 */
public class CompanyHBaseUtils {
    HBaseAdmin admin = null;
    Configuration conf = null;
    /**
     * 私有构造方法:加载一些必要的参数
     */
    private CompanyHBaseUtils(){
        conf = new Configuration();
        conf.set("HBase.zookeeper.quorum","hadoop:2181");
        conf.set("HBase.rootdir","hdfs://hadoop:9000/HBase");
        try {
            admin = new HBaseAdmin(conf);
        } catch(IOException e){
            e.printStackTrace();
        }
    }
    private static CompanyHBaseUtils instance = null;
    public static synchronized CompanyHBaseUtils getInstance(){
        if(null == instance){
            instance = new CompanyHBaseUtils();
        }
        return instance;
    }
    /**
```

```java
     * 根据表名获取 HTable 实例
     */
    public HTable getTable(String tableName){
        HTable table = null;
        try {
            table = new HTable(conf,tableName);
        } catch(IOException e){
            e.printStackTrace();
        }
        return table;
    }
    /**
     * 根据表名和输入条件获取 HBase 的记录数
     */
    public Map<String,Long> query(String tableName,String condition)
        throws Exception {
        Map<String,Long> map = new HashMap<>();
        HTable table = getTable(tableName);
        String cf = "info";
        String qualifier = "step_count";
        Scan scan = new Scan();
        // 使用列前缀过滤器
        Filter filter = new PrefixFilter(Bytes.toBytes(condition));
        scan.setFilter(filter);
        ResultScanner rs = table.getScanner(scan);
        for(Result result:rs){
            String row = Bytes.toString(result.getRow());
            long companyStepCount = Bytes.toLong(result.getValue
                (cf.get Bytes(),qualifier.getBytes()));
            map.put(row,companyStepCount);
        }
        return map;
    }
}
```

（7）类 CompanyStepCount 定义，实现公司内员工步数统计，代码实现如下：

```java
package com.study.sparkweb.domain;
import org.Springframework.stereotype.Component;

@Component
public class CompanyStepCount {
    private String name;
    private long value;
    public String getName(){
        return name;
```

```
    }
    public void setName(String name){
        this.name = name;
    }
    public long getValue(){
        return value;
    }
    public void setValue(long value){
        this.value = value;
    }
}
```

(8) 类 Dao 层 CompanyStepCountDAO 定义,代码实现如下:

```
package com.study.sparkweb.dao;

import com.study.sparkweb.domain.CompanyStepCount;
import com.study.sparkweb.until.CompanyHBaseUtils;
import org.Springframework.stereotype.Component;
import java.util.ArrayList;
import java.util.List;
import java.util.Map;
@Component
public class CompanyStepCountDAO {
    /**
     * 根据天查询
     */
    public List<CompanyStepCount> query(String day)throws Exception {
        List<CompanyStepCount> list = new ArrayList<>();
        // 去 HBase 表中根据 day 获取每个公司对应的步数
        Map<String,Long> map = CompanyHBaseUtils.getInstance().query
            ("companySportAnalysis","20180607");
        for(Map.Entry<String,Long> entry:map.entrySet()){
            CompanyStepCount model = new CompanyStepCount();
            model.setName(entry.getKey());
            model.setValue(entry.getValue());
            list.add(model);
        }
        return list;
    }
// 进行单元测试
public static void main(String[] args)throws Exception{
    CompanyStepCountDAO dao = new CompanyStepCountDAO();
    List<CompanyStepCount> list = dao.query("20180607");
    for(CompanyStepCount model:list){
        System.out.println(model.getName()+ ":" + model.getValue());
```

```
        }
    }
}
```

（9）公司实时查询展示，代码实现如下：

```java
package com.study.sparkweb.spark;

import com.study.sparkweb.dao.CompanyStepCountDAO;
import com.study.sparkweb.domain.CompanyStepCount;
import org.Springframework.beans.factory.annotation.Autowired;
import org.Springframework.web.bind.annotation.RequestMapping;
import org.Springframework.web.bind.annotation.RequestMethod;
import org.Springframework.web.bind.annotation.ResponseBody;
import org.Springframework.web.bind.annotation.RestController;
import org.Springframework.web.servlet.ModelAndView;
import java.util.HashMap;
import java.util.List;
import java.util.Map;
@RestController
public class CompanyStepStatApp {
    private static Map<String,String> company = new HashMap<>();
    static {
        company.put("1","A 公司");
        company.put("2","B 公司");
        company.put("3","C 公司");
        company.put("4","D 公司");
        company.put("5","E 公司");
        company.put("6","F 公司");
    }
    @Autowired
    CompanyStepCountDAO companyStepCountDAO;
    @RequestMapping(value = "/company_step_count",method = Request Method.
        POST)
    @ResponseBody
    public List<CompanyStepCount> companyStepCount()throws Exception {
        List<CompanyStepCount> list = companyStepCountDAO.query
            ("20180607");
        for(CompanyStepCount model:list){
            model.setName(company.get(model.getName().substring(9)));
        }
        return list;
    }
    @RequestMapping(value = "/echart",method = RequestMethod.GET)
    public ModelAndView echarts(){
        return new ModelAndView("echart");
    }
}
```

（10）页面调用数据并进行展示，代码实现如下：

```html
<!DOCTYPE html>
<html lang="en">
<head>
    <meta charset="UTF-8"/>
    <title>公司智能终端运动大数据实时展现</title>
    <!-- 引入 ECharts 文件 -->
    <script src="/js/echarts.min.js"></script>
    <!-- 引入 JQuery 文件 -->
    <script src="/js/JQuery.js"></script>
</head>
<body>
<div id="main" style="width:600px;height:400px;position:absolute;top:
    50%;left:50%;margin-top:-200px;margin-left:-300px"></div>
<script type="text/javascript">
    // 基于准备好的dom,初始化echarts实例
    var myChart = echarts.init(document.getElementById('main'));
    var option = {
        title:{
            text:'公司智能终端运动实时步数统计',
            subtext:'公司步数',
            x:'center'
        },
        tooltip:{
            trigger:'item',
            formatter:"{a} <br/>{b}:{c}({d}%)"
        },
        legend:{
            orient:'vertical',
            left:'left'
        },
        series:[
            {
                name:'公司步数',
                type:'pie',
                radius:'55%',
                center:['50%','60%'],
                data:(function(){ //<![CDATA[
                    var datas = [];
                    $.ajax({
                        type:"POST",
                        url:"/company_step_count",
                        dataType:'json',
```

```
                async:false,
                success:function(result){
                    for(var i=0;i<result.length;i++){
                        datas.push({"value":result[i].value,
                            "name":result[i].name})
                    }
                }
            })
            return datas;
            //]]>
        })(),
        itemStyle:{
            emphasis:{
                shadowBlur:10,
                shadowOffsetX:0,
                shadowColor:'rgba(0,0,0,0.5)'
            }
        }
    ]
};
// 使用刚指定的配置项和数据显示图表
myChart.setOption(option);
</script>
</body>
</html>
```

(11) 打开页面 localhost: 8080/echart, 统计可视化界面如图 7-12 所示。

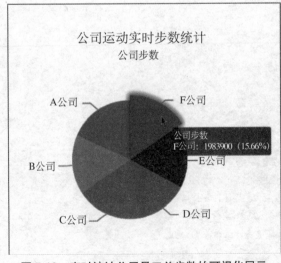

图 7-12　实时统计公司员工总步数的可视化展示

7.4 项目小结

本章介绍了实时处理服务的开发方法，从 Flume 采集体检数据文件后，输出 Kafka 上，之后消息主题发送给 SparkStreaming 中间件，进行数据处理和统计，最后数据统计结果进入 HBase 数据库。这是实时计算的典型流程，可以推广到其他业务上。最后借助 echart 可视化工具实现了智能终端运动数据统计，这是目前常用的数据展示方式，还有很多主流的大数据可视化工具，请读者自行学习，不在本章赘述。

第 8 章

大数据智能分析微服务引擎

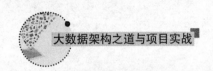

架构之道分享之八：孙子兵法的《虚实篇》提出了善于应用多种方式调动敌人，使敌人力量分散或者出现薄弱点，然后集中优势兵力进行避实击虚。运用到项目上，数据集构建和特征选择是提升分析准确率的有效手段，机器学习专家要善于工程处理和算法设计，灵活应用算法框架来提升敏感度和特异性。

本章学习目标

★ 掌握机器学习的技术原理和常用分类算法理论知识

★ 掌握分类算法在项目中的实战方法

★ 掌握 Spark 架构原理和 Mlib 库在机器学习中的实战技巧

8.1 核心需求分析和优秀解决方案

大数据智能分析服务引擎是大数据技术和物联网等业务领域结合的必然产物，给各机构和用户带来了不同程度的价值提升，不仅可以指导用户实现个性化锻炼和合理饮食，而且帮助专业机构按需实现数据共享和经营辅助决策。国内众多顶级互联网公司参与了物联网项目，致力于医疗、健康、养老和车联网等典型应用，建设了大数据智能分析服务，让社会的优质资源达到了合理分配和高效利用。近几年可穿戴设备和大数据的迅猛发展，催生了互联网+物联网的一系列应用和大数据数据中心，这些数据中心的涵盖了异构多源的物联网大数据，具备了海量、多变性、时效性、真实性等诸多特征，如何借助工程服务和机器学习算法，构建一个大数据智能分析服务引擎，快速孵化各种新型应用是本章要阐述的内容。在物联网行业，常见的大数据智能分析服务引擎包括健康疾病预测、体检花费预测、慢病趋势分析、关键影响因子发现和协同过滤推荐等。本章针对一些物联网大数据进行用户体检费用预测、疾病预测、药品推荐等服务构建，有效指导相关机构实现资源优质化配置。

8.2 服务引擎的技术架构设计

大数据智能分析服务引擎包括 5 个核心模块，如图 8-1 所示，每一个模块的设计和实现中，不仅要考虑数据源的数据格式灵活转换，而且要保证分析服务的通用性和可复用，具体说明如下。

（1）核心模块一：介绍机器学习基本原理，详细阐述逻辑回归、SVM、决策树等分类算法原理。

（2）核心模块二：介绍基于 RDD 的 Spark 架构和原理，深入 Mlib 算法库讲解。

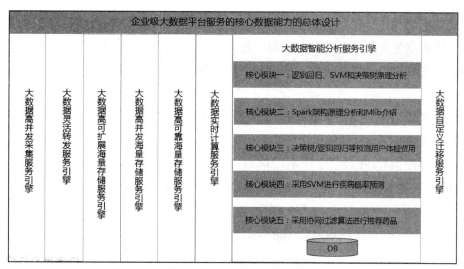

图 8-1 大数据智能分析服务引擎的模块化设计

（3）核心模块三：借助分类算法，针对体检数据集的 323 个特征，采用决策树、逻辑回归和随机森林算法预测体检费用。

（4）核心模块四：借助 SVM 分类算法，针对体检数据样本，预测某种疾病的患病概率。

（5）核心模块五：借助协同过滤推荐算法，针对用药数据样本，实现辅助药品推荐。

8.3 核心机器学习算法讲解和应用

机器学习是一门交叉学科，涉及概率论学、算法复杂度、工程学、计算机科学和数据挖掘等多门学科，它也是人工智能领域的一个重要分支，其原理是能够从历史数据中提取关键特征并进行推理预测。机器学习分为监督学习、非监督学习、半监督学习和强化学习。监督学习是给定了一组带分类标签的样本集，学习出一个函数，当新的数据到来后，可以根据已知函数预测出新数据的分类标签，常用的监督学习算法包括回归和分类。无监督学习是有一组没有带分类标签的样本集，通过机器学习得到数据分类，然后对正确分类行为进行激励，常用的无监督学习算法如聚类等。半监督学习就是对少量已标注样本和大量未标注样本进行训练和分类，提升学习能力。强化学习就是一种以环境反馈作为输入的方法，学习对象以周围环境的反馈为依据做出判断，常用的强化学习如机器人控制等。其中分类算法应用占据了全部机器学习算法的半壁江山，接下来重点介绍分类器的原理和实践。

8.3.1 逻辑回归的原理分析

回归分析是数据挖掘中的一种重要方法，是对具有因果关系的影响因素（自变量）

和预测对象（因变量）所进行的数理统计分析处理，旨在确定两种或两种以上变量间相互依赖的定量关系，主要用于预测、分类和因素分析。回归分析的基本原理是找到反映输入变量和输出变量间关系的回归方程，利用回归方程完成预测、分类和因素分析的任务。

逻辑回归（Logistic Regression）是回归分析的一种，主要用来做分类，适用于二分类问题，可以推广到多分类问题。

设 $\{x_1, x_2, \cdots, x_n, y\}$ 为一个样本，$\{x_1, x_2, \cdots, x_n\}$ 是样本输入，$y \in \{0,1\}$ 是样本所属类别。逻辑回归的输出是 $(0,1)$ 区间上的一个值，表示样本属于某个类别的概率。

$$P\{y=1 | x_1, x_2, \cdots, x_n\} \in (0,1) \tag{1}$$

引入单调、任意阶可导函数 Sigmoid，如图 8-2 所示。

$$S(x) = \frac{1}{1+e^{-x}} \tag{2}$$

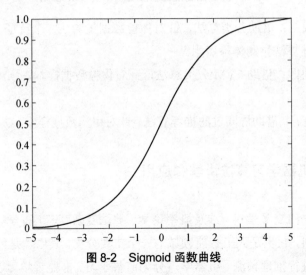

图 8-2　Sigmoid 函数曲线

当 x 由 $-\infty \to +\infty$ 时，$S(x)$ 由 $0 \to 1$。

$$P\{y=1 | x_1, x_2, \cdots, x_n\} = \frac{1}{1+e^{-z}} \tag{3}$$

进一步对 z 进行线性回归：

$$i = 1, 2, \cdots, N$$

其中 $\omega_i, i = 0, 1, \cdots, n$ 为权值。

得到回归模型：

$$P\{y=1 | x_1, x_2, \cdots, x_n\} = \frac{1}{1+e^{-(\omega_0 + \omega_1 x_1 + \cdots + \omega_n x_n)}} \tag{4}$$

通过对对数似然函数

$$\sum_{i=1}^{K}[(1-y^{(i)})\ln(1-P\{y^{(i)}=1\,|\,x^{(i)}_{1},x^{(i)}_{2},\cdots,x^{(i)}_{n}\})+y^{(i)}\ln(P\{y^{(i)}=1\,|\,x^{(i)}_{1},x^{(i)}_{2},\cdots,x^{(i)}_{n}\})]$$

最大化，得到权值 $\omega_i, i=0,1,\cdots,n$。其中 K 是样本总数，$y^{(i)} \in \{0,1\}$ 是第 i 个样本的类别，$\{x^{(i)}_{1}, x^{(i)}_{2}, \cdots, x^{(i)}_{n}\}$ 是第 i 个样本的输入。

推广到多分类问题，一种解决方法是对每个类别 c 都建立一个二分类回归模型，用来计算样本属于类别 c 和不属于类别 c 的概率，最终比较样本属于各类别的概率，取最大的即可。

8.3.2 支持向量机原理分析

支持向量机是由 Vapnik 等人提出的一种机器学习算法，在解决小样本、非线性及高维模式识别中表现出独特的优势，并逐步推广到函数拟合等其他机器学习问题中。其主要思想是将低维空间的样本通过非线性变换映射到高维空间，从而将低维空间的线性不可分问题转化为高维空间的线性可分问题。具体实现是用低维空间中满足一定条件的核函数实现高维空间中的内积运算，从而构造高维空间中的最优分类超平面，达到分类目的。

支持向量机是扩展的线性分类器，如图 8-3 所示，线性分类器是用一个超平面将不同类别的样本分开，图中各超平面都能将两类样本分开，但中间的超平面鲁棒性最好，泛化能力最强。支持向量机是通过最优化的方法，找到那个鲁棒性最强、泛化能力最好的超平面。

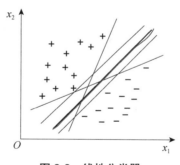

图 8-3 线性分类器

如图 8-4 所示，要找到两个相互平行的分隔超平面，使得类别 1 中的点到它的距离大于等于 1，类别 –1 中的点到它的距离小于等于 –1，两个类别中最靠近分隔超平面的点称为支持向量。

对于线性不可分的分类问题，采用的方法是将样本从原始空间映射到一个更高维的特征空间，使样本在这个特征空间内线性可分。通过选择恰当的核函数，实现高维空间中向量间的内积运算，进一步求得分隔超平面。具体过程如下：

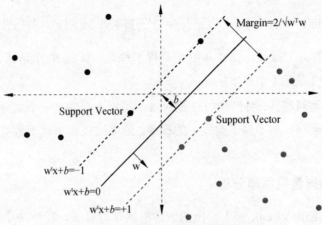

图 8-4　支持向量机原理图示

设样本向量为 $S_i, i=1,2,\cdots,N$，$S_i \in R^{3n}$，N 为样本个数。每个样本属于 ω_1, ω_2 之一，用 c_i 标识其类别，若 $S_i \in \omega_1$，则 $c_i = -1$；若 $S_i \in \omega_2$，则 $c_i = 1$。

（1）输入样本向量及其所属类别 (S_i, c_i)，$i=1,2,\cdots,N$，用以建立分类模型。

（2）指定核函数 $K(S_1, S_2)$ 类型。

（3）得到判别函数，通过计算函数系数 a^* 和 b^*，就可以确定这个超平面的方程：

$$f(S) = \sum_{i=1}^{N} c_i a_i^* K(S_i, S_j) + b^*$$

其中 a_i^* 通过解下列约束优化问题得到：

$$\max H(a) = \sum_{i=1}^{N} a_i - \frac{1}{2} \sum_{i=1}^{N} \sum_{j=1}^{N} c_i c_j a_i a_j K(S_i, S_j)$$

$$\text{subject to} \quad \sum_{i=1}^{N} c_i a_i = 0, a_i \geq 0, i = 1, \cdots, N$$

（4）利用二次规划方法求解上述优化问题，得到最优 Lagrange 乘子 a^*。

（5）利用样本库中的一个支持向量 S_k 和判别函数，可得到偏差值 b^*：

$$b^* = f(S_k) - \sum_{i=1}^{N} c_i a_i^* K(S_i, S_j)$$

利用生成的判别函数可以进行样本的分类。

8.3.3　决策树原理分析

决策树是用于分类和预测的一种树结构，决策树方法常用于类别属性的数据集的分类和预测，对于数值属性的数据集，可以经过数据预处理后使用。

决策树的建立是基于样本的递归的学习过程，每个样本都是具有确定的属性的数据，决策树就是基于样本的各属性建立起来的。决策树分为根节点、内部节点和叶节点。从根节点出发，自顶向下，构建分支和内部节点，在内部节点进行属性值的比较，并根据属性的不同取值确定从该节点向下的分支,最终在叶节点给出类别。整棵树就是一个规则集,任意样本都可以从根节点出发,根据样本各属性的取值,找到对应的树中的分支(规则),从而得到样本的类别。

具体方法是:选择一个属性置于根节点,根据这个属性的不同取值将测试样本集划分成多个子集,一个子集对应一个属性取值,然后在每个分支上递归地重复这个过程,直到在一个节点上的所有样本具有相同的类别,即到达叶子结点。

在上述构建方法中,关键的问题在于如何选择在根节点、中间节点进行划分的属性。一个基本的原则是构建较小的决策树,使得上述递归的过程尽早停止,即尽早完成分类。要达到上述目的,需要度量每个节点对应的样本子集的纯度,即样本子集中类别分布的情况,类别数越少,各类别分布越不均匀,纯度越高。信息熵(entropy)是度量集合纯度的最常用的一种指标,定义如下:

设当前样本集合 D 中第 k 类样本所占比例为 p_k,则 D 的样本熵定义为

$$Ent(D) = -\sum_{k=1}^{|D|} p_k \log_2 p_k$$

$Ent(D)$ 的值越小,则 D 的纯度越高。

在选择划分属性时,分别计算上一级节点对应的样本子集的信息熵和选择各属性划分后该样本子集的信息熵,二者差值(信息增益)最大划分的对应的属性即为划分属性。

决策树构建流程如下:

输入:训练集 $D = \{(x_1, y_1), (x_2, y_2), \cdots, (x_m, y_m)\}$

属性集 $A = \{a_1, a_2, \cdots, a_k\}$

生成树:$TreeGenerate(D, A)$

(1)生成节点 $node$。

(2)If D 中样本全属于同一类别 C

将 $node$ 标记为 C 类叶节点

end if

(3)If $A = \varnothing$ or D 中样本在 A 上取值相同,then

将 $node$ 标记为叶节点,其类别标记为 D 中样本数最多的类

end if

(4)从 A 中选择最优划分属性 a_{opt}

for a_{opt} 的每一个属性值 a_{opt}^v do

为 $node$ 生成一个分支,D_v 表示 D 中在 a_{opt} 上取值为 a_{opt}^v 的样本集合

If D_v 为空,then

将分支节点标记为叶节点,其类别标记为 D 中样本最多的类

else

以 $TreeGenerate(D_v, A \setminus \{a_{opt}\})$ 为分支节点

  end if
 end for
输出：以 node 为根节点的一棵决策树。

8.3.4 聚类算法原理分析

  聚类算法通常按照中心点或者分层的方式对输入数据进行归并。聚类算法都试图找到数据的内在结构，以便按照最大的共同点将数据进行归类。常见的聚类算法包括 k-Means 算法以及期望最大化算法 EM。聚类算法属于非监督式学习，通常被用于探索性的分析，就是"物以类聚"的原理，将本身没有类别的样本聚集成不同的组，一组数据对象的集合叫作簇，并且对每一个这样的簇进行描述的过程。它的目的是使得属于同一簇的样本之间应该彼此相似，而不同簇的样本应该足够不相似，应用案例包括用户画像、用户行为分析、价值评估。Mlib 目前支持广泛使用的 k-Means 聚类算法。案例：导入上班族的智能终端运动行为训练数据集，使用 k-Means 对象来将数据聚类到多个类簇当中，所需的类簇个数会被传递到算法中，然后计算集内均方差总和，可以通过增加类簇的个数 k 来减小误差。k-Means 算法中最简单的就是基于距离的聚类算法，采用距离作为相似性的评价指标，即认为两个对象的距离越近，其相似度就越大。该算法认为类簇是由距离靠近的对象组成的，因此把得到紧凑且独立的簇作为最终目标。核心思想是：通过迭代寻找 k 个类簇的一种划分方案，使得用这 k 个类簇的均值来代表相应各类样本时所得的总体误差最小。k 个聚类具有以下特点：各聚类本身尽可能的紧凑，而各聚类之间尽可能的分开。k-Means 算法基础是最小误差平方和准则。

8.3.5 关联规则算法原理分析

  关联规则学习通过寻找最能够解释数据变量之间关系的规则，来找出大量多元数据集中有用的关联规则。常见算法包括 Apriori 算法和 Eclat 算法等。我们使用 FP-growth 算法来高效发现频繁项集，方法是将数据集存储在一个特定的称作 FP 树的结构之后发现频繁项集或者频繁项对，即常在一块出现的元素项的集合 FP 树，这种做法使算法的执行速度要快于 Apriori，通常性能要好两个数量级以上。FP-growth 算法的工作流程是：首先构建 FP 树，然后利用它来挖掘频繁项集。为构建 FP 树，需要对原始数据集扫描两遍。第一遍对所有元素项的出现次数进行计数，去掉不满足最小支持度的元素项，生成这个头指针表。数据的第一遍扫描用来统计出现的频率，而第二遍扫描中只考虑那些频繁元素。

8.3.6 协同过滤原理分析

  协同过滤 Collaborative Filtering 就是给用户推荐相似的物品或者人，协同过滤算法

又分为基于用户的协同过滤算法和基于物品的协同过滤算法。用户偏好具有相似性，即用户可分类。这种分类的特征越明显，推荐准确率越高。物品之间具有相似性，即偏好某物品的人，都很可能也同时偏好另一件相似物品。CF 常用方法有 3 种，分别是欧式距离法（计算每两个点的距离）、皮尔逊相关系数法和余弦相似度法。皮尔逊相关系数是指：两个变量之间的相关系数越高，从一个变量去预测另一个变量的精确度就越高，这是因为相关系数越高，就意味着这两个变量的共变部分越多，所以从其中一个变量的变化就可越多地获知另一个变量的变化。如果两个变量之间的相关系数为 1 或-1，那么你完全可由变量 X 去获知变量 Y 的值。当相关系数为 0 时，X 和 Y 两变量无关系。相关系数的绝对值越大，相关性越强，相关系数越接近于 1 和-1。余弦相似度法是指：两个向量有相同的指向时，余弦相似度的值为 1；两个向量夹角为 90°时，余弦相似度的值为 0；两个向量指向完全相反的方向时，余弦相似度的值为-1。算法思路：首先构建物品的同现矩阵，然后构建用户对物品的评分矩阵，最后通过矩阵计算得出推荐结果，其中推荐结果就是要户评分矩阵与同现矩阵的乘积。

8.4　Spark 架构原理与数据预测

Spark 是基于弹性分布式数据集 RDD 的采用内存计算的并行计算框架，是对 Hadoop MapReduce 框架改进升级后的迭代计算框架。MapReduce 在处理复杂数据处理任务时会遇到严重的磁盘频繁读写性能瓶颈，Spark 逐步成为了 MapReduce 的数据处理优秀替代者，得益于遵循摩尔定律的内存高速发展。在与 Hadoop 结合方面，可以兼容 Hadoop 的 HDFS、HBase、Hive 等分布式存储系统。具备特性说明如下。

（1）优秀的数据批处理框架：在进行 MapReduce 数据批处理时，作业任务需要读取 HDFS 文件作为数据输入进行聚合，而统计输出的结果也要存储到 HDFS 上。如果是一次数据处理需要运行多个 MapReduce 作业，其中间结果通过 HDFS 保存与传递，如果是多次 HDFS 读写操作，会产生 I/O 读写效率低和处理时间长的瓶颈。但是，如果采用 Spark 进行数据批处理时，替代多个 MapReduce 作业任务的是一个 Spark 作业程序，不仅可以缩短作业的申请、资源分配过程，而且把作业执行时的中间结果可保存于内存中，减少 HDFS 的读写次数，从而减少了磁盘读写开销，大幅缩短数据处理时间，提高了数据处理效率。

（2）高可扩展的编程接口：相比 MapReduce 编程模型，Spark 提供了更为灵活的 DAG 编程模型。DAG 编程模型不仅包含了 map、reduce 接口，还增加了 filter、flatMap、union 等操作接口，使得编写 Spark 程序更为方便。Spark 提供了编程语言 Java、Scala、Python、R 的 API，以及 SQL 的支持，支持开发者编写 Spark 程序。同时还提供了 Spark Shell

以支持用户进行交互式编程。

（3）多源异构数据处理：Spark 支持数据批处理，还支持流式数据处理、复杂分析（包括机器学习、图计算）、交互式数据查询（包括 SQL）。Spark 可以运行，或者在 Hadoop Yarn 集群管理器，兼容 Hadoop 已有的各种数据类型，支持多种数据源，如 HDFS、Hive、HBase、Parquet 等。

Spark 应用程序是由一个驱动程序 Driver Program 和多个任务组成的，它们都运行在一组进程上，这些进程分布在 Worker 节点上并被 Driver Program 的进程协调管理，一个 Spark 应用由一个驱动程序和多个执行器 Executor 构成，驱动程序中的 SparkContext 对象负责协调管理多个 Executor，而一个 Executor 可以执行多个任务 Task，并可以将数据保存在缓存中。每个 Spark 应用所拥有的 Executor 进程是独立的，这些执行器进程会随着 Spark 应用的运行而运行，并且通过多线程的方式运行任务 Task，如图 8-5 所示。

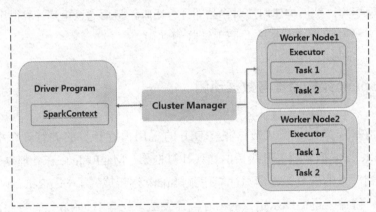

图 8-5 Spark 应用程序执行流程

Spark 应用程序的执行流程如下：首先当一个应用提交后，系统启动一个驱动程序 Driver Program，并创建一个 SparkContext 对象。然后 SparkContext 对象连接集群管理器（不同模式集群的管理器不同，如 Standalone 模式、Yarn 模式、Mesos 模式）分配资源，连接之后分配到多个用来计算和存储数据的执行器 Executor。接着 SparkContext 对象将用户提交的程序代码（JAR 或 Python 文件）发送给执行器 Executor，最后在执行器 Executor 上运行从 SparkContext 上发送来的任务 Task。

使用分布式内存存放待处理数据和优化数据处理过程是 Spark 数据处理效率得到提高的主要因素，称为未雨绸缪。在 Spark 中分布式内存数据进行了抽象，称为弹性分布式数据集 RDD。RDD 是 Spark 中核心的数据结构，所有数据处理操作都围绕 RDD 进行。数据操作即是对 RDD 的转换处理，把一个 RDD 转换为另一个新的 RDD。一个 Spark 应用的数据处理过程即是多个 RDD 的转换过程，Spark 对数据处理过程提出了优化方法：首先把数据处理过程中使用到的 RDD 和 RDD 之间的转换过程构成有向无环图 DAG；其次根据 RDD 之间的转换类型划分阶段 Stage，同一阶段 Stage 中的 RDD 转换操作在不同的节点上可以并行执行；最后执行划分好阶段的 DAG 图。

有向无环图的阶段划分如图 8-6 所示，图中的标注 A、B、C、D、E、F 和 G 是 RDD，小方块代表分别在不同节点上，并标明了 RDD 之间的转换关系，如 RDD A 到 RDD B 为 groupBy 操作、RDD C 到 RDD D 为 map 操作等，这些 RDD 根据它们之间的转换关系构成了一个有向无环图 DAG。虚线框为阶段 Stage，图中分 3 个阶段：阶段 Stage 1、阶段 Stage 2 和阶段 Stage 3。阶段的划分依据为 RDD 的转换是否能合并与并行进行，如某个节点上从 RDD C 到 RDD D 的转换和 RDD D 到 RDD F 的转换可以合并执行，不需要等待其他节点的转换结果；而 RDD A 到 RDD B 的转换需要涉及所有相关节点，故被划分到两个阶段。最后依次执行各阶段算出 RDD G。

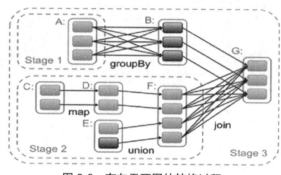

图 8-6　有向无环图的转换过程

8.4.1　YARN 运行架构工作原理

（1）Spark 在 0.6 版本之后，增加了 Yarn 工作模式，如图 8-7 所示介绍原理之前先说明几个名词概念。

① Driver：和 ClusterManager 通信，进行资源申请、任务分配并监督其运行状况等。

② ClusterManager：指 YARN。

③ DAGScheduler：把 spark 作业转换成 Stage 的 DAG 图。

④ TaskScheduler：把 Task 分配给具体的 Executor。

⑤ ResourceManager：负责整个集群的资源管理和分配。

⑥ ApplicationMaster：YARN 中每个 Application 对应一个 AM 进程，负责与 RM 协商获取资源，获取资源后告诉 NodeManager 为其分配并启动 Container。

⑦ NodeManager：每个节点的资源和任务管理器，负责启动/停止 Container，并监视资源使用情况。

⑧ Container：YARN 中的抽象资源。

（2）spark on yarn 有两种模式，一种是 cluster 模式；另一种是 client 模式。

① 执行命令 ./spark-shell --master yarn，默认运行的是 client 模式。

② 执行"./spark-shell --master yarn-client"或者"./spark-shell --master yarn --deploy-mode client"，运行的也是 client。

③ 执行"./spark-shell--master yarn-cluster"或者"./spark-shell--master yarn--deploy-mode cluster"，运行的是 cluster 模式。

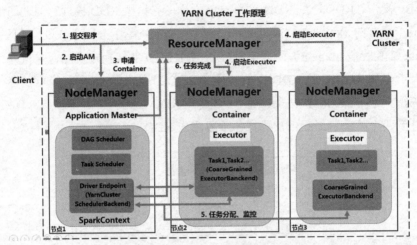

图 8-7　Yarn Cluster 模式工作原理

（3）client 和 cluster 模式的主要区别在于，前者的 driver 是运行在客户端进程中，后者的 driver 是运行在 NodeManager 的 ApplicationMaster 之中。具体工作流程如图 8-8 所示。

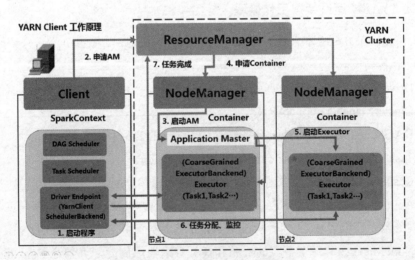

图 8-8　Yarn Client 模式工作原理

① Spark Yarn Client 向 YARN 的 ResourceManager 申请启动 Application Master。同时在 SparkContent 初始化中将创建 DAGScheduler 和 TASKScheduler 等，由于我们选择的是 Yarn-Client 模式，程序会选择 YarnClientClusterScheduler 和 YarnClientSchedulerBackend。

② ResourceManager 收到请求后，在集群中选择一个 NodeManager，为该应用程序分配第一个 Container，要求它在这个 Container 中启动应用程序的 ApplicationMaster，与 YARN-Cluster 的区别是在该 ApplicationMaster 中不运行 SparkContext，只与 SparkContext 进行联系进行资源的分派。

③ Client 中的 SparkContext 初始化完毕后，与 ApplicationMaster 建立通信，向 ResourceManager 注册，根据任务信息向 ResourceManager 申请资源（Container）。

④ ApplicationMaster 申请到资源（也就是 Container）后，便与对应的 NodeManager 通信，要求它在获得的 Container 中启动 CoarseGrainedExecutorBackend，CoarseGrainedExecutorBackend 启动后会向 Client 中的 SparkContext 注册并申请 Task。

⑤ Client 中的 SparkContext 分配 Task 给 CoarseGrainedExecutorBackend 执行，CoarseGrainedExecutorBackend 运行 Task 并向 Driver 汇报运行的状态和进度，以让 Client 随时掌握各个任务的运行状态，从而可以在任务失败时重新启动任务。

⑥ 应用程序运行完成后，Client 的 SparkContext 向 ResourceManager 申请注销并关闭自己。

⑦ 任务完成，回收资源。

8.4.2 Spark Mlib 核心技术

1. Mlib 数据类型介绍

（1）本地变量的基类是 Vector：支持密集向量 Dense Vector 和稀疏向量 Sparse Vector，scala 实现如下：

val dv：Vector = Vector.dense (5.0, 6.0, 7.0)

val sv：Vector = Vector.sparse (3, Array (0, 2), Array (1.0, 3.0))

（2）标点类型 LabeledPoint：由一个标签和本地向量组成，标签可以是 Int 型或者 Double 型。scala 实现如下：

val pos=LabledPoint(1.0, Vector.dense (1.0, 0.0, 3.0))

val pos=LabledPoint(0.0, Vector.sparse (3, Array (0,2), Array (1.0, 3.0)))

（3）稀疏数据：如 LibSVM 格式，label index1: value1 index2: value2 index3: value3。

（4）本地矩阵：基类是 Matrix，Mlib 提供了 DenseMatrix 实现。

MLlib 是 Spark 对常用的机器学习算法的实现库，同时包括相关的测试和数据生成器。Spark 的设计初衷就是为了支持一些迭代的 Job，这正好符合很多机器学习算法的特点。MLlib 目前支持 4 种常见的机器学习问题：分类、回归、聚类和协同过滤。分类算法属于监督式学习，使用类标签已知的样本建立一个分类函数或分类模型，应用分类模型，能把数据库中的类标签未知的数据进行归类。分类在数据挖掘中是一项重要的任务，目前在商业上应用最多，常见的典型应用场景有流失预测、精确营销、客户获取、个性偏好等。MLlib 目前支持的分类算法有逻辑回归、支持向量机、朴素贝叶斯和决策树。回归算法属于监督式学习，每个个体都有一个与之相关联的实数标签，并且我们希望在给出用于表示这些实体的数值特征后，所预测出的标签值可以尽可能接近实际值。MLlib 目前支持的回归算法有线性回归、岭回归、Lasso 和决策树。

2. 评价数据模型需要评价指标，如灵敏度和特异度等，具体概念说明如下：

- TP：true（预测是正确），positive（预测为正样本），表示预测为正样本预测正确。
- FN：false（预测是错误），negative（预测为负样本），表示预测为负样本预测错误。
- TN：true（预测是正确），negative（预测为负样本），表示预测为负样本预测正确。
- FP：false（预测是错误），positive（预测为正样本），表示预测为正样本预测错误。
- P（实际为正样本）=TP+FN，表示包括预测为正样本预测正确，预测为负样本预测错误。
- N（实际为负样本）=TN+FP，表示包括预测为负样本预测正确，预测为正样本预测错误。
- 正确率（accuracy）=TP+TN/P+N，表示正负样本而言，预测是正确的概率。
- 错误率（error rate）=FN+FP/P+N，表示正负样本而言，预测是错误的概率。
- 灵敏度（敏感度，召回率 recall，查全率 sensitive）=TP/P =TPR，表示对所有正样本而言，预测是正确的概率。
- 特异度（特效度 specificity）=TN/N，表示对所有负样本而言，预测是正确的概率。
- 精度（查准率，准确率 precision）=TP/TP+FP，表示预测为正样本而言，预测是正确的概率。

8.4.3 Spring Maven 工程构建

构建 pom.xml 文件，spark-mllib_2.11 就是机器学习类库包，具体如下：

```
<?xml version="1.0" encoding="UTF-8"?>
<project xmlns="http://maven.apache.org/POM/4.0.0"
      xmlns:xsi="http://www.w3.org/2001/XMLSchema-instance"
      xsi:schemaLocation="http://maven.apache.org/POM/4.0.0 http://
         maven.apache.org/xsd/maven-4.0.0.xsd">
   <modelVersion>4.0.0</modelVersion>
   <groupId>BSR</groupId>
   <artifactId>Bigdata</artifactId>
   <version>1.0-SNAPSHOT</version>
```

```xml
<properties>
    <scala.version>2.11.8</scala.version>
    <hadoop.version>2.7.4</hadoop.version>
    <spark.version>2.0.2</spark.version>
</properties>
<dependencies>
    <dependency>
        <groupId>org.scala-lang</groupId>
        <artifactId>scala-library</artifactId>
        <version>${scala.version}</version>
    </dependency>
    <dependency>
        <groupId>org.apache.spark</groupId>
        <artifactId>spark-core_2.11</artifactId>
        <version>${spark.version}</version>
    </dependency>
    <dependency>
        <groupId>org.apache.hadoop</groupId>
        <artifactId>hadoop-client</artifactId>
        <version>${hadoop.version}</version>
    </dependency>
    <!-- https://mvnrepository.com/artifact/org.apache.spark/spark-mllib -->
    <dependency>
        <groupId>org.apache.spark</groupId>
        <artifactId>spark-mllib_2.11</artifactId>
        <version>${spark.version}</version>
    </dependency>
    <dependency>
        <groupId>mysql</groupId>
        <artifactId>mysql-connector-java</artifactId>
        <version>5.1.38</version>
    </dependency>
    <dependency>
        <groupId>org.apache.spark</groupId>
        <artifactId>spark-sql_2.11</artifactId>
        <version>2.0.2</version>
    </dependency>
    <dependency>
        <groupId>org.apache.spark</groupId>
        <artifactId>spark-hive_2.11</artifactId>
        <version>2.0.2</version>
    </dependency>
    <dependency>
        <groupId>org.apache.spark</groupId>
```

```xml
            <artifactId>spark-streaming_2.11</artifactId>
            <version>2.0.2</version>
        </dependency>
        <dependency>
            <groupId>org.apache.spark</groupId>
            <artifactId>spark-streaming-flume_2.11</artifactId>
            <version>2.0.2</version>
        </dependency>
        <dependency>
            <groupId>org.apache.spark</groupId>
            <artifactId>spark-streaming-kafka-0-8_2.11</artifactId>
            <version>2.0.2</version>
        </dependency>
    </dependencies>
    <build>
        <sourceDirectory>src/main/java</sourceDirectory>
        <testSourceDirectory>src/test/java</testSourceDirectory>
        <plugins>
            <plugin>
                <groupId>org.apache.maven.plugins</groupId>
                <artifactId>maven-shade-plugin</artifactId>
                <version>2.3</version>
                <executions>
                    <execution>
                        <phase>package</phase>
                        <goals>
                            <goal>shade</goal>
                        </goals>
                        <configuration>
                            <filters>
                                <filter>
                                    <artifact>*:*</artifact>
                                    <excludes>
                                        <exclude>META-INF/*.SF</exclude>
                                        <exclude>META-INF/*.DSA</exclude>
                                        <exclude>META-INF/*.RSA</exclude>
                                    </excludes>
                                </filter>
                            </filters>
                            <transformers>
                                <transformer implementation="org.apache.
                                    maven.plugins.shade.resource.
                                    ManifestResourceTransformer">
                                    <mainClass></mainClass>
                                </transformer>
```

```
                </transformers>
            </configuration>
        </execution>
    </executions>
</plugin>
        </plugins>
    </build>
</project>
```

8.4.4 决策树预测体检费用

(1) 体检数据集说明：323 个特征值，包括用户年龄，性别，住址，籍贯，既往病史和检查结果等，目标特征是分类变量 1~9，5 表示体检花费 5.0（千元），需要转换为 LabledPoint 类型的稀疏数据格式，具体如下：

```
**********output all features and label by String in the line beforeprocessing
******************
1,1,1,30,2201,0,1,20,210105,51,3,5,0,0,0,0,0,0,0,0,0,0,0,0,0,0,0,
0,0,0,0,0,0,0,0,0,0,0,0,0,0,0,0,0,0,0,0,0,0,0,0,0,0,0,3.2,0,3.
2,3.2,3.2,3.2,0,1,0,1,1,0,10,4680,1,1,80,0,1,2,2,5,0,1,2101,1,1,2101,
3,1,1,1,1,1,1,0,0,0,0,1,300117,300102,300133,300133,13,1,300102,900704,
0,0,0,0,58176,8907.53,1,1,1,1,1,1,1,1,1,1,1,0,0,1,1,0,0,1,0,1,0,0,
0,0,0,0,0,0,0,0,0,0,0,0,0,0,0,0,0,0,0,0,0,0,0,0,0,0,0,0,0,0,0,0,
0,0,0,0,0,0,0,0,0,0,0,0,0,0,0,0,0,0,0,0,0,0,0,0,0,0,0,0,0,0,0,0,
0,0,0,0,0,0,0,0,0,0,0,0,0,0,0,0,0,0,0,0,0,0,0,0,0,0,0,0,0,0,0,0,
0,0,0,0,0,0,0,0,0,0,0,0,0,0,0,0,0,0,0,0,0,0,0,0,0,0,0,0,0,0,0,0,
0,0,0,0,0,0,0,0,0,0,0,0,0,0,0,0,0,0,0,0,0,0,0,5
********output all features and label by LabeledPoint in the line after
processing***************
(5.0,(323,[0,1,2,3,4,5,6,7,8,9,10,11,12,13,14,15,16,17,18,19,20,21,22,
23,24,25,26,27,28,29,30,31,32,33,34,35,36,37,38,39,40,41,42,43,44,
45,46,47,48,49,50,51,52,53,54,55,56,57,58,59,60,61,62,63,64,65,66,
67,68,69,70,71,72,73,74,75,76,77,78,79,80,81,82,83,84,85,86,87,88,
89,90,91,92,93,94,95,96,97,98,99,100,101,102,103,104,105,106,107,
108,109,110,111,112,113,114,115,116,117,118,119,120,121,122,123,
124,125,126,127,128,129,130,131,132,133,134,135,136,137,138,139,
140,141,142,143,144,145,146,147,148,149,150,151,152,153,154,155,
156,157,158,159,160,161,162,163,164,165,166,167,168,169,170,171,
172,173,174,175,176,177,178,179,180,181,182,183,184,185,186,187,
188,189,190,191,192,193,194,195,196,197,198,199,200,201,202,203,
204,205,206,207,208,209,210,211,212,213,214,215,216,217,218,219,
220,221,222,223,224,225,226,227,228,229,230,231,232,233,234,235,
236,237,238,239,240,241,242,243,244,245,246,247,248,249,250,251,
252,253,254,255,256,257,258,259,260,261,262,263,264,265,266,267,
268,269,270,271,272,273,274,275,276,277,278,279,280,281,282,283,
284,285,286,287,288,289,290,291,292,293,294,295,296,297,298,299,
```

```
300,301,302,303,304,305,306,307,308,309,310,311,312,313,314,315,
316,317,318,319,320,321,322],[1.0,1.0,1.0,30.0,2201.0,0.0,1.0,20.
0,210105.0,51.0,3.0,5.0,0.0,0.0,0.0,0.0,0.0,0.0,0.0,0.0,0.0,0.
0,0.0,0.0,0.0,0.0,0.0,0.0,0.0,0.0,0.0,0.0,0.0,0.0,0.0,0.0,0.0,0.
0,0.0,0.0,0.0,0.0,0.0,0.0,0.0,0.0,0.0,0.0,0.0,0.0,0.0,0.0,0.0,0.
0,0.0,0.0,0.0,3.2,0.0,3.2,3.2,3.2,3.2,0.0,1.0,0.0,1.0,1.0,0.0,10.0,4680.
0,1.0,1.0,80.0,0.0,1.0,2.0,2.0,5.0,0.0,1.0,2101.0,1.0,1.0,2101.0,
3.0,1.0,1.0,1.0,1.0,1.0,1.0,0.0,0.0,0.0,0.0,1.0,300117.0,300102.0,
300133.0,300133.0,13.0,1.0,300102.0,900704.0,0.0,0.0,0.0,0.0,58176.
0,8907.53,1.0,1.0,1.0,1.0,1.0,1.0,1.0,1.0,1.0,1.0,1.0,0.0,0.0,1.0,
1.0,0.0,0.0,1.0,0.0,0.0,0.0,0.0,0.0,0.0,0.0,0.0,0.0,0.0,0.0,0.0,
0.0,0.0,0.0,0.0,0.0,0.0,0.0,0.0,0.0,0.0,0.0,0.0,0.0,0.0,0.0,0.0,
0.0,0.0,0.0,0.0,0.0,0.0,0.0,0.0,0.0,0.0,0.0,0.0,0.0,0.0,0.0,0.0,
0.0,0.0,0.0,0.0,0.0,0.0,0.0,0.0,0.0,0.0,0.0,0.0,0.0,0.0,0.0,0.0,
0.0,0.0,0.0,0.0,0.0,0.0,0.0,0.0,0.0,0.0,0.0,0.0,0.0,0.0,0.0,0.0,
0.0,0.0,0.0,0.0,0.0,0.0,0.0,0.0,0.0,0.0,0.0,0.0,0.0,0.0,0.0,0.0,
0.0,0.0,0.0,0.0,0.0,0.0,0.0,0.0,0.0,0.0,0.0,0.0,0.0,0.0,0.0,0.0,
0.0,0.0,0.0,0.0,0.0,0.0,0.0,0.0,0.0,0.0,0.0,0.0,0.0,0.0,0.0,0.0,
0.0,0.0,0.0,0.0,0.0,0.0,0.0,0.0,0.0,0.0,0.0,0.0,0.0,0.0,0.0]))
```

（2）更多体检数据集可以在本书指定的网络上下载，决策树预测体检花费的代码实现如下：

```scala
import org.apache.spark.mllib.linalg.Vectors
import org.apache.spark.mllib.regression.LabeledPoint
import org.apache.spark.mllib.tree.DecisionTree
import org.apache.spark.mllib.tree.configuration.Algo
import org.apache.spark.mllib.tree.impurity.Entropy
import org.apache.spark.{SparkConf,SparkContext}
/**
  * Created by changyaobin
  */
object DecisionTreeTest {
  def main(args:Array[String]):Unit = {
    val conf = new SparkConf().setAppName("DesionTrain").setMaster
      ("local[2]")
    val sc = new SparkContext(conf)
    // 加载数据
    val data = sc.textFile("f://physicalCheck.csv").map(lines => {
      val fields = lines.split(",")
      val lable = fields(fields.length - 1).toDouble
      val features = fields.slice(1,fields.length - 1).map(x => x.
        toDouble)
      LabeledPoint(lable,Vectors.dense(features))
```

```
})
val labe = data.map(_.label)
// 了解决策树的参数
val model = DecisionTree.train(data,Algo.Classification,Entropy,5,9)
val predictionAndLabel = data.map { point =>
  val score = model.predict(point.features)
  (score)
}
println(predictionAndLabel.collect().toBuffer)
val acc = labe.zip(predictionAndLabel).filter(x => {
  x._1.equals(x._2)
}).count()/ labe.count().toDouble
println(acc)
  }
}
```

（3）预测结果如图 8-9 所示。

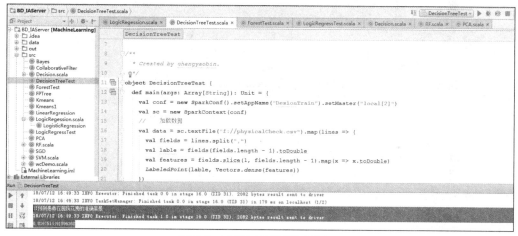

图 8-9　决策树的数据预测展示

8.4.5　逻辑回归预测体检费用

（1）更多体检数据集可以在本书指定的网络地址下载，逻辑回归预测体检花费的代码实现如下：

```
import org.apache.spark.mllib.classification.{LogisticRegressionWith LBFGS,
  LogisticRegressionWithSGD}
import org.apache.spark.mllib.linalg.Vectors
import org.apache.spark.mllib.regression.LabeledPoint
import org.apache.spark.{SparkConf,SparkContext}
/**
  * Created by changyaobin.
  */
object LogicRegressTest {
```

```scala
def main(args:Array[String]):Unit = {
  val conf = new SparkConf().setMaster("local[2]").setAppName
    ("LogisticRegression")
// 设置环境变量
  val sc = new SparkContext(conf)
  val data=sc.textFile("f://physicalCheck.csv").map(lines=>{
    val fields=lines.split(",")
    val lable=fields(fields.length-1).toDouble
    val features=fields.slice(1,fields.length-1).map(x=>x.toDouble)
    LabeledPoint(lable,Vectors.dense(features))
  })
  val lable=data.map(_.label)
  val model = new LogisticRegressionWithLBFGS()
    .setNumClasses(9)
    .run(data)
  val predictionAndLabels =data.map { case LabeledPoint(label,features)=>
    val prediction = model.predict(features)
    (prediction)
  }
  predictionAndLabels.foreach(x=> println(x))
  val acc=lable.zip(predictionAndLabels).filter(x=>{
    x._1.equals(x._2)
  }).count()/lable.count().toDouble
  println("LR 预测患者在医院花费的准确率是")
  println(acc)
 }
}
```

（2）预测结果如图 8-10 所示。

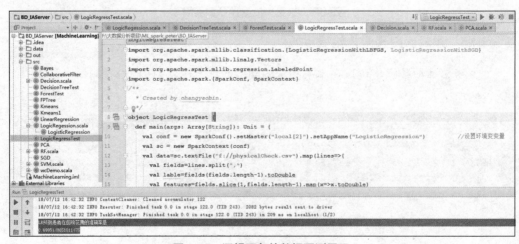

图 8-10 逻辑回归的数据预测展示

8.4.6 随机森林预测体检费用

（1）更多体检数据集可以在本书指定的网络地址下载，随机森林预测体检花费的代码实现如下：

```scala
import org.apache.spark.mllib.linalg.Vectors
import org.apache.spark.mllib.regression.LabeledPoint
import org.apache.spark.mllib.tree.RandomForest
import org.apache.spark.mllib.tree.configuration.Algo
import org.apache.spark.mllib.tree.impurity.Entropy
import org.apache.spark.{SparkConf,SparkContext}
/**
 * Created by changyaobin.
 */
object ForestTest {
  def main(args:Array[String]):Unit = {
    val conf=new SparkConf().setAppName("DesionTrain").setMaster("local[2]")
    val sc=new SparkContext(conf)
    // 加载数据
    val data=sc.textFile("f://physicalCheck.csv").map(lines=>{
      val fields=lines.split(",")
      val lable=fields(fields.length-1).toDouble
      val features=fields.slice(1,fields.length-1).map(x=>x.toDouble)
      LabeledPoint(lable,Vectors.dense(features))
    })
    val labe=data.map(_.label)
    // 配置决策树的参数
    val model= RandomForest.trainClassifier(data,9,Map[Int,Int](),20,
      "auto", "entropy",30,300)
    val predictionAndLabel = data.map { point =>
      val score = model.predict(point.features)
      (score)
    }
    predictionAndLabel.foreach(x=> println(x))
    // 测试准确率
    val acc=labe.zip(predictionAndLabel).filter(x=>{
      x._1.equals(x._2)
    }).count()/labe.count().toDouble
    println("Forest 预测患者在医院花费的准确率是")
    println(acc)
  }
}
```

（2）预测结果如图 8-11 所示。

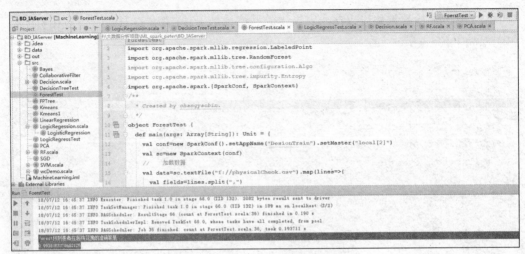

图 8-11　随机森林的数据预测展示

8.4.7　支持向量机预测疾病概率

（1）以下数据是代表患者化验检查的部分结构，数据集的数据格式是 lable 是患病的标签，后面是特征值，表示年龄和化验结果等。20 条数据集如下：

```
0 1:77 2:2 3:49.4 4:2 5:1
0 1:36 2:1 3:57.2 4:1 5:1
0 1:61 2:2 3:168 4:2 5:1
1 1:58 2:3 3:98 4:4 5:3
1 1:55 2:3 3:80 4:3 5:4
0 1:71 2:1 3:94.4 4:2
0 1:38 2:1 3:76 4:1 5:1
0 1:42 2:1 3:240 4:3 5:2
0 1:40 2:1 3:74 4:1 5:1
0 1:58 2:3 3:63.6 4:2 5:2
0 1:78 2:3 3:132.8 4:4 5:2
1 1:25 2:2 3:94.6 4:4 5:3
0 1:52 2:1 3:56 4:1 5:1
0 1:31 2:1 3:49.8 4:2 5:1
1 1:36 2:3 3:31.6 4:3 5:1
0 1:42 2:1 3:66.2 4:2 5:1
1 1:14 2:3 3:138.6 4:3 5:3
0 1:32 2:1 3:114 4:2 5:3
0 1:35 2:1 3:40.2 4:2 5:1
1 1:70 2:3 3:177.2 4:4 5:3
```

（2）针对以上数据集构建数据模型，代码实现如下：

```
import org.apache.spark.mllib.classification.SVMWithSGD
```

```scala
import org.apache.spark.mllib.evaluation.MulticlassMetrics
import org.apache.spark.mllib.linalg.Vectors
import org.apache.spark.mllib.regression.LabeledPoint
import org.apache.spark.mllib.util.MLUtils
import org.apache.spark.{SparkContext,SparkConf}
object DiseasesPredict {
  def main (args: Array[String]) {
    val conf = new SparkConf ()                                //创建环境变量
      .setMaster ("local")                                     //设置本地化处理
      .setAppName ("SVMTest")                                  //设定名称
val sc = new SparkContext (conf)                               //创建环境变量实例
val data = MLUtils.loadLibSVMFile (sc,"c: //svm.txt")          //获取数据集
val splits = data.randomSplit (Array (0.7,0.3),seed = 11L)     //对数据集切分
val parsedData = splits (0)                                    //分割训练数据
val parseTtest = splits (1)                                    //分割测试数据
val model = SVMWithSGD.train (parsedData,50)                   //训练模型
val predictionAndLabels = parseTtest.map {                     //计算测试值
  case LabeledPoint (label,features) =>                        //计算测试值
    val prediction = model.predict (features)                  //计算测试值
    (prediction,label)                                         //存储测试和预测值
}
val metrics = new MulticlassMetrics (predictionAndLabels)      //创建验证类
val precision = metrics.precision                              //计算验证值
println ("Precision = " + precision)                           //打印验证值
val patient = Vectors.dense (Array (70,3,180.0,4,3))           //计算患者可能性
if (patient == 1) println ("患者的患某种疾病概率很大。")          //做出判断
else println ("未见异常。")                                     //做出判断
  }
}
```

（3）对一个新患者（70，3，180.0，4，3）的预测结果为未见异常，如图 8-12 所示。

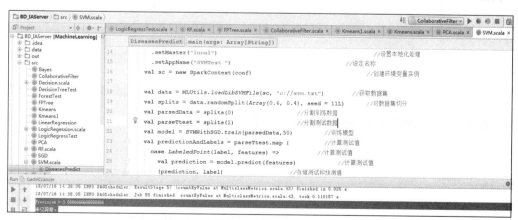

图 8-12 支持向量机的疾病预测结果展示

8.4.8 协同过滤推荐药品

（1）以下是用户的数据信息，数据集的数据格式是：lable 表示用户类型（1-4），后面是特征值，表示药品类型（10～20）和疗效（0～5）。20 条数据集如下：

```
1 17 2,1 12 3,1 14 1,1 18 0,1 11 1,2 11 1,2 12 2,2 17 1,2 14 3,2 15 1,
2 13 5,2 10 1,3 17 4,3 12 2,3 18 1,3 14 3,3 11 1,3 19 5,4 16 1,
4 13 2, 4 12 2,4 11 1,4 14 3,4 11 1,4 18 2,4 12 1,4 14 4
```

（2）对如上数据集进行转换，将数据集转化为专用 Rating，调用 recommendProducts 函数进行推荐，代码实现如下：

```scala
import org.apache.spark._
import org.apache.spark.mllib.recommendation.{ALS,Rating}
object CF {
  def main(args:Array[String]){
    val conf = new SparkConf().setMaster("local").setAppName("CF ")
                                              //设置环境变量
    val sc = new SparkContext(conf)           //实例化环境
    val data = sc.textFile("c://cf.txt")      //设置数据集
    val ratings = data.map(_.split(' ')match { //处理数据
      case Array(user,item,rate)=>            //将数据集转化
        Rating(user.toInt,item.toInt,rate.toDouble)
                                              //将数据集转化为专用Rating
    })
    val rank = 2                              //设置隐藏因子
    val numIterations = 2                     //设置迭代次数
    val model = ALS.train(ratings,rank,numIterations,0.01) //进行模型训练
    var rs = model.recommendProducts(2,2)     //为用户2推荐一个药品
    rs.foreach(println)                       //打印结果
  }
}
```

（3）为用户 2 推荐一个药品，结果如图 8-13 所示。

图 8-13 协同过滤推荐药品的结果展示

8.5 项目小结

本章针对物联网体检大数据进行了分类分析,实现了体检费用预测、疾病预测、药品推荐等应用,采用决策树、逻辑回归、SVM 等机器学习算法,这些算法使用的都是通用的数据训练和预测方法,我们工程师只有在真正理解算法工作原理和应用场景之后,才能解决实际分析问题,尤其是提升预测准确率方面。其中数据集的生成是主要工作,工作占比大概在 60%以上。LabeledPoint 和 LibSVM 格式是主流的 Spark Mlib 的数据加载常用格式。下面总结各个知识点:

1. 理论知识点

(1)逻辑回归(Logistic Regression)是回归分析的一种,主要用来做分类,适用于二分类问题,可以推广到多分类问题。主要工作为了求出回归系数,并建立方程,为后续新向量预测提供函数。

(2)支持向量机是扩展的线性分类器,线性分类器是用一个超平面将不同类别的样本分开。其主要思想是将低维空间的样本通过非线性变换映射到高维空间,从而将低维空间的线性不可分问题转化为高维空间的线性可分问题。计算函数系数 a^* 和 b^*,就可以确定这个超平面的方程

$$f(S) = \sum_{i=1}^{N} c_i a_i^* K(S_i, S_j) + b^*$$

(3)决策树的建立是基于样本的递归的学习过程,每个样本都是具有确定的属性的数据,决策树就是基于样本的各属性建立起来的。决策树分为根节点(信息增益最大的节点)、内部节点和叶节点。从根节点出发,自顶向下,构建分支和内部节点,在内部节点进行属性值的比较,并根据属性的不同取值确定从该节点向下的分支,最终在叶节点给出类别。

(4)聚类算法通常按照中心点或者分层的方式对输入数据进行归并。聚类算法都试图找到数据的内在结构,以便按照最大的共同点将数据进行归类。

(5)关联规则学习通过寻找最能够解释数据变量之间关系的规则,来找出大量多元数据集中有用的关联规则。常见算法包括 Apriori 算法和 FP-growth 算法。

(6)协同过滤就是给用户推荐相似的物品或者人,协同过滤算法又分为基于用户的协同过滤算法和基于物品的协同过滤算法。

2. 实战知识点

(1)标点类型 LabeledPoint:由一个标签和本地向量组成,标签可以是 Int 型或者 Double 型。

（2）稀疏数据：如 LibSVM 格式 label index1: value1 index2: value2 index3: value3。

（3）把数据集转成 LabeledPoint 格式的通用方法，Scala 实现如下：

```
//1.加载数据
val data = sc.textFile("f://physicalCheck.csv").map(lines => {
  val fields = lines.split(",")
  val lable = fields(fields.length - 1).toDouble
  val features = fields.slice(1,fields.length - 1).map(x => x.toDouble)
//2.转换数据
  LabeledPoint(lable,Vectors.dense(features))
})
```

（4）不同分类算法的建模函数不同，但是预测方法都一样，如 val prediction = model.predict (features)。

第 9 章

大数据自定义迁移微服务引擎

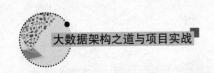

架构之道分享之九：孙子兵法的《始计篇》论述了庙算，即出兵前在庙堂上比较敌我的各种条件，估算战事胜负的可能性，并制订作战计划。大数据自定义迁移要考虑和选择服务对象及数据格式之后，针对从不同数据库迁移到集群环境的不同存储位置，为后续数据分析和数据处理奠定坚实基础。

本章学习目标

★ 掌握 Sqoop 引擎迁移数据到 HDFS、HBase、Hive 中的实战技巧
★ 掌握 MapReduce 框架的数据处理方法
★ 掌握 HBase 的数据库设计和通过模板类实现数据管理
★ 掌握数据迁移微架构服务构建方法

9.1 核心需求分析和优秀解决方案

大数据自定义迁移引擎是把物联网数据资源池的有价值的数据抽取到集群中，可以方便后续智能分析和构建数据模型使用，本章讨论的物联网数据资源池中有价值的数据包括用户信息、智能终端运动和体检业务数据等，随着物联网设备增多和体检数据不断积累，需要按照日期把历史数据迁移到 HBase 或者 Hive 数据仓库中。数据迁移要考虑几个核心需求：一是建立统一和标准化的迁移工具抽取数据到 HBase，并考虑数据清洗；二是考虑抽取和迁移性能，使用主流的 Sqoop 引擎实现批量处理迁移；三是验证迁移后的数据是否具备完整性和一致性；四是考虑迁移工具的可扩展性，采用灵活配置实现模块化处理。

9.2 服务引擎的技术架构设计

大数据自定义迁移引擎包括 5 个核心模块，如图 9-1 所示，每一个模块设计都要考虑模块化和工具化两个关键因素，具体说明如下。

（1）核心模块一：基于 Sqoop 引擎，把一个关系型数据库（如 MySQL、Oracle、SQLServer 等）中的用户和业务数据导入 Hadoop 的 HDFS 中。

（2）核心模块二：设计 HBase 的表 Rowkey 和列簇，并进行数据清洗，然后把 HDFS 的用户和业务数据迁移到 HBase 库。

（3）核心模块三：设计和实现 HBaseTemplete 模板类，进行 Dao 层接口定义和实现。

（4）核心模块四：按照时间和用户等条件查询在 HBase 中的迁移数据，验证数据统一性和数据完整性。

（5）核心模块五：考虑另外的迁移方式实现离线统计，即把历史数据迁移到 Hive。

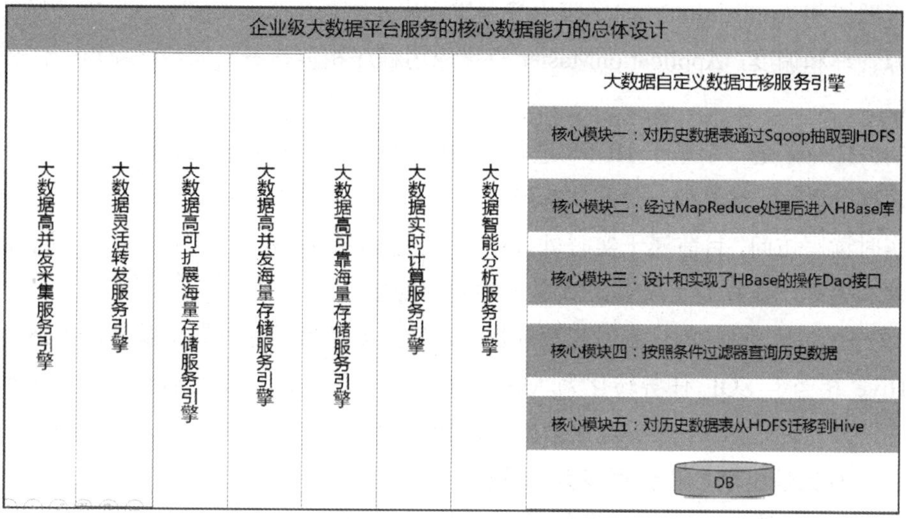

图 9-1 大数据自定义数据迁移服务模块化设计

9.3 核心技术讲解及模块化实现

9.3.1 Hadoop 生态的核心组件

Hadoop 是 Apache 的一个开源项目，是一个高可靠、高可扩展和高性能的分布式并行计算框架和技术生态，主要由分布式文件存储系统 HDFS 和分布式并行计算框架 MapReduce 组成。本书项目的运行环境是 Hadoop V2.7.3，构建 Hadoop V2.7.3 集群环境文档和项目数据可以在本书指定位置下载。Hadoop 主要解决的是分布式环境下的数据存储、并行计算和资源协同管理问题，核心价值是能高效组织一大批廉价机器构建分布式环境来协同存储和处理海量数据。

（1）HDFS 是分布式文件系统，具备易扩展、廉价、高吞吐量、高可靠性的特性，一个文件被分片为多个文件进行分布式存储，一个文件默认为 128MB。HDFS 系统由主节点 NameNode 和多个从节点 DataNode 组成。主节点 NameNode 的职责是管理文件所在目录及对应的 block 位置，DataNode 负责实际数据的存储，其理论基础是数据复制和数据分片。

（2）MapReduce 是一个分布式并行计算框架，实现了在集群环境下的并行任务分解处理的 Map 作业和结果汇总的 Reduce 作业，采用了"分而治之"的作战思想。MapReduce 框架解决了并行数据处理下的分布式存储、任务调度、负载均衡、容错机制、容错处理等核心技术问题。

（3）YARN 是新的多级调度资源管理系统，它是 MapReduce 的升级版本，核心思想是将 MapReduce 中 JobTracker 的资源管理和作业调度两个职责分别由 ResourceManager 资源管理器和 ApplicationMaster 应用管理进程来实现。ResourceManager 负责整个集群的资源管理和调度，ApplicationMaster 负责应用程序相关事务，如任务调度、任务监控和容错等。

（4）HBase 是一个基于 HDFS 进行 NoSQL 存储模型的列式 KV 数据库，具备高可伸缩、高可靠性、高性能、分布式等特点。它通过高效的 Rowkey 行键实现了对海量数据的实时读写访问，目前属于实时处理中非常主流的数据库之一，可以采用 MapReduce 来存储和处理，让存储实现真正意义上的并行处理。

（5）Hive 是 FaceBook 公司开源项目，设计初衷是为了处理海量数据的离线统计问题。Hive 相当于 SQL 任务转化为 Mapdeduce 并行任务，可以大幅度减少 Mapdeduce 的程序开发，通过 SQL 来实现数据处理，如过滤、清洗、编码、整合等任务。它属于 SQL 的并行处理引擎，适合离线数据分析。

（6）Flume 是 CLoudera 开源的日志收集系统，具备分布式、高可靠性、高容错性、高可扩展的特性。对实时数据实现了从接收、传输、处理、写入过程的数据流处理，可以接收多源异构数据源，对数据流实现数据整合，如过滤、清洗和编码等。

（7）HBase 集群采用 Master/Slave 的一主多从架构设计，一个主节点服务 HMaster，多个从节点服务 HRegionServer，ZooKeeper 通过 HMaster 对集群进行调度，是集群环境的协调器。

9.3.2　HBase 工作原理

HBase 集群是由 HMaster 和 HRegionServer 等服务组成的。HMaster 是 HBase 集群的管理者，负责管理多个 HRegionServer，以及对其上的表和区域 Region 的管理、对用户数据请求的响应。集群工作主要是客户端和集群交互进行数据文件的读写，由客户端直接和 HRegionServer 通信，当出现故障后 HMaster 负责集群的故障切换、HRegion 拆分、管理操作接口，因此 HMaster 出现问题后需要尽快恢复。HBase 集群可以实现高可用，至少两个 HMaster 节点，最多 10 个 HMaster 节点，可以避免 HMaster 发生单点故障。通过 ZooKeeper 的 Master Election 选举机制来保证任何时刻集群内只有一个 HMaster 在运行，如图 7-2 所示。

HMaster 的具体任务如下：

（1）负责将分区 Region 分配给 HRegionServer。

（2）负责 HRegionServer 的负载均衡。

（3）负责监控集群的从节点状态，发现失效的 HRegionServer 并重新分配其上的 Region。

（4）负责维护表和 Region 的元数据。

（5）负责管理用户对表的增、删、改、查操作。

（6）处理 Schema 更新请求，GFS 上的垃圾文件回收。

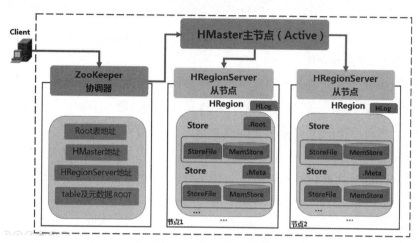

图 9-2　HBase 的工作原理

1. HRegion 和 HRegionServer 原理介绍

在存储表数据时，有唯一行键 rowkey 将表分成很多块进行存储，每一块称为一个 HRegion。每个 HRegion 由一个或多个 Store 组成，每个 Store 保存表中一个列族的数据，每个 Store 由一个 MemStore 和若干个 StoreFile 组成，MemStore 保存在内存中，默认配置是 128MB，StoreFile 的底层实现则是以 HFile 的形式存储在 HDFS 上。

（1）HRegionServer 对 HRegion 的管理：负责管理本服务器上的 HRegion，处理对 HRegion 的 I/O 请求。一台服务器上一般只运行一个 HRegionServer，每个 HRegionServer 管理多个 HRegion。每台 HRegionServer 上有一个 HLog 文件，记录该 HRegionServer 上所有 HRegion 的数据更新备份。

（2）Table 的 HRegion 存储机制：每个 table 先只有一个 HRegion，HRegion 是 HBase 表数据存储分配的最小单位，一张 table 的所有 HRegion 会分布在不同的 HRegionServer 上，但一个 HRegion 内的数据只会存储在一个 HRegionServer 上。当客户端进行数据更新操作时，先连接有关的 HRegionServer，由其向 HRegion 提交变更，提交的数据会首先写入 HLog 和 MemStore，其中的数据累计到设定的阈值时，HRegionServer 会启动一个单独的线程，将 MemStore 中的内容保存到磁盘上，生成一个新的 StoreFile。当 StoreFile 文件的数量增长到设定值后，就会将多个 StoreFile 合并成一个 StoreFile，合并时会进行版本合并和数据删除。说明一下，HBase 平时一直在增加数据，所有的更新和删除操作都是在后续的合并过程中进行的。当 HRegion 中单个 StoreFile 大小超过设定的阈值时，HRegionServer 会将该 HRegion 拆分为两个新的 HRegion，并且报告给主服务器，由 HMaster 来分配由哪些 HRegionServer 存放新产生的两个 HRgion，最后旧的 HRegion

不再需要时会被删除。反过来，当两个 HRegion 足够小时，HBase 也会将它们合并。

（3）Client 端有访问 HBase 的接口，并通过缓存来加快对 HBase 的访问。它使用 HBase 的 RPC 机制与 HMaster 和 HRegionServer 进行通信：对于管理类操作，Client 与 HMaster 进行通信；对于数据读写类操作，Client 与 HRegionServer 进行通信。

2. ZooKeeper 原理介绍

ZooKeeper 是一个开放源码的分布式集群协调器，主要用于解决分布式应用中的统一命名服务、状态同步服务、集群管理、配置项管理等问题。HBase 安装包中含有内置 ZooKeeper，也可以使用独立安装的 ZooKeeper，主要有如下作用。

（1）解决 HMaster 的单点故障问题：HBase 中可以启动多达 10 个 HMaster，通过 ZooKeeper 的 Master Election 机制保证任何时刻只有一个 HMaster 在运行。

（2）解决实时监控 HRegionServer 在线问题：监控 HRegionServer 的上、下线信息并及时通知 HMaster，若有 HRegionServer 崩溃可以通过 ZooKeeper 来进行分配协调。

（3）解决快速 Region 寻址问题：ZooKeeper 中存储了-ROOT 表的地址、HMaster 的地址、HRegionServer 地址、HBase 的 Schema 和表的元数据，当 Client 连接到 HBase 时，需要首先访问 ZooKeeper 以获取这些核心数据。

3. 元数据的原理介绍

用户表被按行键分隔成不同的 HRegion 来保存，用户表的 HRegion 元数据被存储在.META 表中，该表在 HBase 中也以 HRegion 的形式来进行存储。随着.META 表中数据增多后，它也会被拆分成多个 HRegion 来保存，.META 表中各个 HRegion ID 及其映射信息组成了 HBase 的-ROOT 表，由 ZooKeeper 来记录-ROOT 表的位置信息。-ROOT 表永远不会被分割且只有一个 HRegion，这样可以保证经过 3 次跳转就可以定位到任意一个 HRegion：客户端访问用户数据时，首先访问 ZooKeeper 获得-ROOT 表的位置，然后访问-ROOT 表获得.META 表的位置，最后根据.META 表中的信息确定用户数据存放的位置。

9.3.3 Sqoop 工作原理

Sqoop 是负责把 Hadoop 和关系型数据库中的数据相互转移的工具，可以将一个关系型数据库中的数据导入 Hadoop 的 HDFS 中，也可以将 HDFS 的数据导入关系型数据库中。对于某些 NoSQL 数据库，它也提供了连接器。Sqoop 使用元数据模型来判断数据类型，并在数据从数据源转移到 Hadoop 时确保类型安全的数据处理。Sqoop 专为大数据批量数据迁移设计，能够分割数据集并创建 Hadoop 任务来处理每个区块。

9.3.4 MapReduce 工作原理

MapReduce 是一种分布式计算框架，由 Google 公司提出，主要用于搜索领域，以解决海量数据的计算问题。MapReduce 由两个阶段组成：Map 阶段和 Reduce 阶段，用户只需实现 map 和 reduce 两个函数就可以实现分布式计算。有两个工作线程，分别是 JobTracker 和 TaskTracker。JobTracker 进程负责资源管理和任务调度，TaskTracker 进程服务任务处理，它们在集群中协同工作。具体工作原理如下：

（1）建立通信。客户端提交一个作业，JobClient 与 JobTracker 服务进行通信，JobTracker 返回一个 JobID。

（2）复制作业资源文件。JobClient 复制作业资源文件到集群的 HDFS，包括 MapReduce 程序打包的 JAR 文件、配置文件和客户端的输入划分信息。这些文件都存放在 JobTracker 进程为该作业创建的文件夹中，文件夹名为该作业的 Job ID。

（3）提交作业任务。开始提交任务，任务的描述信息包括 jobid、jar 存放的位置、配置信息等。

（4）初始化任务。创建作业对象，JobTracker 进程接收到作业后，将其放在一个作业队列中，等待作业调度器对其进行调度。

（5）分配任务。对 HDFS 上的资源文件进行分片，每一个分片对应一个 MapperTask，当作业调度器根据自己的调度算法调度到该作业时，会根据输入划分信息为每个划分创建一个 map 任务，并将 map 任务分配给 TaskTracker 执行。

（6）建立心跳机制。TaskTracker 进程会向 JobTracker 进程返回一个心跳信息（任务的描述信息），根据心跳信息，分配任务每隔一段时间会给 JobTracker 发送一个心跳，汇报当前 map 任务完成的进度等信息。当 JobTracker 收到作业的最后一个任务完成信息时，便把该作业设置为"成功"。当 JobClient 查询状态时，它将得知任务已完成，便显示一条消息给用户。

（7）任务分配。TaskTracker 进程从 HDFS 上获取作业资源文件，对于 map 和 reduce 任务，TaskTracker 根据主机核的数量和内存的大小有固定数量的 map 槽和 reduce 槽。map 任务分配有一个就近原则，将 map 任务分配给含有该 map 处理的数据块的 TaskTracker 上，同时将程序 JAR 包复制到该 TaskTracker 上来运行，在分配 reduce 任务时并不考虑数据本地化。

（8）任务执行。TaskTracker 进程启动一个 child 进程来执行具体任务。

9.3.5 Sqoop 抽取历史数据到 HDFS

基于 Sqoop 引擎把一个关系型数据库（如 MySQL、Oracle、SQLServer 等）中的用户数据、智能终端运动、体检数据导入 Hadoop 的 HDFS 中，具体步骤如下：

（1）MySQL 的 usertbl 表和 sports 表导入 Hadoop 的 HDFS，通过 Sqoop 操作 Windows 的 MySQL 的 usertbl 表和 sports 表。在 SQL yog 下执行给 Sqoop 授权后，才保证在集群 Linux 环境下可以操作 Windows 的 MySQL 的所有表，192.168.106.1 地址是开发环境 Eclipse 所在 Windows 的本地网卡地址，192.168.106.111 地址是伪分布式集群 Linux 环

境下的 Master 地址，只有同一个网络内，可以实现集群环境操作本地 MySQL 数据库。在 SQL yog 下执行命令如下：

```
GRANT ALL PRIVILEGES ON *.* TO 'root'@'192.168.106.111' IDENTIFIED BY 'root' WITH GRANT OPTION;FLUSH PRIVILEGES;
```

（2）在伪分布式集群 Linux 环境中执行语句，数据导入 aggregate /usertbl，结果见图 9-3 所示，执行命令如下：

```
sqoop import --connect jdbc:mysql://192.168.106.1:3306/aggregate --username root --password root --table usertbl --target-dir '/aggregate/usertbl/' --fields-terminated-by '\t'
```

（3）查看进入 HDFS 的 usertbl 文件的数据格式命令如下：

```
Hadoop fs-cat/aggregate/usertbl/part-m-00000
```

数据导入 HDFS 成功，数据如下：

```
1001 0526939 p939 null AppA;AppB;AppC;AppD 1 caompany939 test939 null null null 0 2018-05-05 14:58:13 null null
```

（4）在虚拟机中执行语句，数据导入 aggregate / sports，执行命令如下：

```
sqoop import --connect jdbc:mysql://192.168.106.1:3306/aggregate --username root --password root --table sports --target-dir '/aggregate/sports/' --fields-terminated-by '\t'
```

（5）查看进入 HDFS 的 usertbl 文件的数据格式命令如下：

```
Hadoop fs-cat/aggregate/usertbl/part-m-00000
```

数据导入 HDFS 成功，数据格式按照'\t'分割如下：

```
398 p155 0526155 stepDetail AppA;AppB;AppC;AppD No8-2 0 2018-07-01 09:33:00 2018-07-01 2018-07-01 09:53:16 null [{"snp5":"768,3328,5888,8448,11008,13568,16128,18688,21248,23808,26368,0"},{"knp5":"3,103,459,303,403,759,603,959,803,903,1259,1359"},{"level2p5":"10,10,10,10,10,10,10,10,10,10,10,0"},{"level3p5":"0,0,0,0,0,0,0,0,0,0,0"},{"level4p5":"0,0,0,0,0,0,0,0,0,0,0"},{"yuanp5":"0,0,0,0,0,0,0,0,0,0,0"},{"hour":"2"},{"measureTime":"2018-07-01"}] 1 0 0 0
```

（6）创建表 usertbl 和列簇 info，命令如下：

```
create 'usertbl, 'info'
```

（7）创建表 usertbl 和列簇 info，命令如下：

```
create 'sports, 'data'
```

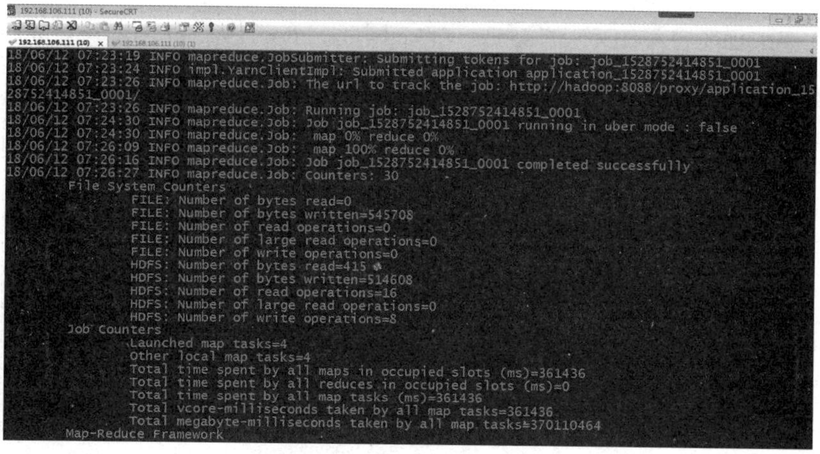

第 9 章 大数据自定义迁移微服务引擎

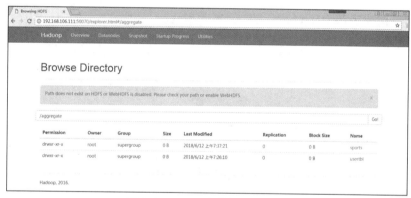

图 9-3　MR 任务执行工程

（8）查看已经导入 HDFS 的文件（网址：http://192.168.106.111:50070/），如图 9-4 所示。

图 9-4　HDFS 上数据目录展示

（9）查看已经导入 HDFS 的 usertbl 文件（网址：http://192.168.106.111:50070/explorer.html#/aggregate/usertbl），如图 9-5 所示。

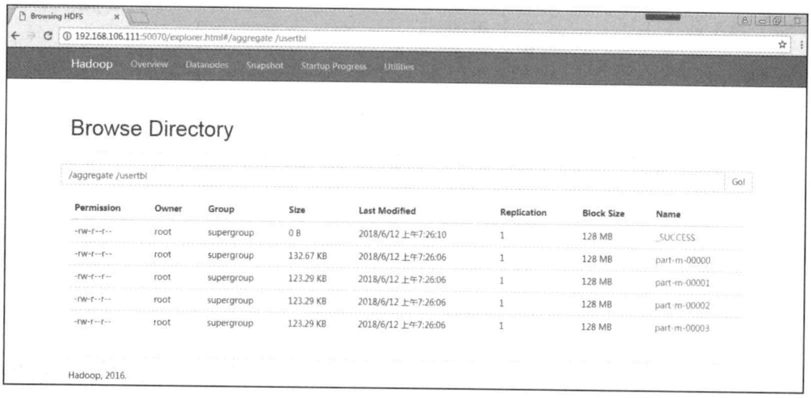

图 9-5　HDFS 上用户数据表分片情况

· 363 ·

（10）查看已经导入 HDFS 的 sports 文件（网址：http://192.168.106.111:50070/explorer.html#/aggregate/sports），如图 9-6 所示。

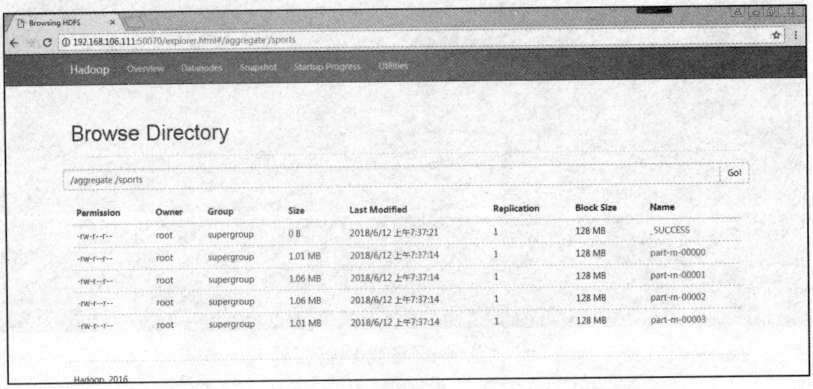

图 9-6　HDFS 上智能终端运动数据表分片情况

（11）按照如上方法，把 sport 的简要包和详细包导入 Hadoop。

9.3.6　构建工程 BD_CustomTransfer_Maven

数据导入微服务工程架构如图 9-7 所示。

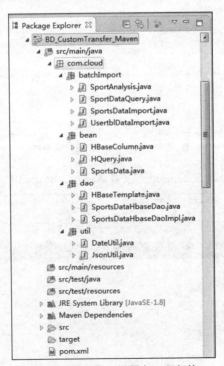

图 9-7　数据导入微服务工程架构

（1）构建工程后导入数据，把 usertbl 表的数据从 HDFS 导入 HBase 的 usertbl 表的 info 列簇，代码实现如下：

```java
package com.cloud.batchImport;

import org.apache.hadoop.conf.Configuration;
import org.apache.hadoop.HBase.client.Put;
import org.apache.hadoop.HBase.mapreduce.TableOutputFormat;
import org.apache.hadoop.HBase.mapreduce.TableReducer;
import org.apache.hadoop.HBase.util.Bytes;
import org.apache.hadoop.io.LongWritable;
import org.apache.hadoop.io.NullWritable;
import org.apache.hadoop.io.Text;
import org.apache.hadoop.mapreduce.Counter;
import org.apache.hadoop.mapreduce.Job;
import org.apache.hadoop.mapreduce.Mapper;
import org.apache.hadoop.mapreduce.lib.input.FileInputFormat;
import org.apache.hadoop.mapreduce.lib.input.TextInputFormat;
/*数据格式如下：
 * sqoop import --connect jdbc:mysql://192.168.106.1:3306/aggregate
   --username root --password root --table usertbl --target-dir
   '/aggregate/ usertbl/' --fields-terminated-by '\t'
 * usertbl 数据格式如下：
 * 1001 0526939 p939 null AppA;AppB;AppC;AppD 1 caompany939 test939 null
   null null 0 2018-05-05 14:58:13 null null
 */
public class UsertblDataImport {
    static class UsertblDataImportMapper extends
            Mapper<LongWritable,Text,LongWritable,Text> {
        Text v2 = new Text();
        protected void map(LongWritable key,Text value,Context context)
                throws java.io.IOException,InterruptedException {
            final String[] splited = value.toString().split("\t");
            try {
                String rowKey = splited[1]+splited[2];
                v2.set(rowKey + "\t" + value.toString());
                context.write(key,v2);
            } catch(NumberFormatException e){
                final Counter counter = context.getCounter
                    ("Usertbl DataImport","ErrorFormat");
                counter.increment(1L);
                System.out.println("出错了" + splited[0] + " " + e.
                    getMessage());
            }
        }
    };
}
    /**
```

```
 * obsDatetime,value,conceptId,patientId,conceptName
 *
 * @author changyaobin
 *
 */
static class UsertblDataImportReducer extends
        TableReducer<LongWritable,Text,NullWritable> {
    protected void reduce(LongWritable key,
            java.lang.Iterable<Text> values,Context context)
            throws java.io.IOException,InterruptedException {
        for(Text text:values){
            final String[] splited = text.toString().split("\t");
            final Put put = new Put(Bytes.toBytes(splited[0]));
            put.add(Bytes.toBytes("info"),Bytes.toBytes("Id"),
                    Bytes.toBytes(splited[1]));
            put.add(Bytes.toBytes("info"),Bytes.toBytes("deviceId"),
                    Bytes.toBytes(splited[2]));
            put.add(Bytes.toBytes("info"),Bytes.toBytes("patientID"),
                    Bytes.toBytes(splited[3]));
            put.add(Bytes.toBytes("info"),Bytes.toBytes("deviceType"),
                    Bytes.toBytes(splited[4]));
            put.add(Bytes.toBytes("info"),Bytes.toBytes("appType"),
                    Bytes.toBytes(splited[5]));
            put.add(Bytes.toBytes("info"),Bytes.toBytes
              ("device UseFlag"),
                    Bytes.toBytes(splited[6]));
            put.add(Bytes.toBytes("info"),Bytes.toBytes("company"),
                    Bytes.toBytes(splited[7]));
            put.add(Bytes.toBytes("info"),Bytes.toBytes("pname"),
                    Bytes.toBytes(splited[8]));
            put.add(Bytes.toBytes("info"),Bytes.toBytes("email"),
                    Bytes.toBytes(splited[9]));
            put.add(Bytes.toBytes("info"),Bytes.toBytes("teamName"),
                    Bytes.toBytes(splited[10]));
            put.add(Bytes.toBytes("info"),Bytes.toBytes
              ("company Name"),
                    Bytes.toBytes(splited[11]));
            put.add(Bytes.toBytes("info"),Bytes.toBytes
              ("isActivate"),
                    Bytes.toBytes(splited[12]));
            put.add(Bytes.toBytes("info"),Bytes.toBytes("laseTime"),
                    Bytes.toBytes(splited[13]));
            put.add(Bytes.toBytes("info"),Bytes.toBytes
              ("modifyTime"),
                    Bytes.toBytes(splited[14]));
```

```java
                put.add(Bytes.toBytes("info"),Bytes.toBytes("ywId"),
                    Bytes.toBytes(splited[15]));
                context.write(NullWritable.get(),put);
            }
        }
    }
    public static void main(String[] args)throws Exception {
        final Configuration configuration = new Configuration();
        // 1.设置 ZooKeeper
        configuration.set("HBase.zookeeper.quorum","hadoop");
        // 2. 设置 HBase 表名称
        configuration.set(TableOutputFormat.OUTPUT_TABLE,"usertbl");
        // 3. 将该值改大,防止 client 连接 ZK 超时退出 s
        configuration.set("dfs.socket.timeout","180000");
        final Job job = new Job(configuration,"UsertblDataImport");
        job.setMapperClass(UsertblDataImportMapper.class);
        job.setReducerClass(UsertblDataImportReducer.class);
        // 4.设置 map 的输出,不设置 reduce 的输出类型
        job.setMapOutputKeyClass(LongWritable.class);
        job.setMapOutputValueClass(Text.class);
        job.setInputFormatClass(TextInputFormat.class);
        // 5.不再设置输出路径,而是设置输出格式类型 s
        job.setOutputFormatClass(TableOutputFormat.class);
        FileInputFormat.setInputPaths(job,"hdfs://192.168.106.111:9000/
            aggregate/usertbl");
        // 6.提交作业任务到集群
        job.waitForCompletion(true);
    }
}
```

（2）导入数据，把 sprots 表的数据从 hdfs 导入 HBase 的 sports 表的 data 列簇，代码实现如下：

```java
package com.cloud.batchImport;

import org.apache.hadoop.conf.Configuration;
import org.apache.hadoop.HBase.client.Put;
import org.apache.hadoop.HBase.mapreduce.TableOutputFormat;
import org.apache.hadoop.HBase.mapreduce.TableReducer;
import org.apache.hadoop.HBase.util.Bytes;
import org.apache.hadoop.io.LongWritable;
import org.apache.hadoop.io.NullWritable;
import org.apache.hadoop.io.Text;
import org.apache.hadoop.mapreduce.Counter;
import org.apache.hadoop.mapreduce.Job;
```

```java
import org.apache.hadoop.mapreduce.Mapper;
import org.apache.hadoop.mapreduce.lib.input.FileInputFormat;
import org.apache.hadoop.mapreduce.lib.input.TextInputFormat;
import com.cloud.util.DateUtil;
/*
 * sqoop import --connect jdbc:mysql://192.168.106.1:3306/aggregate
   --username root --password root --table sports --target-dir
   '/aggregate/ sports/' --fields-terminated-by '\t'
 * 数据格式如下:
 * 398 p155 0526155 stepDetail AppA;AppB;AppC;AppD No8-2 0
 * 2018-07-01 09:33:00 2018-07-01 2018-07-01 09:53:16 null
 * [{"snp5":"768,3328,5888,8448,11008,13568,16128,18688,21248,23808,
   26368,0"},
 * {"knp5":"3,103,459,303,403,759,603,959,803,903,1259,1359"},
 * {"level2p5":"10,10,10,10,10,10,10,10,10,10,10,0"},
 * {"level3p5":"0,0,0,0,0,0,0,0,0,0,0,0"},{"level4p5":"0,0,0,0,0,0,0,
   0,0,0,0,0"},
 * {"yuanp5":"0,0,0,0,0,0,0,0,0,0,0,0"},{"hour":"2"},{"measureTime":
   "2018-07-01"}]
 *       1      0      0      0
 *
 */
public class SportsDataImport {
    static class SimpleDataImportMapper extends
            Mapper<LongWritable,Text,LongWritable,Text> {
        Text v2 = new Text();
        protected void map(LongWritable key,Text value,Context context)
                throws java.io.IOException,InterruptedException {
            final String[] split = value.toString().split("\t");
            try {
                String phone = split[2].trim();
                String deviceId = split[3].trim();
                String dataType = split[4].trim();
                String time = split[7].trim();
                String receiveTime = DateUtil.dateToString(time);
                String rowKey = phone + "_" + deviceId + "_" + dataType
                    + "_"+ receiveTime;
                String val = value.toString();
                v2.set(rowKey + "\t" + value.toString());
                context.write(key,v2);
            } catch(Exception e){
                final Counter counter = context.getCounter
                    ("SimpleData Import","ErrorFormat");
                counter.increment(1L);
```

```java
                System.out.println("出错了" + split[0] + " " + e.
                    getMessage());
            }
        };
    }
}
/**
 * obsDatetime,value,conceptId,patientId,conceptName
 *
 * @author ulove
 *
 */
static class SimpleDataImportReducer extends
        TableReducer<LongWritable,Text,NullWritable> {
    protected void reduce(LongWritable key,
            java.lang.Iterable<Text> values,Context context)
            throws java.io.IOException,InterruptedException {
        for(Text text:values){
            final String[] splited = text.toString().split("\t");
            final Put put = new Put(Bytes.toBytes(splited[0]));
            put.add(Bytes.toBytes("data"),Bytes.toBytes("Id"),
                Bytes.toBytes(splited[1]));
            put.add(Bytes.toBytes("data"),Bytes.toBytes("phone"),
                Bytes.toBytes(splited[2]));
            put.add(Bytes.toBytes("data"),Bytes.toBytes("deviceId"),
                Bytes.toBytes(splited[3]));
            put.add(Bytes.toBytes("data"),Bytes.toBytes("dataType"),
                Bytes.toBytes(splited[4]));
            put.add(Bytes.toBytes("data"),Bytes.toBytes("appType"),
                Bytes.toBytes(splited[5]));
            put.add(Bytes.toBytes("data"),Bytes.toBytes("pname"),
                Bytes.toBytes(splited[6]));
            put.add(Bytes.toBytes("data"),Bytes.toBytes("sendFlag"),
                Bytes.toBytes(splited[7]));
            put.add(Bytes.toBytes("data"),Bytes.toBytes
                ("receive Time"),
                Bytes.toBytes(splited[8]));
            put.add(Bytes.toBytes("data"),Bytes.toBytes("realTime"),
                Bytes.toBytes(splited[9]));
            put.add(Bytes.toBytes("data"),Bytes.toBytes("sendTime"),
                Bytes.toBytes(splited[10]));
            put.add(Bytes.toBytes("data"),Bytes.toBytes
                ("device Type"),
                Bytes.toBytes(splited[11]));
            put.add(Bytes.toBytes("data"),Bytes.toBytes (
                "dataValue"),
```

```java
                    Bytes.toBytes(splited[12]));
            put.add(Bytes.toBytes("data"),Bytes.toBytes("AppA
                _flag"),Bytes.toBytes(splited[13]));
            put.add(Bytes.toBytes("data"),Bytes.toBytes("AppB_
                flag"),
                    Bytes.toBytes(splited[14]));
            put.add(Bytes.toBytes("data"),Bytes.toBytes("AppC_
                flag"),
                    Bytes.toBytes(splited[15]));
            put.add(Bytes.toBytes("data"),Bytes.toBytes("AppD_
                flag"),
                    Bytes.toBytes(splited[16]));
            context.write(NullWritable.get(),put);
        }
    };
}
public static void main(String[] args)throws Exception {
    final Configuration configuration = new Configuration();
    // 1.设置 ZooKeeper
    configuration.set("HBase.zookeeper.quorum","hadoop");
    // 2.设置 HBase 表名称
    configuration.set(TableOutputFormat.OUTPUT_TABLE,"sports");
    // 3.将该值改大,防止 client 连接 ZK 超时退出 s
    configuration.set("dfs.socket.timeout","180000");
    final Job job = new Job(configuration,"SportsDataImport");
    job.setMapperClass(SimpleDataImportMapper.class);
    job.setReducerClass(SimpleDataImportReducer.class);
    // 4.设置 map 的输出,不设置 reduce 的输出类型
    job.setMapOutputKeyClass(LongWritable.class);
    job.setMapOutputValueClass(Text.class);
    job.setInputFormatClass(TextInputFormat.class);
    // 5.不再设置输出路径,而是设置输出格式类型 s
    job.setOutputFormatClass(TableOutputFormat.class);
    FileInputFormat.setInputPaths(job,"hdfs://192.168.106.111:9000/
        aggregate/sports");
    job.waitForCompletion(true);
}
}
```

(3) 导入数据,把 sprots 表的数据从 HDFS 导入 HBase 的 sports 表的 sort 列簇,为后续统计步数做好准备。代码实现如下:

```java
package com.cloud.batchImport;

import org.apache.hadoop.conf.Configuration;
```

```java
import org.apache.hadoop.HBase.client.Put;
import org.apache.hadoop.HBase.mapreduce.TableOutputFormat;
import org.apache.hadoop.HBase.mapreduce.TableReducer;
import org.apache.hadoop.HBase.util.Bytes;
import org.apache.hadoop.io.LongWritable;
import org.apache.hadoop.io.NullWritable;
import org.apache.hadoop.io.Text;
import org.apache.hadoop.mapreduce.Counter;
import org.apache.hadoop.mapreduce.Job;
import org.apache.hadoop.mapreduce.Mapper;
import org.apache.hadoop.mapreduce.lib.input.FileInputFormat;
import org.apache.hadoop.mapreduce.lib.input.TextInputFormat;
import com.cloud.util.DateUtil;
import com.cloud.util.JsonUtil;
import net.sf.json.JSONArray;
/*
 * sqoop import --connect jdbc:mysql://192.168.106.1:3306/aggregate
   --username root --password root --table sports --target-dir
   '/aggregate/ sports/' --fields-terminated-by '\t'
 * 数据格式如下:
355 p898 0526898 stepCount AppA;AppB;AppC;AppD No8-1 0 2018-07-01 09:33:00
   2018-07-01 09:33:00 2018-07-01 09:53:06 null
[{"stepSum":"526899000"},{"calSum":"526899000"},
{"distanceSum":"3161394"},{"yxbsSum":"0"},{"weight":"70"},
{"stride":"70"},{"degreeOne":"52220"},{"degreeTwo":"52220"},
{"degreeThree":"52220"},{"degreeFour":"52220"},{"uploadType":"1"},
{"measureTime":"2018-07-01 09:33:00"}]    1    0    0    0
 */
public class SportAnalysis {
    static class SportAnalysisMapper extends Mapper<LongWritable,Text,
LongWritable,Text> {
        Text rowData = new Text();
        protected void map(LongWritable key,Text value,Context context)
            throws java.io.IOException,InterruptedException {
            final String[] split = value.toString().split("\t");
            try {
                String stepSum = "";
                String phone = split[2].trim();
                String deviceId = split[3].trim();
                String dataType = split[4].trim();
                String time = split[7].trim();
                String receiveTime = DateUtil.dateToString(time);
                String rowKey = phone + "_" + deviceId + "_" + dataType
                    + "_" + receiveTime;
                String stepData = split[11].trim();
```

```java
                    if("stepCount".equals(split[3].trim())){
                        JSONArray ja = JSONArray.fromObject(stepData);
                        stepSum = JsonUtil.getJsonParamterString
                          (ja.getJSONObject(0),"stepSum");
                        rowData.set(rowKey + "\t" + value.toString()+ "\t" +
                          stepSum);
                        System.out.println(rowData);
                        context.write(key,rowData);
                    }
            } catch(Exception e){
                final Counter counter = context.getCounter
                  ("Simple DataImport","ErrorFormat");
                counter.increment(1L);
                System.out.println("出错了" + split[0] + " " + e.
                  getMessage());
            }
        }
    };
}
/**
 * obsDatetime,value,conceptId,patientId,conceptName
 * @author changyaobin
 *
 */
static class SportAnalysisReducer extends TableReducer<LongWritable,
  Text,NullWritable> {
    protected void reduce(LongWritable key,java.lang.Iterable<Text>
      values,Context context)
            throws java.io.IOException,InterruptedException {
        for(Text text:values){
            final String[] splited = text.toString().split("\t");
            final Put put = new Put(Bytes.toBytes(splited[0]));
            put.add(Bytes.toBytes("sort"),Bytes.toBytes("Id"),
              Bytes.toBytes(splited[1]));
            put.add(Bytes.toBytes("sort"),Bytes.toBytes("phone"),
              Bytes.toBytes(splited[2]));
            put.add(Bytes.toBytes("sort"),Bytes.toBytes("deviceId"),
              Bytes.toBytes(splited[3]));
            put.add(Bytes.toBytes("sort"),Bytes.toBytes("dataType"),
              Bytes.toBytes(splited[4]));
            put.add(Bytes.toBytes("sort"),Bytes.toBytes("appType"),
              Bytes.toBytes(splited[5]));
            put.add(Bytes.toBytes("sort"),Bytes.toBytes("pname"),
              Bytes.toBytes(splited[6]));
            put.add(Bytes.toBytes("sort"),Bytes.toBytes("sendFlag"),
              Bytes.toBytes(splited[7]));
```

```java
            put.add(Bytes.toBytes("sort"),Bytes.toBytes
                ("receive Time"),Bytes.toBytes(splited[8]));
            put.add(Bytes.toBytes("sort"),Bytes.toBytes("realTime"),
                Bytes.toBytes(splited[9]));
            put.add(Bytes.toBytes("sort"),Bytes.toBytes("sendTime"),
                Bytes.toBytes(splited[10]));
            put.add(Bytes.toBytes("sort"),Bytes.toBytes
                ("device Type"),Bytes.toBytes(splited[11]));
            put.add(Bytes.toBytes("sort"),Bytes.toBytes("dataValue"),
                Bytes.toBytes(splited[12]));
            put.add(Bytes.toBytes("sort"),Bytes.toBytes("AppA_
                flag"),Bytes.toBytes(splited[13]));
            put.add(Bytes.toBytes("sort"),Bytes.toBytes("AppB_
                flag"),Bytes.toBytes(splited[14]));
            put.add(Bytes.toBytes("sort"),Bytes.toBytes("AppC_
                flag"),Bytes.toBytes(splited[15]));
            put.add(Bytes.toBytes("sort"),Bytes.toBytes("AppD_
                flag"),Bytes.toBytes(splited[16]));
            put.add(Bytes.toBytes("sort"),Bytes.toBytes("stepSum"),
                Bytes.toBytes(splited[17]));
            context.write(NullWritable.get(),put);
        }
    };
}
public static void main(String[] args)throws Exception {
    final Configuration configuration = new Configuration();
    // 1. 设置 ZooKeeper
    configuration.set("HBase.zookeeper.quorum","hadoop");
    // 2. 设置 HBase 表名称
    configuration.set(TableOutputFormat.OUTPUT_TABLE,
      "sports Analysis");
    // 3. 将该值改大,防止 client 连接 ZK 超时退出 s
    configuration.set("dfs.socket.timeout","180000");
    final Job job = new Job(configuration,"SportAnalysis");
    job.setMapperClass(SportAnalysisMapper.class);
    job.setReducerClass(SportAnalysisReducer.class);
    // 4. 设置 map 的输出,不设置 reduce 的输出类型
    job.setMapOutputKeyClass(LongWritable.class);
    job.setMapOutputValueClass(Text.class);
    job.setInputFormatClass(TextInputFormat.class);
    // 5. 不再设置输出路径,而是设置输出格式类型 s
    job.setOutputFormatClass(TableOutputFormat.class);
    FileInputFormat.setInputPaths(job,"hdfs://192.168.106.111:
      9000/aggregate/sports");
```

```
            job.waitForCompletion(true);
    }
}
```

9.3.7 智能终端运动数据从 MySQL 数据迁移到 Hive

（1）创建用户表 usertbl 在 Hive 中，代码实现如下：

```
create table usertbl(id int,deviceID String,patientID String,deviceType
   String,appType String,deviceUseFlag String,company String,pname
   String, email String,teamName String,companyName String,isActivate
   String,lastTime String,modifyTime String,ywId String)row format
   delimited fields terminated by '\t' stored as textfile;
```

（2）加载数据到 Hive 中，代码实现如下：

```
hive->load data local inpath '/aggregate/sports' overwrite into table
   sports
hive->load data local inpath '/aggregate/usertbl' overwrite into table
   usertbl
hive->dfs -ls /usr/local/data
hive-->select * from sports;
```

（3）迁移体检数据到 HDFS 之后，再导入 Hive 中，代码实现如下：

```
sqoop import --connect jdbc:mysql://192.168.106.1:3306/storage --username
   root --password root --table physical_check --target-dir '/aggregate/
   physical_check/' --fields-terminated-by '\t'
   create table physical_check
(
   CHECK_ID            int,
   CARD_NO             String,
   REG_DATE            String,
   CLINIC_CODE         String,
   SEX_CODE            String,
   PAYKIND_CODE        String,
   PACT_CODE           String,
   PACT_NAME           String,
   REGLEVL_CODE        String,
   REGLEVL_NAME        String,
   DEPT_CODE           String,
   DEPT_NAME           String,
   DOCT_CODE           String,
   DOCT_NAME           String,
   YNSEE               String,
   IN_SOURCE           String,
   ADDRESS_CODE        String,
   ADDRESS_NAME        String,
```

```
        AGE                     String,
        DIAGNOSE                String,
        TIMES                   int,
        Y00000010435            String,
        Y00000010770            String,
        Y00000010408            String,
        Y00000010428            String,
        Y00000010902            String,
        Y00000010388            String,
        Y00000010403            String,
        Y00000010966            String,
        Y00000010990            String,
        Y00000010391            String,
        Y00000010865            String,
        Y00000010867            String,
        Y00000010816            String,
        Y00000010406            String,
        Y00000010837            String,
        Y00000010833            String,
        Y00000010682            String,
        Y00000010863            String,
        Y00000010889            String,
        Y00000010678            String,
        Y00000010179            String,
        F00000038623            String,
        F00000038500            String,
        F00000038457            String,
        F00000038487            String,
        F00000038462            String,
        F00000038624            String,
        F00000038541            String,
        F00000038413            String,
        F00000038414            String,
        F00000038627            String,
        F00000038297            String,
        F00000038597            String,
        F00000038493            String,
        F00000038536            String,
        F00000038465            String,
        F00000038497            String,
        F00000038412            String,
        F00000038495            String,
        F00000039159            String,
        F00000038299            String,
        F00000038496            String,
```

```
    F00000039158        String,
    F00000038608        String,
    F00000038508        String,
    F00000038498        String,
    F00000038503        String,
    F00000038467        String,
    F00000038499        String,
    SUM_OWN             int,
    SUM_PUB             int,
    SUM_OVER            int,
    SUM_DOWN            int,
    SUM_DPUB            int,
    SUM_DOVER           int,
    SUM_OWN_P           int,
    SUM_OWN_U           int,
    SUM_OWN_UZ          int,
    SUM_OWN_UC          int,
    SUM_OWN_CLASS       String,
    SUM_PUB_CLASS       String,
    SUM_OVER_CLASS      String,
    SUM_OWN_CLASS_P     String,
    SUM_OWN_CLASS_U     String,
    SUM_OWN_CLASS_UZ    String,
    SUM_OWN_CLASS_UC    String,
    DRUG_PERCENT        String,
    LABEL               String,
    LATENT1             String,
    LATENT2             String,
    COST_PREDICT        String,
    RESULT              String,
    COST_RESULT         String
) row format delimited fields terminated by '\t' stored as textfile;
```

（4）在 root 目录执行，加载 HDFS 的文件命令：

```
load data inpath '/aggregate/ physical_ check' overwrite into table
  physical_check
```

加载 Linux 本地的文件命令：

```
load data local inpath '/aggregate/ physical_check' overwrite into table
  physical_check
```

启动 MySQL，存放了 Hive 文件的元数据，命令如下：

```
mysql -uroot -proot
```

查询数据，命令如下：

```
select * from physical_check;
```

9.4 项目小结

自定义数据迁移服务引擎的主流大数据工具就是 Sqoop 引擎，它从关系型数据库（如 MySQL、Oracle、SQLServer 等）中的用户和业务数据导入 Hadoop 的 HDFS 中之后，设计好 HBase 的表 Rowkey 和列簇，并进行数据清洗后迁移到 HBase 库，入库采用了 HBaseTemplete 主流类，核心是编写 Dao 层接口定义和实现，最后讲解了历史数据迁移到 Hive，为数据智能分析和数据建模做好准备。

1. 理论知识点

（1）Hadoop 主要解决的是分布式环境下的数据存储、并行计算和资源协同管理问题，核心价值是能高效组织一大批廉价机器构建分布式环境来协同存储和处理海量数据。

（2）HDFS：主节点 NameNode 的职责是管理文件所在目录及对应的 block 位置，DataNode 负责实际数据的存储。

（3）MapReduce 框架解决了并行数据处理下的分布式存储、任务调度、负载均衡、容错机制、容错处理等核心技术问题。

（4）YARN 核心思想是将 MapReduce 中的 JobTracker 的资源管理和作业调度两个职责分别由 ResourceManage 资源管理器和 ApplicationMaster 应用管理进程来实现。

（5）HBase 是一个基于 HDFS 进行 NoSQL 存储模型的列式 KV 数据库，具备高可伸缩、高可靠性、高性能、分布式等特点。要理解 HMaster 和 HRegionServer 的职责。

（6）Hive 是为了处理海量数据的离线统计而生。Hive 相当于 SQL 任务转化为 Mapdeduce 并行任务。

（7）ZooKeeper 是一个开放源码的分布式集群协调器，主要用于解决分布式应用中的统一命名服务、状态同步服务、集群管理、配置项管理等问题。要理解 ZooKeeper 的职责。

（8）要理解 MapReduce 的工作流程。

（9）Sqoop 是负责把 Hadoop 和关系型数据库中的数据相互转移的工具。

2. 实战知识点

（1）Sqoop 引擎把一个关系型数据库（如 MySQL、Oracle、SQLServer 等）中的用户数据、运动、体检数据导入 Hadoop 的 HDFS 中。

（2）借助 MR 工具，把用户数据、运动、体检数据从 HDFS 导入 HBase 中。

（3）HBase 数据库建立和 Rowkey 的创建方法。

（4）借助 Hive 工具，把用户数据、运动、体检数据从 HDFS 导入 Hive 中。